Y0-BDX-956

Maps and Statistics

Peter Lewis

Maps and Statistics

A Halsted Press Book

John Wiley & Sons, New York

First published 1977 by Methuen & Co Ltd
11 New Fetter Lane, London EC4P 4EE

Printed in Great Britain at the
University Printing House, Cambridge

ISBN 0 416 65370 7 *hardback*
ISBN 0 416 65380 4 *paperback*

Library of Congress Cataloging in Publication Data

Lewis, Peter, 1938–
Maps and statistics.

Bibliography: p. 310
Includes index.
1. Maps, Statistical. I. Title.
GA109.8.L48 1977 001.4'226 77-1184
ISBN 0-470-99094-5

To my parents

Contents

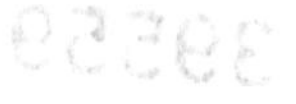

Categorization of test procedures

PROPOSITION	MEASUREMENT: NOMINAL *Class or category*	MEASUREMENT: DISCRETE QUANTITY *Position in sequence, ranked*
RANDOMNESS	Probability distributions Binomial (1.9, 2.2.1) Multinomial (2.2.2) Hypergeometric (2.2.3, 2.3.1) Poisson (2.2.4) Number of runs (2) (3.1.1) Number of runs (1) (3.1.2) Number of runs, linked pairs (3.1.3) Wald-Wolfowitz (3.4.1)	Cox and Stuart (3.3.1) Kolmogorov (3.5.1)
INDEPENDENCE DEPENDENCE ASSOCIATION	Fisher's exact (2.3.1) Chi-square (2.3.2, 2.5, 2.6, 3.5.2) Cramer's coefficient (2.4) Tschuprow's coefficient (2.4) Pearson's contingency coefficient (2.4) Kendall's phi coefficient (2.4) Cross-product ratio (2.4) Chi-square ($r \times c$) (2.5) Chi-square ($r \times c \times l$) (2.6) Log-linear models (2.6)	Chi-square ($r \times 2$) subsets (2.5) Chi-square ($r \times 2$) regression (2.5) Chi-square ($r \times c \times l$) (2.6) Log-linear models (2.6)
EQUALITY OF LOCATION IDENTITY OF DISTRIBUTION GOODNESS-OF-FIT	Sign test (1.9) Wald-Wolfowitz (3.4.1) Westenberg-Mood median test (3.4.2) Chi-square (goodness-of-fit) (3.5.2) McNemar's test (Ex. 3.7)	Wilcoxon rank-sum (2.7.1) Mann-Whitney, U (2.7.2) Kruskal-Wallis (2.7.3) Friedman (2.7.4) Matched-pairs sign test (3.4.2) Wilcoxon signed-rank (3.4.3)

Point symbol maps are discussed in section 2. Line symbol maps and angular measurements are discussed in section 3. Numbers in brackets indicate the article in which the test statistic is discussed in detail.

SCALE

CONTINUOUS QUANTITY *Distribution-free or not-normal*	*Von Mises (angular)* *Normal*
Runs up and down, Edgington (3.2.1) Periodicity in runs, Noether (3.2.2) Kolmogorov (3.5.1)	
Cox and Stuart (3.3.1) Spearman's rho (3.3.2) Daniel's test for trend (3.3.2) Kendall's tau (3.3.2) Kendall's concordance, *W* (3.3.2)	*t* for *b* coefficient *t* for Pearson's *r* coefficient
Smirnov test (3.4.4) Kolmogorov test (3.5.1) Lilliefors test (3.5.1) Kuiper's test (angular data) (3.6.1) Mardia's uniform scores test (angular data) (3.6.2)	*t* in 1 and 2 sample cases *F* in multi-sample cases Goodness-of-fit not needed by assumption Watson and Williams, Rayleigh for angles

Figures, tables and maps

Acknowledgements

I should like to thank the following for permission to reproduce the tables in appendix 2: Addison-Wesley Publishing Company for table 9; Biometrika Trustees for tables 4, 6, 7, 12 and 17; Department of Statistics, Florida State University, for table 14; Institute of Mathematical Statistics, Hayward, California, for tables 10 and 15; *Journal of the American Statistical Association* for tables 3(b), 8, 11 and 16; *Journal of the Royal Statistical Society* for table 18; *Statistica Neerlandica* for table 13; and Stanley Thornes (Publishers) Ltd for tables 1, 2, 3(a) and 5.

Implicit in any book is the debt owed to the people whose work appears in the references. It is a pleasure to offer my thanks to those people who have contributed to the development of this book. Like the other authors in this series I have benefited from the suggestions made so pleasantly by Bill Morgan, the series editor, and by Janice Price, general editor for Methuen. I should like to thank Toby Lewis of Hull University and Joyce Snell of Imperial College, London, for reading the manuscript and making valuable recommendations, and I hope my response to those suggestions has been adequate. My thanks, too, to the technical and secretarial staff of the Department of Geography, Birkbeck College, for producing the fair copies of the illustrations and for typing a difficult manuscript. My deepest thanks happily go to Rosemary and Simon and Catherine for providing the environment for writing this book, for giving so much encouragement during its preparation, and for helping on so many of the jobs that arose at all stages of its formation.

Section 1/ Maps, measurement and probability

Maps, measurement and probability

Maps are scales for measuring the property *location*. Location is that property of objects which geographers consider central to their study and central to the problems of understanding which interest them. Although maps may show objects with respect to attributes other than location, their principal purpose is to depict objects in terms of their locational property. We recognize objects and we refer to them in terms of their properties. All properties are relational. We increase our understanding of objects by defining their properties and then comparing objects in terms of one or more shared properties. In order to compare objects in terms of any property, that property has to be measured in some way. Measurement entails assigning numerals to things with respect to a particular property according to some determinative, non-degenerate rule. A rule is determinative if the same numerals are given to the same things under the same conditions. It is non-degenerate if different numerals are given to different things under the same conditions or to the same things under different conditions.

Rules that satisfy these conditions are referred to as measurement scales. These conditions are the minimum restrictions placed upon measurement and, as numeral assignments can be made which satisfy more restrictions, scales of measurement can be classified in terms of these increasing restrictions (Ellis 1966). The imposition of certain restrictions implies that the numerals, assigned as measures of a property, themselves have certain properties which can be used in comparing objects. It is useful to distinguish three important classes of measurement scale. First we recognize NOMINAL scales, which correspond to names of classes. For example we might distinguish days as being wet days or dry days, or we might classify plants according to whether they survive or die in a particular environment. Measurement on a nominal scale consists of allocating objects to classes in terms of one or more properties. The properties involved are what many people would refer to as *qualities.* In distinction to qualities, we denote as *quantities* those properties which objects can possess to different degrees. Measurement of quantities is taken to mean that objects are given numbers according to the degree to which they possess a particular property. It is useful to make a distinction between DISCRETE-QUANTITY scales in which *order* is recognized and CONTINUOUS-QUANTITY scales in which the number sequence is dense over the interval of measurement so that the size of two or more measurements can be related. We might distinguish buildings according to their rank in a sequence of increasing degrees of dilapidation. In this case there is no necessary implication that a building with a measure of 6 is twice as dilapidated as a building with a measure of 3. Continuous-quantity scales correspond most closely to our usual idea of measurement of, say, length or time, when, if our measuring instrument is sufficiently refined, we can give a number of any accuracy to our measurement. In the case of length a value of 6 units does entail that the object is twice as long as another object with a value of 3 units. Properties can be measured on one or more of these scale types. Location can be measured on scales of all three types and in articles 1.1, 1.2 and 1.3 some of the implications of restricting the measurement scale are discussed.

Two important consequences can be anticipated. In the case of both sorts of quantity scale, the numbers given as a result of measurement correspond to the degree to which the objects have the property. There is a correspondence between number order and property order such that when the numbers are arranged in a sequence of increasing values the objects are also arranged in a sequence according to the degree to which they have a particular property. This is a linear or one-dimensional ordering relationship. It is the increase in information that accompanies this order which is exploited by the mathematical procedures applied to the numbers we get as measurements of quantities. Some geographers may feel that location cannot be characterized satisfactorily by this restriction to measurement scales which entail one-dimensional ordering relationships. The importance of the map is that it can provide a two-dimensional scale for ordering this property of location, and yet its locational information can be characterized on the conventional, more restricted quantity scales or on the least restricted nominal scales which have no such linear-order property. Two-dimensional ordering relationships can be characterized numerically, but they have a weaker order property. Some future work in geography will be concerned with the algebras and number systems of such relationships. Secondly, the use of a particular scale for measuring location determines the statistical inferential procedures which we can use to decide if a particular assertion, made about the objects in terms of location and one or more other of their properties, is acceptable or not. Articles 1.4 and 1.9 discuss this aspect thoroughly.

1.1 The concept of location class

The least restricted scales of measurement are referred to as NOMINAL scales and, as the name implies, such scales refer to classes or to categories of a property which are distinct enough to be named. The property and its classes have to be defined sufficiently precisely to be able to allocate an object unequivocally to a particular class. A common requirement is that the categories must be mutually exclusive and exhaustive. The first condition ensures that each object can be allocated to just one category in terms of the particular classification. The second condition ensures that all the objects to be classified can be put into at least one category. Measurement on a nominal scale implies either that objects can be said to be equal or that they can be said to be different with respect to some property, and with respect to some criteria for subdividing that property.

There can be two or more categories. In the simplest case of two categories the measurement involves recognizing two cases: is, is not. It is common practice to associate the object and the property category: thus we say, a day is wet, a factory has survived, a site is occupied. Such qualitative classification may not seem to be what is customarily intended by the term measurement. In all such cases of qualitative classification the names of the categories can be replaced by numerals, usually integers, and, so long as their assignment is made according to a determinative, non-degenerate rule, it is nominal measurement. If those two conditions are not met, the literal or the numeral designations have no useful purpose, and when the conditions are met the literal and the numeral descriptions are equivalent.

The literal description is often retained because it evokes the particular property state more readily than the numeral description, but the numeral description is often a more convenient shorthand when the number of classes or the number of properties is large. In order to make some probability judgement of an assertion about the particular property categories we need to analyse the objects statistically in terms of the property. To do this we need to specify a random variable (defined in article 1.8); the desirable and usable properties of random variables are characterized as numbers. Literal or numeral designations of categories do not carry much information about the property and it is in the sense of satisfying so few conditions that nominal measurements are said to be the least restricted scales. Indeed any permutation transformation can be made without loss of information. This means that the numeral or literal designations can be rearranged amongst the categories without losing the information contained in the measurement.

This must not be seen as a weakness without benefit, as often this lack of restriction accurately reflects the degree of confidence that a geographer feels in making an assertion about a set of objects in terms of some particular property. For example, a geographer might consider that fig. 1.1 (a) justified an assertion that the objects with location class A_1 had a greater propensity to survive than objects not in location class A_1. At the same time the geographer may doubt that an assertion which demanded any more restricted measurement of either attribute, that is, location class or survival propensity, would be appropriate. Such an assertion might be that the likelihood of survival diminished with increasing rank-distance from a coal pithead. All four representations (a) (b) (c) (d) in fig. 1.1 show that 9 of the 15 objects have the same location class and that 8 of the 15 have the same survival class.

Maps like fig. 1.1 (a) which show objects in terms of their location class are common. Any preference for fig. 1.1 (a) over the other representations reflects a belief that the irregular shape of location class A_1 corresponds to the *correct location* of, in this case, the coalfield. A belief in a correct location set implies that either (a) locations are objects, or (b) that location is a property of objects and that there is exactly one scale for measuring location such that the locational relationships of the objects are then defined uniquely. The difficulties of (a) are apparent as soon as an attempt is made to determine what properties a location has.

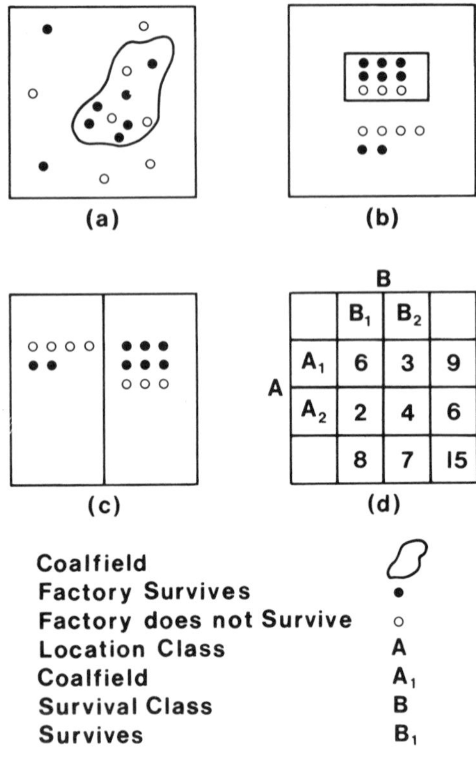

1.1 Objects and location as a class.

In the second situation the belief loses credibility because it leads to a contradiction in the following way. If location is a property of objects, then, as with any property, it is meaningful only to speak of an object's location in relationship to the same property of another object. This property is characterized in terms of a measurement, but as there is a choice of measurement scale (article 1.3) we can arrive at different measures of this property, as we can of any property, by exercising this choice of scale. Consequently by changing the way in which we measure location we can change the way in which we characterize the locational relationships between a set of objects. We shall see in articles 1.2 and 1.3 that this is a quite sensible statement of what happens in geographic analysis; it is only contrary to our preconceptions under either (a) or (b) above.

Early attempts to map things, in the sense of providing a scale to measure their locational relations, were experimental. These early maps were often an amalgam of fear of the unknown, of difficulty, and of time as well as of directed distance. These attempts emphasize both the arbitrary nature of the measurement scale and that location is a composite property: indeed it has been suggested that the quantity names used generally in science are such cluster concepts (Gasking 1960). The need for a standard scale and an agreed origin is plain. This standard should have certain properties of arithmetic convenience, of simplicity and of consistency in one, two and three dimensions. Distance has these characteristics and has been accepted as the standard linear scale for maps because of them. The study of map projections makes geographers accept that it is impossible to retain in a two-dimensional map all the components of the locational relationships of objects on a sphere in terms of a distance measurement scale. Consequently geographers accept that two-dimensional maps are wrong in the sense that some aspect of the relationship in three dimensions *has* to be lost. However it seems at least as important to accept that even if it were possible to retain the distance properties of three dimensions in a two-dimensional map there would still not be a uniquely correct location set.

Let us by all means express objects' locations on a standard scale so long as we recognize that such a map is not uniquely correct. In the cases depicted in fig. 1.1 all the maps are equally correct because location is measured on a nominal scale, and the information actually used is shown correctly on them all. Fig. 1.1 (a) contains more information than is required by the assertion.

It is often the case that location-class names result from identifying one set of objects, say factories, with another object that is both extensive in terms of a standard distance scale from an origin, and comparatively stable over time. Reference objects such as vegetation types, mineral deposits and physical features are convenient shorthand descriptions and are the basis of many elementary geographic assertions. Often, however, the usefulness of location classes based upon such objects depends upon other properties that relate more closely to the specific property studied but which are less readily measured. We are free to define homogeneous location classes in terms of any measurement scale; it is convenience not necessity that makes it customary to define location classes in terms of a standard distance scale. When we do so we discard some of the information implicit in such a distance scale.

1.2 Using location as a quantity

The idea of measurement can be extended if we use the order property of numbers, such that if we say two objects are not equal with respect to a particular property then we can recognize whether one object has more or less of that property than the other object. Inequality in this sense implies an order that is not entailed by the qualitative inequality of nominal measurement. Properties which are considered to have gradations, such that the numerals that are given to those gradations reflect the magnitude of the property, are called quantities. Thus, if we have two objects, A and B, which have a quantity in common we can say that if A is not equal in that property to B, then either A is greater in that property than B, or A is less in that property than B.

Location is often treated as a quantity, such that any one of those three relations can hold, but only one of them does hold in a particular comparison of two objects. Consider fig. 1.2 in which we have a map using a standard distance scale which we can use to measure the location of four objects in relation to a coalpit. We agree to measure location on a standard distance scale. We say A is not equal in location to B and write $A \neq_L B$. Similarly we can write $A \neq_L C$ and $B \neq_L C$, and $A =_L D$.

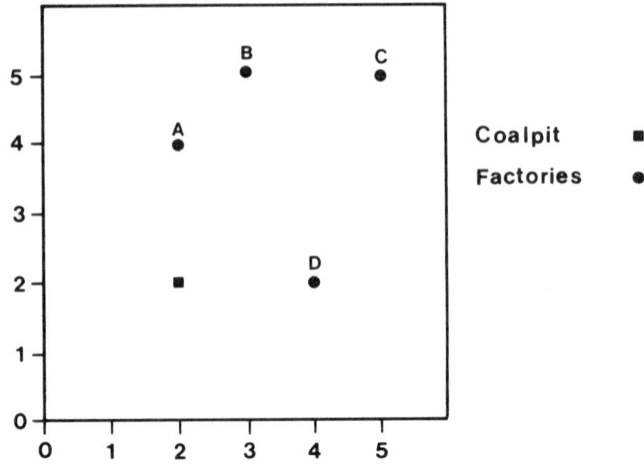

1.2 Objects and location as a quantity.

A and *D* have the same location in precisely the same way that in fig. 1.1 (a) we said that the objects on the coalfield had the same location. In fig. 1.2 we are saying *A*, *D* have the same location (measure), whereas in fig. 1.1 we said two objects had the same location (class), in terms of a set of conventions for measuring the locational relations of these objects. Under these conventions we can say that $A <_L B$, $A <_L C$ and $B <_L C$. The objects can be ordered in terms of their property location when it is measured on a standard distance scale to give

$$A = D < B < C$$

In many cases geographers obviously have believed that there is a crude order of location relationships which is matched by such a measurement scale. Indeed distance has become the principal quantity for measuring location in assertions about relationships between location and non-location properties of objects.

Because it is customary to use standard scales such as the metric or imperial scales for measuring distance in such assertions, it must not be assumed that it is necessary to measure distance in such a way. In article 1.3 we shall consider alternative scales and note that they arise equally naturally as measures of location. Nor are we saying that location is equivalent to distance measured on whatever scale, but that for the particular proposition under investigation we are agreeing to measure location in a way that admits of gradations to which we can assign numerals such that these numbers discriminate those gradations.

In common with all discrete–and continuous–quantity measurement we are associating a quantity with a linear order. Location is the property that provides the ordering relationship and we wish to find a measure that assigns numerals to that property such that they correspond to the order of the objects with respect to that property. We recognize that location is a composite property, usually thought of as two-dimensional, which we are prepared to treat as a one-dimensional quantity in order to relate it to other properties. For many purposes this restriction to a linear order has been convenient and useful. If we find that the benefits of this restriction are accompanied by too great a loss in information about locational relationships then we shall have to increase our exploration of other ways of measuring location.

Of course on standard maps using the same distance scale on each axis location is taken to be equivalent to position and is defined uniquely as a vector quantity. Vectors are not numbers in our ordinary sense, because they consist of ordered pairs of numbers in two dimensions and of ordered triples of numbers in three dimensions. It is because vectors are ordered pairs (triples, etc.) of numbers that it makes no sense to speak of a linear-order relationship. There are procedures for discussing weaker order relations and these are basic to exciting aspects of progress in geography in location analysis. Vectors have direction and length and are sometimes referred to as directed distances. To date, geographers have felt that distance is preferable to direction as the linear-ordering relationship for the composite property of location. This is evident from the theoretical developments as well as from the infrequent use of maps showing direction consistently. Although either direction or distance could be combined in geographic assertions relating location and other attributes, the justification for preferring one must be found in the success which its use yields in establishing such relationships. Because of the preference for distance measures in geographic theory and practice, the weight of the statistical inferential procedures covered in this text is biased towards that measure. However a number of tests designed specifically for angular measurements are included. It is often unwise to use standard tests for angular measurements as is illustrated in article 3.6.

The use of quantity measures and linear-ordering relationships seems to reflect a tacit acceptance amongst geographers that our progress in understanding should follow a path similar to that which led to the development of numerical laws in the physical sciences. Unless logical grounds for the falsity of such an assumption are produced there is good reason to explore the possibility of establishing such laws in geography by comparable procedures. The numerical laws of physical science express relationships between quantities. These quantities are properties of objects and are measured on scales of particular sorts. The results of those measurements are numbers and the relationship contained in the numerical law is a relationship between numerals assigned on continuous-quantity scales. Formal attempts to provide similar relations in geography between location and other properties have attempted to use similar quantitative scales. In these attempts location has been measured on a standard distance scale.

Weaker assertions than these are made often in human geography, and the reduced restriction reflects a feeling amongst geographers, first, that distance is not a complete measure of location and, secondly, that its refinement as a measuring scale is greater than needed for the relationships proposed because the discrimination implicit in the continuous distance scale is greater than the distance discrimination that occurs in the arrangement of each object in response to other objects. Geographers are often more confident in saying that things are arranged in some order than in saying that they are arranged at specific distances from each other. Order is acknowledged whereas interval is not. Assertions embodying order rather than interval arise frequently in studies relating to *location theory.* It is still common to find elementary assertions of industrial location such that a factory is market-oriented, that an iron foundry is raw-material-oriented or that an electronics factory is footloose in the sense that its location is indiscriminate with respect to market or to raw material. In such blunt forms these statements mean that in a particular industrial production sequence from P_0, the source of raw material, to P_n, the ultimate market, a factory in the *i*th stage, P_i, will have a particular locational relationship to the P_nth or the P_0th stage, or that it will be located indiscriminately with respect to P_n and P_0. This is an ordinal relation of location, and uses a discrete-quantity scale. This scale is more refined than a nominal scale because it allows for ordered gradations of location. In such an assertion something is stated about the position in a sequence of events of a

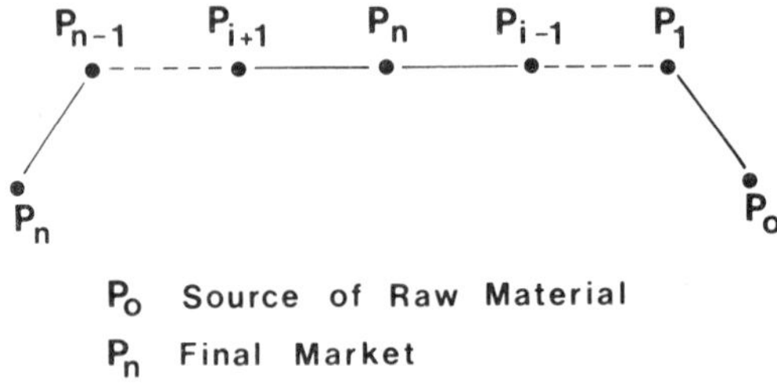

1.3 Location as position in a sequence.

factory and its corresponding position in a distance sequence between the two end points. The statement asserts that the factory will be (i) nearer, (ii) as near, (iii) further from an end point than another similar object at a different stage in the same production sequence with respect to the same end points. It is not saying that the object will be any particular distance from either end point or from any other object in the production sequence. The use of a discrete-quantity scale rather than a continuous-quantity scale is important because it affects the choice of statistical tests that are appropriate.

1.3 Scale variations in location measurement

Even if we assume that location can be represented satisfactorily by distance, there is no need to assume that distance can be measured in only one way, because we have accepted simply that measurement entails giving numbers to a property according to a scale. There is no confusion between the use of the word scale in geography to refer to representative fractions and its more general use in measurement. A scale for measuring length involves a determinative, non-degenerate rule for giving numerals to things in respect of this property. For two railway stations we may say that the distance between them is 1 mile. When we give the number 1 we mean the distance is 1 unit on a particular scale. In this case it is the Imperial scale and the unit name is a mile. The same relationship can be represented on a different length scale, say the metric scale, whose unit name is a kilometre, but in this case we should use a different number, 1·609, to express their locational relationship. Similarly on a map representing the stations we measure length on a different scale, say A', such that $A' = f(A)$. We could decide to make $A' = \frac{1}{63360}(A)$ and so get a typical representative fraction, *RF*, of cartography.

We note that if we put the objects in order of the numerals we put them in the order of the quantity. The numbers are different, but the order is not. This is all quite acceptable because it is familiar and because the relationship between the scales is linear, as we can see in fig. 1.4(a). However *order* can be retained by relationships that are non-linear. For example we may take pairs of numbers such as $(X, \sqrt{X})$, (X, X^2), $(X, \log X)$ as in fig. 1.4 (b) (c) (d) as the functional relationship between the scales. When we use distance as our quantity for measuring location relationships we use a standard scale such as the metric or Imperial. In such a scale if we give a value to one object, A, of x units and of $2x$ units to another object, B, then we say that B is twice as far as A from a reference object.

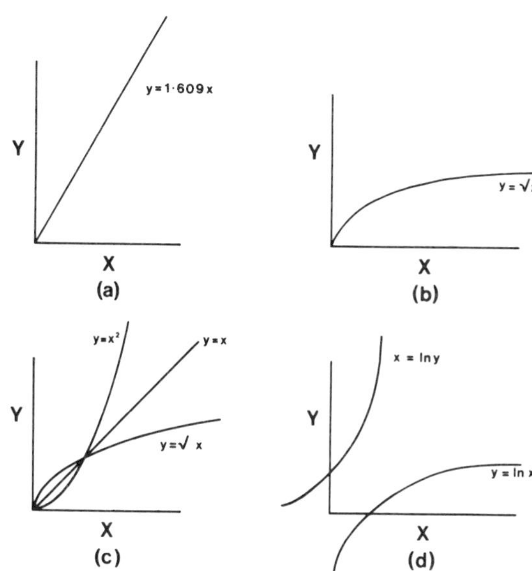

1.4 Location as a quantity on different scales.

This applies to all scales that are linear functions of our standard Imperial scale even though the numbers given to the objects are different. In such scales, magnitudes are additive. However, for scales which are non-linear, monotonic* functions of a standard scale, the order property is retained but magnitudes are not additive. A widely quoted example of such a scale in a geographic context is

* A function such that $f(A) < f(B)$ for $A < B$ is said to be monotone increasing; a function such that $f(A) > f(B)$ for $A > B$ is said to be monotone decreasing. A function which is either of these is said to be monotonic.

the one used by Hägerstrand (Hägerstrand 1957). Further consideration of this problem is available in Tobler's work (Tobler 1961).

This illustration is useful because it emphasizes that measurement is not necessarily to be seen as a question of counting units. For something to be measured as twice as far from another thing as a third thing is a reference to the relationship between the numeral assignments made on a particular scale. Now this seems to be exactly the situation in geography when we consider objects whose location relationships reflect human response. It is well established that human response to many stimuli—grey tones, area, time interval, value and length—are not linear but are monotonic transformations of some linear scale and consequently the order remains the same. If we accept that a quantity is defined by its order then we must accept that all three scales are scales for measuring the same quantity and are simply different scales. To the extent that distance is an acceptable component of location to characterize it satisfactorily for some assertions, then we accept that geographers can experiment with scales to find ones that give relations among the numeral assignments that correspond to the relations in human response. That being so we should find nothing unacceptable about a map such as Hägerstrand's because it could be argued that such a map shows location more accurately in terms of correspondence of numeral assignments and response than a map using an additive distance scale. This would then be an important conceptual step towards accepting that there is no proper set of location relations because the choice of measurement for location is arbitrary. The basis of the judgement for using one scale rather than another is then shifted to an argument of convenience. A scale could be convenient because it led to relationships between properties being discerned more readily than on a standard scale, or because it had arithmetically desirable properties. For many people it is both conceptually and arithmetically more convenient to work with scales that are additive like our standard length scales. Man seems to locate himself and his activities on non-linear scales. As there is an identity between unit and scale and as scale identifies quantity, maps show location with respect to a scale: this scale is usually additive. Non-linear scales corresponding to response seem sensible bases for map construction, but as we are concerned with response we must recognize that response can be modified by learning and numeral assignments may not be stable. Indeed the use of a response scale could yield different assignments to the same objects at different times or to similar objects at the same time but in different countries. These cases are contrary to the criteria for measurement, of rules that are determinative and non-degenerate. Cost scales raise this sort of problem in theories of the location of economic activities.

As we have changed the scale for measuring location, we could consider some of the implications of changing our quantity from distance to some other quantity name as an equally acceptable substitute for the composite property location. Human response to location of objects is a composite of various quantities each with its one-dimensional scale, because our judgements of good location are made on variables that are measured in one dimension such as time and value. Expressions such as time-distance or cost-distance, when cost is a specified scale for measuring value, are met frequently in geography. It is not clear immediately what such expressions mean. Time-distance and cost-distance could be quantity names for the property accessibility which we may suppose is a composite of the two properties time and distance or of value and distance. It is apparent that in the first case we give a property name, and a scale name in the second case. If time or cost scales are monotonic transformations of distance scales, then they all yield the same order and should be taken to be equivalent quantities. If they do not yield the same order then we are dealing with two quantities, time and distance, or value and distance for which the new property name, accessibility, may be recommended. Alternatively we may suppose that accessibility incorporates more of the sense of location than either quantity separately. Perhaps a more precise definition of the property and quantity involved would be a useful preliminary to studies of accessibility. On the argument that location is the name for the property of two-dimensional order relations, which is approximated by convenient, but inadequate, one-dimensional quantity scales, of which distance is the most common, we may expect all attempts to capture this two-dimensional order to be included under the general property name location, whether we invent other names for these subsets or not.

Whether cost and time orders of objects are the same as distance orders for those objects is an empirical question, not a question of logic. We may suppose that deciding whether two orders are the same is a straightforward matter in the sense that if the sequence of objects is identical then they are in the same order and we have the same quantity. However, when we are considering human response to a property such as distance, we do not expect that our judgements will give an order exactly the same as by some more objective procedure. We would not suppose that the occurrence of some inversions of order would imply that we were dealing with two distinct properties. A decision as to whether in any particular case there is evidence for one or two properties could be made on the basis of their similarity. Some of the ways of measuring relationships are discussed in article 3.3.2. Studies that are essentially correlations of scales are a common element of recent geography. The use of various one-dimensional scales as approximations to the two-dimensional location property is expected to yield strong correlations, but with some inversions in order (exercise 1.29).

Maps, then, are scales for measuring the property location of objects. They are usually two-dimensional scales with the same scale type and the same unit of measurement on both axes. The property location is a composite property and because it is usually considered in two dimensions it has no unique order relation. It can be measured on scales of all classes of restriction and within each class different sorts of scale form can be used. Maps are the bases of such measurement experiments. Other properties of the same objects can be measured on a map and there are conventions for accomplishing this increase in information (articles 2.1, 2.3, 2.5, 2.6). These scales are also subject to restrictions. These restrictions must be given due consideration in the interpretation of the objects' relationships. In order to increase the properties of the numerals assigned to the objects' attributes, restrictions are imposed on the measurement scales. This means that the composite-property location is characterized by a single quantity such as distance or angle and the measurements made of such a quantity carry an order relation on either a discrete or continuous scale. If the scale is continuous further restrictions are often assumed, and all of these restrictive assumptions must be recognized in testing the likelihood of an assertion made about the relationships depicted in the map.

1.4 Establishing conventions of map inference

We have argued that maps are scales for measuring the locational property of objects, although other attributes may be shown as well. Some maps are designed as data stores for locational information, but the maps that concern us in this book and which are used commonly by geographers have a more directed purpose than that of data storage. This purpose is implied by the discussion of (a) the relational nature of location and (b) experiments with numeral assignments on different scales for measuring this relation. Although the use of maps to depict objects and their properties is important, geographers use maps as the bases of assertions about those objects and their properties. These assertions arise from the problems that interest geographers and which relate to particular classes of problem and to the theories that have been established to provide frameworks for the study of such problems. Maps are used to stimulate or engender assertions as well as to provide evidence for judging whether a particular statement is acceptable. In order to reduce the risk of making an unjustifiable assertion as well as to increase the reliability of the judgement, certain conventions of map construction have been established. Rules and procedures to reduce bias have become an integral part of cartographic training with the result that maps conforming to such standards present a common, regulated basis for both assertion and judgement. Modifications to these procedures continue to be made, encouraged by the questions raised about the objectivity of maps as well as by the construction and analysis of maps on computers. Two important modifications are those of symbol density introduced as *graphical rational patterns*, *GRP* (Bachi 1968) and contour algorithms (Heap 1972). All such procedures designed to standardize map symbolism are intended to make inference less arbitrary.

The next logical development is to establish conventions of map inference which enable geographers to continue to use maps for their traditional and established purposes but with more objective criteria for judgement. I shall use the terms assertion and proposition as equivalent terms referring to a statement made about one or more properties of the objects depicted on a map. These statements arise from questions that interest geographers and the judgement refers to the acceptability of the assertion with respect to the particular map. A map is used as evidence for a proposition. Propositions are not always either true or false. They may be likely or unlikely and in terms of their likelihood they are judged to be acceptable or unacceptable. In other words we envisage putting assertions in order on the basis of their likelihood, and probability is the scale on which propositions are ordered. The need to make probability judgements on the basis of evidence is not confined to maps, but occurs so widely that many procedures have been devised. Such inductive procedures are part of the field of probability and statistics. Our use of probability is simply the introduction of a logical procedure to give greater precision and more objectivity to the measurement of the acceptability or likelihood of an assertion about a map. Statistical inference denotes a judgement about an assertion in terms of the likelihood of that assertion and the reliability of the inference. A proposition which is the subject of statistical inference is called an *hypothesis*, the inductive procedures used are termed *tests* and the judgement of the acceptability of the proposition is made as a result of applying a *test statistic* to a particular hypothesis. The outcome of the test is a number assigned on the probability scale and its value lies between zero for impossible and one for certainty. This number expresses the likelihood of the hypothesis under the conditions of the test statistic. The judgement, based on this number, is determined by the individual's willingness to make the assertion at that particular probability of its being correct.

Unlike the scales of measurement we have met so far, assignments on the probability scale cannot always be made unequivocally, because assignments made on the same scale by different inductive procedures do not result necessarily in the same value. The distinction between a proposition and a hypothesis illustrates that a proposition is put in a form suitable for being evaluated on the probability scale by a particular test statistic and the assumptions entailed by that procedure have been met. The tests are more specific than the proposition. Often there is a number of tests for the general idea contained in a proposition as we shall discover when we consider, for example, assertions about randomness or relationship, and the tests of these assertions put as specific hypotheses. Each test is constructed with respect to particular assumptions about the information that is available. The choice of the appropriate inductive procedure is made on the basis of the following considerations:

a) The sort of proposition.
b) The number of attributes or samples involved.
c) The measurement scale used to characterize the attributes.

These criteria are used in the classification of the test statistics used in this book. Each test is given its customary name and put in the category that seems most appropriate. A summary table is given for reference on the inside cover.

1.4.1 The proposition

A substantial proportion of the statements we wish to make about maps is covered by a comparatively small number of propositions. An assertion basic to many maps is that a particular property of objects is in some sense random. For example such an assertion could be made about the locations of the homes of undergraduate students at Birkbeck (fig. 2.14) or about the levels of pollution along the river Trent (fig. 3.13). Judging whether a proposition is acceptable in such cases depends on defining the idea of randomness more exactly as a hypothesis and in terms of a particular test statistic. The choice of the test statistic could be made as a result of specifying an alternative if the hypothesis of randomness is judged to be unacceptable. For example we might be more interested in any evidence of increasing pollution downstream rather than a simple indication of non-randomness in the sequence of values.

Propositions of independence and of association are combined in this classification. It seems sensible to put *measures* of association with *tests* of independence because it is possible to determine, on the basis of a test that leads us to reject a hypothesis of independence, that dependence is likely without providing an estimate of the association. Certainly many maps are described more satisfactorily by a measure of association linked to an assertion of dependence than by an assertion that merely rejects independence.

The third class of proposition refers to equality of location and to identity of distribution. Both terms are used in their statistical sense. *Distribution* refers to the frequency of all the values that occur in measuring the property that has given rise to the proposition. *Location* refers to a measure of an average value in this collection of all values; this average is usually the arithmetic mean or the median (article 1.8.2), which is the name given to the half-way value. Both terms have geographical connotations that we have used earlier, and some initial confusion is unavoidable between their geographical and their statistical usage. Both usages are so well established that replacement terms would cause even more confusion. An indication of the statistical usage is gained by reference to the map showing pollution measures along the river Trent at two separate times. Our assertion might be that pollution levels had not changed and we might base our acceptance of this assertion on a test that determined whether the two sets of values were essentially the same, or that their two distributions had a common median (article 1.8.1).

1.4.2 The attributes

In many maps assertions can be made about

a) location
b) location and one or more other properties
c) non-locational attributes.

We shall see that (c) is an inevitable consequence of certain types of map and such assertions form a natural part of geographic enquiry (article 2.6). As the number of attributes increases or as the number of sets of information about one attribute increases, so the amount of information to be accommodated in the test statistic increases. There is usually a distinction between test statistics that are designed for one set of measurements and those for two or more sets of measurements. Occasionally tests are designed for just two sets of measurements. The proposition is general to one or more attributes, although the hypothesis need not be.

1.4.3 The measurement

All the tests are directed to some property and this is characterized as a set of values resulting from a measuring procedure. These measurements are made on a scale which varies from the least restricted nominal scale, through the ordinal scales to the continuous scales. These restrictions to frequencies in category, to position in a sequence and to compactness in some interval are all involved explicitly in the derivation of the test statistic. Further restrictions can be placed on the measurements by specifying the form of the pool of all possible such measurements. This pool is usually referred to as the population of such measurements. Characteristics of the population are termed parameters; typical parameters are the mean, that is the arithmetic average, and the variance which describes the variability, or dispersion of the measurement values (article 1.8.2). The form of the population is usually used to refer to the shape of the graph of all such measurements when the vertical axis of the graph plots the probability of each measurement value, and the horizontal axis shows particular values. Plainly these are strong restrictions; if they can be met then test statistics based on these restrictions should be used because the values contain a great deal of information that is exploited by the test statistic. The appropriate test statistics are known as the t test statistic and the F test statistic. Together they form the basis of a large part of traditional statistical inference and are of great importance. They are discussed very well in many textbooks, together with their assumptions.

1.5 Measurement and probability judgements

Our probability judgements about a proposition, or hypothesis, are always made on the basis of the information available to us. This information is contained in three distributions:

a) The actual observations and their characteristics: a frequency distribution.
b) The population of all such observation-characteristic measurements: an assumed probability distribution.
c) The test statistic: a probability distribution.

The test statistic enables us to compare an actual result from (a) with the result under the specified conditions of (c). It is important to ensure as well as possible that these conditions are satisfied. The most important conditions relate to restrictions in (b). In general, the more stringent the assumptions the less often they are satisfied, consequently there is least certainty when assumptions are made about the form of the probability distribution of the measurements. It is usually possible to say that the population of measurements is continuous, as with distance measures of location, or that it is continuous in the interval that contains some specified parameter such as the median. If this is not the case then it may be possible to say that the measurements were made on an ordinal scale, or on a nominal scale. As the scales become less restricted the numerals carry less information and this reduction in information must be matched by changes in the assumptions of the test statistic. We can exercise control over the test statistics and these are devised to correspond to the various combinations shown in the summary classification. Test statistics also use information from (a); often the measures in (a) are summarized by straightforward sample-linked characteristics such as frequency, rank, or position in a sequence rather than by magnitudes. The test statistics refer to these sample-linked characteristics rather than to some infinite, unknown population of magnitudes.

1.5.1 Distribution-free and non-parametric test statistics

Tests which do not require the distribution of the population of measurements to be specified are termed *distribution-free* procedures. In this sense they are different from tests which require the distribution function to be known. These distribution-free procedures use characteristics of the observed measurements such that the distribution (c) of this observation-characteristic can be derived without the requirement that the distribution of the measurement population (b) is specified. A quantile (article 1.8.1) such as the median is often used as the appropriate characteristic of the sample. Similarly, position in a sequence or the frequency of some category provides the basis for deriving the distribution of the test statistic. Thus, distribution-free tests have classical, parametric analogues but with less restrictive assumptions. These assumptions usually refer to the sort of measuring scale that applies to the attribute.

The term *non-parametric* refers to tests for hypotheses which do not make a statement about the value of a parameter and these are strictly non-parametric tests. Such hypotheses are important in geography and particularly in maps for which we often wish to make an assertion of randomness or test for trend. Goodness-of-fit tests for hypotheses which result from an assertion about the form of the population of measurements from which the observed measurements are allegedly taken are important and arise often.

The probabilities we assign can vary because we have some choice over the procedures used and the hypotheses selected for a particular assertion. The probability assigned reflects our state of knowledge of the objects and their properties rather than being an apparently variable property of the objects.

The use of statistical procedures to formalize our inference is the natural step after map reading by visual estimation. Just as, when we inspect a map and make an assertion on the basis of our experience, our estimate about its properties may be specific or broad, we can also test a hypothesis against very broad alternatives or against much more restricted ones. Consider a map showing increment in income for farms in each of two location categories (fig. 2.75). For this map we assert that farm income increments in each category are similar (article 2.7). This assertion can be put as the hypothesis that the two distributions of income increment are identical, and this can be tested against the very broad alternative that they are not identical by using the Wald-Wolfowitz test (article 3.4.1) or the Kolmogorov-Smirnov test (article 3.4.4). The alternative could be specified more closely by choosing a test such as the Wilcoxon test (article 2.7.1) that is sensitive to unequal locations, which could lead us to assert that the median increment in one location class was greater than in the other location class. Further work may establish the form of the population of measurements and a test devised which is tailor-made to that population. The Wilcoxon test is particularly sensitive for the logistic distribution and this arises frequently in economic geography.

This illustration emphasizes that a population of measurements can be specified and result in a probability distribution other than the normal distribution. Test statistics appropriate to particular distributions are more sensitive to departures of a sample of actual values from the values expected to arise from that population. In some circumstances a particular non-normal population may be expected to be the relevant one, as in the case of angular measurements. This emphasis on test statistics based on non-normal distributions and on test statistics that are distribution-free is to be treated as a counterweight to the emphasis that customarily is given by geographers to the normal distribution as the appropriate probability distribution for test statistics. The great importance of the normal distribution becomes evident in more advanced and theoretical treatments of inference at which stage the use of the variance also becomes more apparent. There are many advantages in distribution-free test statistics for the conditions and the assertions typical of elementary geography.

Maps prepared in the elementary stages of geographic study often involve measurements and engender assertions that meet the conditions of distribution-free and non-parametric tests more readily than the conditions of classical inference. Test statistics are basic to statistical inference, and these test statistics are based on clear characteristics of the data and can be derived from first principles by using relatively few, straightforward counting rules and school algebra. This means that for a modest investment of effort the geography student can gain a real understanding of probability and inference which can be adapted to many more situations than are dealt with in this text. The ideas are a good basis for tackling normal theory of inference. The procedures are also fairly straightforward to calculate, and with increasing access to electronic calculating machines and to computer programming in undergraduate training, exact probabilities can be given to situations not covered by the published tables.

The remainder of this section establishes the basic material required to deal with the tests and the situations covered in this text. This material is then used to make the procedure of hypothesis-testing quite precise and as such it enables the ideas discussed above to be made formal and specific. Once we have established some basic statistical language we shall return to the task of applying probability to map assertions.

Non-parametric techniques are usually quick to apply and comparatively easy to understand. The properties of the test statistic can be grasped thoroughly because, in most cases, it is a discrete random variable with non-zero probabilities assigned to a finite number of points, and its exact sampling distribution can be determined by direct counting or by simple counting rules.

1.6 Counting rules for probability calculations

Let us consider the occupation of vacant sites in a shopping precinct in which we assume there are only 2 possible outcomes for shop type: food and non-food retail designated F and R respectively. Each time a shop type is occupied we have either an F or an R. If one vacant site is occupied then there are 2 possible outcomes: F or R. If 2 such sites are occupied then there are 4 possible outcomes: FF, FR; RF, RR; with FR meaning that the first site is taken by a food shop and the second by a non-food retail shop. With 3 sites there are 8 possible outcomes: FFF, FFR; FRF, FRR; RFF, RFR; RRF, RRR; in which there are 2 possible outcomes F, R for each of the 4 previous possible outcomes. This doubling of outcomes is general each time the number of sites increases by 1. If there are n possible sites for occupation by 1 of 2 types of shop there are 2^n possible outcomes.

A useful simplification of the idea of probability is to see probability as the ratio of the number of actual cases to the number of possible cases. The task of calculating a probability in a particular situation will often be seen as the problem of counting these cases. The word counting is used in precisely the same enumerative sense that we say 1, 2, 3, . . ., n, but to literally count each outcome separately would often be tedious and time-consuming. Thus we need some counting rules, just as we use multiplication tables to avoid enumeration in elementary arithmetic.

To be consistent with normal statistical terminology we should refer to the occupation of vacant sites as an *experiment* in which each particular site occupation is a *trial.* The possible outcomes of 1 trial, a number of trials or even the occupation of all the vacant sites in 1 shopping precinct, that is the experiment, are referred to as *events.* Of course there may be more than 2 outcomes. Let us extend the shop, site example to the case of 4 possible outcomes, food F, clothing C, household H, and the remaining unspecified retail R. Each time a site is occupied, 1 of these 4 outcomes occurs. Thus on the first trial there are 4 outcomes, on the second there are 4^2, that is 4 outcomes for each of the 4 outcomes of the first trial, on the third there are 4^3, 4 for each of the previous 4^2 outcomes, and so on. If we call the number of outcomes that are possible at each trial r, and the number of trials n, we can specify our first counting rule.

Counting rule 1
If there are r possible outcomes for each of n trials in an experiment then there are r^n possible outcomes of the experiment.

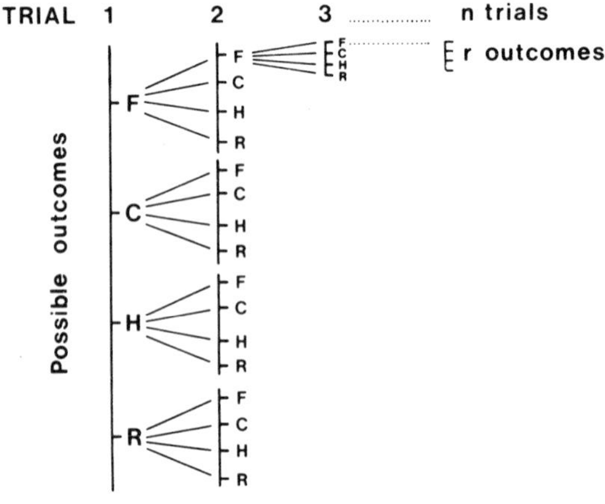

1.5 Counting rule 1.

To understand the second counting rule we can imagine that there are 6 possible sites along a street and that we have 6 distinct shop types to arrange in those sites.

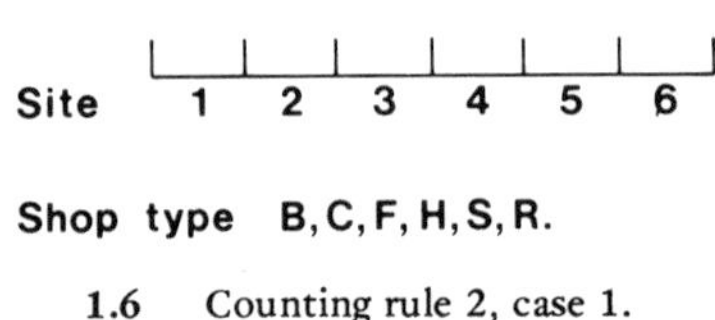

1.6 Counting rule 2, case 1.

Any 1 of the 6 types could occupy site 1, any 1 of the 5 remaining shop types could occupy site 2, and so on. There are 6 ways site 1 could be filled, but only 5 ways for site 2 once site 1 is filled, 4 ways for site 3 and so on until we come to the last site when only 1 type remains. In the case of 6 sites and 6 shop types there are $6 \times 5 \times 4 \times 3 \times 2 \times 1 = 720$ ways of allocating shop types to vacant sites. Let us replace 6 by n then we can write the multiplication as $n(n-1)(n-2)(n-3)(n-4)(n-5=1)$ and if we make n any integer then we can write $n(n-1)(n-2)\ldots(1)$ to mean we intend all the numbers to be multiplied in this fashion. This sort of multiplication is termed *n factorial* and is written as $n!$ We can now write our second counting rule.

Counting rule 2
There are $n!$ ways of arranging n objects in a row.

Such an arrangement along a line or in a time sequence is known as a *permutation*; although permutations are usually seen as linear arrangements they can be imagined in 2 or more dimensions.

If we had 2 food shops, 3 clothing and 1 electrical shop as our 6 shop types there would be 6! ways of arranging them, but those arrangements which simply exchanged the positions of the 2 food shops would not be distinguishable permutations.

F_1	F_2	C	C	C	E
F_2	F_1	C	C	C	E

1.7 Counting rule 2, case 2.

Thus the number of possible permutations is halved if we want to know how many distinguishable permutations for food shops there are, that is 6!/2! Similarly for each of these permutations the 3 indistinguishable clothing shops can be rearranged in 3! ways without affecting the appearance of the arrangement. And the electrical shop can only be rearranged in $1! = 1$ way. So although there are 6! possible arrangements of the 6 shop types the number of distinguishably different arrangements is 6!/2!3!1! giving

$$\frac{6.5.\overset{2}{\not4}.\not3.\not2.1}{\not2.1.\not3.\not2.1.1.} = 60 \text{ ways}$$

We can generalize this to give our third counting rule.

Counting rule 3
For n things of k different kinds with n_1 of the first kind, n_2 of the second kind and n_k of the kth kind there are $n!/n_1!n_2!\ldots n_k!$ distinguishable arrangements.

This sort of calculation occurs often in statistics and a special symbol is used for it. We write

$$\begin{bmatrix} n \\ n_i \end{bmatrix} = \frac{n!}{n_1!\,n_2!\ldots n_i!\ldots n_k!} \tag{1.1}$$

in which n is a positive integer and the n_i are integers such that $0 \leqslant n_i \leqslant n$ and $\sum_{i=1}^{k} n_i = n$. The numbers $\begin{bmatrix} n \\ n_i \end{bmatrix}$ are often called *multinomial coefficients.*

Counting rule 2 is a special case of counting rule 3 when there are as many kinds of object as there are objects so that there is only one object of each kind. As 1! = 1 we have

$$n!/1!1! \ldots 1! = n!$$

You are probably wondering how any two shops can be indistinguishable, when it is plain that J. Harrington the grocer gives personal service and has a very distinctively lettered sign advertising his shop, while F. Brighouse has a grocery self-service. For many investigations shops may be considered by what goods they sell rather than by how they sell them. The fineness of the classification will depend on the purpose of the study. In some cases the distinction between any retail shop may be considered unnecessary because our attention is directed towards detecting any pattern in the incidence of retail, business and services or some such classification. It is in this sense that the word indistinguishable is used.

When just two kinds of object are distinguished, counting rule 3 has a special designation. We write

$$\binom{n}{r} = \frac{n!}{r!\,(n-r)!}$$

in which n is a positive integer and r is an integer such that $0 \leqslant r \leqslant n$. The numbers $\binom{n}{r}$ are often called *binomial coefficients* because they arise as coefficients in the expansion of the binomial expression $(a + b)^n$. If we revert to the earlier distinction of shop types into just 2 kinds, food F, and non-food R, then the arrangement of 6 shops in 6 sites leads to $n!/n_1!n_2!$ distinguishably different arrangements. For fig. 1.7 this is

$$6!/2!4! = 15$$

But as $n = n_1 + n_2$ we can put $n_2 = n - n_1$ and rewrite the equation as

$$n!/n_1!\,(n - n_1)!$$

When only two kinds of attribute are distinguished we have

$$\begin{bmatrix} n \\ n_i \end{bmatrix} = \binom{n}{n_1} = \frac{n!}{n_1!\,(n-n_1)!} = \binom{n}{n-n_1}$$

as the standard symbolism. Sometimes n_1 is replaced by a different letter such as r giving $\binom{n}{r} = \binom{n}{n-r}$. This symbolism is used for the binomial coefficient in this text and in most other texts.

This rule is often used as the rule for counting the number of ways of taking n things r at a time. Suppose we have n objects arranged in a row and we have to take r of them and leave $(n - r)$ of them. Then this situation is just the same as having to arrange the r things in the n possible sites and the $n - r$ things in the remaining $n - r$ sites. This is equivalent to arranging $n - r$ things and then putting the r things in the remaining sites. These three counting rules are enough for progress to the next stage. They will be used throughout the text and some elaborations developed as needed.

1.7 Counting and the probability of an event

In order to demonstrate the use of a counting rule and the idea of probability, let us suppose that there are 6 vacant sites each of which is available to be occupied by a food or non-food retail shop. We are assuming that each site occupation type

SAMPLE SPACE

Points in Sample Space

T_1	T_2	T_3	T_4	T_5	T_6	Event	Counting Rule	Number of Points in Event	
F	F	F	F	F	F	0 R's	$\binom{6}{0}$	1	0·015625
						1 R	$\binom{6}{1}$	6	0·09375
						2 R's	$\binom{6}{2}$	15	0·234375
						3 R's	$\binom{6}{3}$	20	0·3125
						4 R's	$\binom{6}{4}$	15	0·234375
						5 R's	$\binom{6}{5}$	6	0·09375
R	R	R	R	R	R	6 R's	$\binom{6}{6}$	1	0·015625

1.8 The probability of points in a sample space.

is independent of the others and that each type of shop is equally likely. By independent we understand that if any site is occupied by an F this does not affect the likelihood of an adjacent site's being occupied by an F or an R. There are 2 outcomes for each of 6 sites, giving 64 possible outcomes altogether (fig. 1.8). The total number of possible outcomes is known as the *sample space.* Each outcome is 1 point in that sample space. The sequence *FRFFFR* is *1* such outcome. We might be interested in finding out how often just 2 non-food retail shops would occur if there was a 50-50 chance that each site would be an R. In other words we have defined a particular collection of the possible outcomes, or, as it is often expressed, of the points in the sample space. In this particular case we have defined the collection containing $2R$, $4F$. This is our *event.* In general an event is any collection of points in the sample space. An event can be just 1 point, or it can be many points. We can systematically enumerate all of the 64 outcomes as in fig. 1.8. We can attach a number to each outcome. In this case our assumptions lead us to suppose that each of those 64 outcomes is as likely to occur as any other and so we suppose that the likelihood of any outcome is 1/64. Our event contains more than 1 outcome. We can count these from fig. 1.8 and we find that there are 15 outcomes with exactly 2 R's. We say that the probability of an event is the sum of probabilities of all the outcomes that constitute that event, thus the probability of the event *exactly 2 R's* is $15 \cdot \frac{1}{64} = 0{\cdot}234375$, and is quite likely. If we had defined our event as *not more than 2 R's*, that is $R \leqslant 2$, then we should have had to include the 6 outcomes of 1 R, and the 1 outcome of zero R's, giving $15 + 6 + 1 = 22 \cdot \frac{1}{64} = 0{\cdot}34375$ as the probability of the event. This event is even more likely under our assumptions than the event exactly 2 R's. Please note that the outcome *FFFFFF* is neither more nor less likely than *FRFFFR*. If our event had been *all food* or if it had been food shops in sites 1, 3, 4, 5, their probability would have been $\frac{1}{64}$. As our event was less restricted and simply specified 2 R's out of 6 sites, this event is composed of 15 outcomes each equally likely.

For large numbers of outcomes direct enumeration is tedious and error prone. We can get the same result by using counting rule 1 and counting rule 3. We use counting rule 1 to show that there are $2^6 = 64$ outcomes. Our event is $R = 2$ from $n = 6$. Counting rule 3 shows that the number of ways of getting 2 things from 6 things is

$$\begin{aligned} \binom{6}{2} &= \frac{6!}{2!(6-2)!} \\ &= \frac{6.5.4.3.2.1}{2.1.4.3.2.1} \\ &= 15 \text{ ways.} \end{aligned}$$

So our event comprises 15 of the 64 points in the sample space. Similarly, if we wish to count all the points in the event $R \leqslant 2$, then we form the sum from counting rule 3 applied to $R = 0$, $R = 1$, $R = 2$, and denote the fact that we are summing by the use of the Greek letter capital S, pronounced sigma and written Σ. Sigma has subscripts and superscripts to indicate over what interval the sum is to be formed.

$$\sum_{r=0}^{2}\binom{n}{r}\cdot p = \sum_{r=0}^{2}\binom{6}{r}\frac{1}{64}$$

$$= \binom{6}{0}\cdot\frac{1}{64} + \binom{6}{1}\cdot\frac{1}{64} + \binom{6}{2}\cdot\frac{1}{64}$$

$$= \frac{22}{64}$$

We assumed that each outcome was equally likely and gave a probability to each outcome. As there were 64 outcomes the probability we gave to each outcome was $p = 1/64$.

1.8 Random variables

A sample space consists of all the possible outcomes of the property of the objects which is being considered. These outcomes do not have to be numbers. In the shop-site, shop-type example the outcomes are food shops, F, or non-food shops, R. In other geographic situations non-numerical outcomes are common. In studying plant species or factories we might record their presence or absence, whether they survive or die over a given time. Again in observing water flow over a flume we may keep a graphical record and note the pattern of diurnal flow. Visual records such as air photographs are a common source of classificatory information. In all these cases numbers can replace the qualitative record. For example we can give a 1 to survival, a 0 to failure to survive; we can record the maximum and minimum flow off the trace in m^3/sec.

In the shop-site, shop-type example our interest was in the frequency of non-food retail uses, R. The aspect of our sample space that concerns us is the number of outcomes that yield 0 R, 1 R, 2 R up to 6 R. In a sense we have defined a new sample space which consists of numbers given to points in the first sample space. When we give numbers to the points in a sample space we are defining a *random variable.*

Definition. A random variable is a function which gives real numbers to every point in a sample space.

Random variables are denoted by capital letters such as R, X, Y, Z and the real numbers assigned to them are denoted by their lower case equivalents r, x, y, z. In the shop-site, shop-type example the outcomes F, R are not a random variable, but the *number* of non-food retail uses is. Let us designate this random variable as R and the value it takes as r, then $R = r$ is an event in the sample space, it is the event that contains all the outcomes which have the numeral assignment of r. If $r = 2$, then in our example the event contains all the 15 outcomes of rows 8 to 22 in fig. 1.8.

Suppose we record whether days are wet, W, or dry, D. Then for 2 days of observation our sample space consists of the following outcomes:

$$S = \{W, W; W, D; D, W; D, D\}$$

We could define our random variable, X, to be the number of wet days in 2 observations. In this case $X(W, W) = 2$, $X(W, D) = X(D, W) = 1$, $X(D, D) = 0$, and we note that more than 1 outcome can have the same value of the random variable, X. It is the *values* of the random variable that are of interest and the function that specifies the relationship between the outcomes and the numbers is often not discussed. The values often arise directly as the result of a measurement. In the pollution study (article 3.2.1) the chemical evaluation of ammoniacal nitrogen per unit volume determines the random variable directly on a continuous scale. Similarly the water flow across a flume taken off a trace defines the random variable directly again on a continuous scale.

Random variables are not confined to one dimension. Situations where two-dimensional or higher-dimensional random variables arise are common. For example we may evaluate water temperature and the calcium carbonate in solution at various sites along a river. We may measure location with respect to a market and the production capacity of factories.

There can also be more than 1 outcome that is interesting from 1 situation. Consider a map showing the configuration of homes of undergraduate students at Birkbeck College (fig. 2.14). We might be interested in describing the neighbourhood of each student's home or of why the student decided to study at that college. These responses could be converted to appropriate random variables by giving numbers to categories and an event defined on the sample space. Alternatively we might be interested in some numerical characteristics of the map. First we could define our random variable, X, as the number of students whose homes were in any small area of the total map. Secondly we could define Y as the distance between each student's home and its nearest neighbour. Thirdly we could define the minimum distance of a home from an underground station as our random variable Z.

When the random variable can take integer values only, then it is called a *discrete random variable.* In other words it is possible to list the values it can take as $x_1, x_2, x_3, \ldots, x_n$. Plainly the shop-type example is a discrete random variable, so is the example of the number of student homes, because we can list the values that R can take. That is, $R = r$ for $r = 0, 1, 2, \ldots, 6$ or $X = x$ for $x = 0, 1, 2, \ldots, n$ when n is the total number of homes shown on the map (fig. 2.14). When a random variable can take values that are distributed continuously in an interval it is said to be a *continuous random variable.*

1.8.1 Probability distributions

In the shop-type example we assumed that each of the 64 points in the sample space was equally likely and this led us to assign a probability of 1/64 to each outcome. Probabilities are always assigned to every point in the sample space. The random variable also assigns a number to every point in the sample space. Once the probabilities have been assigned to the outcomes in the sample space and once the random variable has been defined on that space, the probability of the random variable, X, is fixed and gives the probability of X taking any value, x_i. For each possible outcome, x_i, a number $p(x_i) = P(X = x_i)$ exists which is called the probability of x_i. These numbers, $p(x_i)$ for $i = 1, 2, \ldots$ must satisfy the conditions of the probability scale that, first, the probability of each outcome must not be less than zero, that is $p(x_i) \geqslant 0$, and, secondly, that the sum of the probabilities of all the outcomes must be 1, that is $\sum_i p(x_i) = 1$. The collection of probabilities over all values, x_i, is termed the *probability distribution of X*. The probability distribution of a discrete random variable is usually written as

$$P_X(x) = P(X = x_i)$$

If the random variable consists of measurements on a continuous scale we say it is a *continuous random variable.* If our random variable is continuous we cannot speak of the ith value of X because the number of values in any interval is not countable and $p(x_i)$ is meaningless. In this case we replace the function $P_X(x)$ defined for $x_1, x_2, \ldots$ by a function f defined for all values x in some interval of measurement. The properties $p(x_i) \geqslant 0$ and $\sum_i p(x_i) = 1$ are replaced by $f(x) \geqslant 0$ and $\int_{-\infty}^{+\infty} f(x)\,dx = 1$, and for any two values a and b of the random variable we define $P(a \leqslant x \leqslant b) = \int_a^b f(x)\,dx$. The sign $\int$ means add-up and is the equivalent for continuous random variables of Σ for discrete ones. In the case of continuous random variables we talk of probability in an interval and make that interval as small as we like. For a continuous random variable the probability distribution is continuous or dense and for this reason it is referred to as the *probability density function* or *pdf.*

The probability distribution for our shop-type example is shown in fig. 1.9. The horizontal axis shows the values that the random variable can take and the vertical axis shows the probability of each of those values. These probabilities and the values of R correspond to the values shown in fig. 1.8.

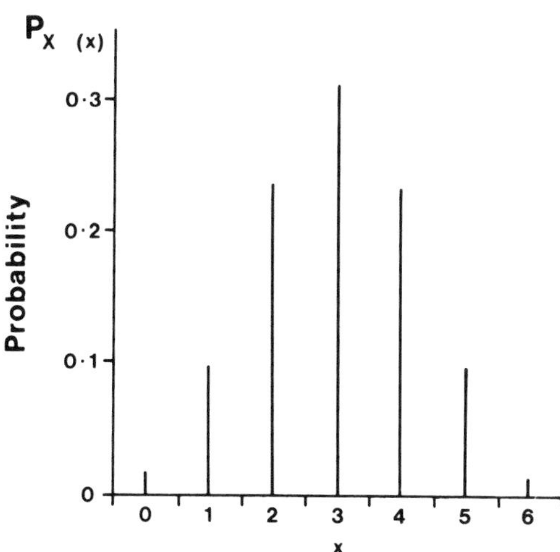

1.9 Probability distribution function of a discrete random variable.

1.8.1.1 Cumulative distribution function

The probabilities in a probability distribution can be summed consecutively for discrete random variables or integrated for continuous random variables to give the cumulative distribution function, *cdf*, of the random variable. This is denoted by

$$F_X(x) = P(X \leqslant x)$$

It is often useful to depict $F_X(x)$ as a graph with the cumulative probability as the vertical axis and the values of the random variable X along the horizontal axis. A random variable is said to be continuous if its *cdf* is continuous and to be discrete if its *cdf* is discrete. Fig. 1.10 is the *cdf* of the values given in fig. 1.9. It is a discrete *cdf* in which the separate steps have been joined to illustrate the accumulation of probabilities and to emphasize that no change in probability occurs between certain pairs of x_i. The change in probability occurs at $x = 2$ and $x = 3$ but not between 2 and 3. The *cdf* illustrates graphically what is meant by the probability that $X \leqslant x$ at a particular value x_i.

The vertical axis can be divided at any value, α, between 0 and 1, such that 100α per cent of the interval lies above it and $100\,(1-\alpha)$ per cent lies below it. For example we may choose $\alpha = 0{\cdot}75$. By continuing a line through α parallel to the horizontal we cut the graph of $F_X(x)$. The value of X that corresponds to this point is such that 100α of the X values are larger than it and $100\,(1-\alpha)$ of them are as small or smaller. In the case of continuous random variables there is a *unique* correspondence between probability values and X values whereas in discrete random variables this may not be the case. This is illustrated in fig. 1.10.

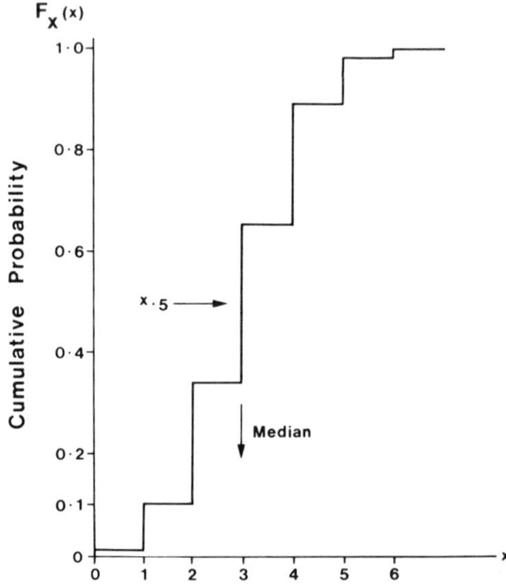

1.10 Cumulative distribution function of a discrete random variable.

We are usually interested in particular values of the *cdf.* The *cdf* can be divided at any value on the vertical axis; these divisions are termed *quantiles.* Some quantiles are commonly used and have been give special names. The 0·5 quantile is termed the *median*, the 0·25 and 0·75 quantiles are called respectively the *lower* and *upper quartiles.* The division of the *cdf* into hundredths gives *percentiles* and the median is often termed the fiftieth percentile. Quantiles can be any fraction of the $F_X(x)$ axis of the *cdf.* The value, x, of the random variable X that corresponds to a particular quantile is referred to as the pth quantile, x_p, of that random variable. In some cases in discrete random variables more than one number can correspond to the pth quantile and the usual convention is to take the average of the highest and lowest of these numbers as the value of the quantile. In the case of fig. 1.10 the median, 0·5 quantile, $x_{.5}$, is 3 because it intersects the vertical line. The $x_{.89}$ quantile is any value from $x = 4$ to $x = 5$ and so the convention would put it as $x_{.89} = (4 + 5)/2 = 4{\cdot}5$.

1.8.1.2 A binomial random variable

Most of the procedures in this text rely on certain well-known probability distributions which have been given special names. We have been using in our example so far one very well-known probability distribution called the *binomial distribution.* That is the probability distribution of a binomial random variable X. We shall use our example to show how the definitions of probability density functions and cumulative distribution functions can be interpreted in relation to this problem and so illustrate how probability statements are expressed.

The binomial distribution is the probability distribution given by the probability distribution function.

$$P_X(x) = P(X = x) = \binom{n}{x} p^x q^{n-x} \qquad x = 0, 1, 2, \ldots, n \qquad (1.8.1)$$

for the random variable X. Let our random variable be the number of non-food shops which we have designated as R. Let the probability, p, of R equal the probability, $q = (1 - p)$ of F so $p = q = 0{\cdot}5$. Then for n independent trials with each trial resulting in either R or F we have

$$P(R = r) = \binom{n}{r} p^r q^{n-r} \qquad (1.8.2)$$

We have just replaced the general designation X by the specific designation R. For $r = 2$ in 6 trials we have

$$\begin{aligned} P(R = r) &= \binom{6}{2} (0{\cdot}5)^2.(0{\cdot}5)^4 \\ &= 15\,(0{\cdot}5)^6 \quad \text{(when } p = q \text{ we can add indices)} \\ &= 15\,(0{\cdot}015625) \\ &= 0{\cdot}234375 \end{aligned}$$

If we were concerned to find the probability that the number of non-food retail shops would be not more than 2 in 6 trials then we are referring to the *cdf.*

$$F_X(x) = P(X \leqslant x) = \sum_{i \leqslant x} \binom{n}{i} p^i q^{n-i} \quad i = 0, 1, 2, \ldots, x$$

giving in this case*

$$F_R(r) = P(R \leqslant r) = \sum_{i \leqslant r} \binom{n}{i} p^i q^{n-i} \quad i = 0, 1, 2,$$

$$= \binom{6}{0} 0{\cdot}5^0 . 0{\cdot}5^6 + \binom{6}{1} 0{\cdot}5^1 . 0{\cdot}5^5 + \binom{6}{2} 0{\cdot}5^2 . 0{\cdot}5^4$$

$$= 0{\cdot}34375$$

Once we have defined our interest to be in this random variable rather than in particular outcomes we are able to appreciate the elegance of this compact form for expressing and calculating the probabilities of the event contained in an assertion. The distribution of a binomial random variable is just one of a number of extremely important discrete random variables which arise in a great many situations. These distributions are the basis of the test statistics referred to throughout this text. We shall continue to use this distribution to illustrate other important notions of inference that occur throughout the subject of probability.

1.8.2 Characteristics of random variables

1.8.2.1 The expected value

In the shop-type example the chance that each site is occupied by an F is the same as the chance that it will be occupied by an R; consequently I imagine that most people would expect to find half the 6 sites occupied by food shops and half by non-food retail shops. Intuitively the number of shops of type R expected out of n sites is the product of the number of sites (trials), n, and the probability, p, of that outcome at each trial. With $n = 6, p = 0{\cdot}5$, the expected value is $np = 3$. As the probability of $F = (1 - p) = 0{\cdot}5 = q$ the expected value of F is also 3. This is indeed the case and a slightly more formal version of this approach is often used to prove that the expected value of a binomial random variable is np. However some people find difficulty in relating this result to the more involved expression of the binomial distribution function. We shall continue with this example to illustrate the general definition of the expected value of a discrete random variable and then use this definition to prove that the expected value of a binomial random variable is np.

The number of non-food retail shops, R, is the random variable. The number of times $r = 0, 1, 2, \ldots, n$ varies and the rule for this number is given by the binomial coefficient $\binom{n}{r}$ which gives us the frequency with which any particular value, r, of R will occur out of n trials. Each of these events has a separate probability and fig. 1.8 gives the probability of the events that R has a value 0, 1, 2, 3, 4, 5, 6.

* The convention that $0! = 1$ is used in this book. Also a number raised to the power 0 is one; a number raised to the power 1 is the number itself. Also see article 2.4.1.

X	$P_X(x)$	$P_X(x).x$	$P_X(x)$		$P_X(x).x$
0	$p^0 q^n = q^n$	0	$0{\cdot}5^0 . 0{\cdot}5^6$	$= 1.(0{\cdot}015625)$	$1.0{\cdot}5^6 .0 = 0.0{\cdot}5^6$
1	$p^1 n.q^{n-1}$	$p^1 n.q^{n-1}$	$0{\cdot}5^1\ 6.0{\cdot}5^5$	$= 6.0{\cdot}5^6$	$6.0{\cdot}5^6 .1 = 6.0{\cdot}5^6$
2	$p^2 \frac{n(n-1)}{1.2} .q^{n-2}$	$p^2 n(n-1)q^{n-2}$	$0{\cdot}5^2 \frac{6.5}{1.2} . 0{\cdot}5^4$	$= 15.0{\cdot}5^6$	$15.0{\cdot}5^6 .2 = 30.0{\cdot}5^6$
3	$p^3 \frac{n(n-1)(n-2)}{1.2.3} .q^{n-3}$	$p^3 \frac{n(n-1)(n-2)}{1.2} .q^{n-3}$	$0{\cdot}5^3 \frac{6.5.4}{1.2.3} . 0{\cdot}5^3$	$= 20.0{\cdot}5^6$	$20.0{\cdot}5^6 .3 = 60.0{\cdot}5^6$
4	$p^4 \frac{n(n-1)(n-2)(n-3)}{1.2.3.4.} .q^{n-4}$	$p^4 \frac{n(n-1)(n-2)(n-3)}{1.2.3} .q^{n-4}$	$0{\cdot}5^4 \frac{6.5.4.3}{1.2.3.4} . 0{\cdot}5^2$	$= 15.0{\cdot}5^6$	$15.0{\cdot}5^6 .4 = 60.0{\cdot}5^6$
5	$p^5 \frac{n(n-1)(n-2)(n-3)(n-4)}{1.2.3.4.5} .q^{n-5}$	$p^5 \frac{n(n-1)(n-2)(n-3)(n-4)}{1.2.3.4} .q^{n-5}$	$0{\cdot}5^5 \frac{6.5.4.3.2}{1.2.3.4.5} . 0{\cdot}5^1$	$= 6.0{\cdot}5^6$	$6.0{\cdot}5^6 .5 = 30.0{\cdot}5^6$
n=6	$p^6 q^0 = p^n$	$p^n \quad n$	$0{\cdot}5^6 \frac{6.5.4.3.2.1}{1.2.3.4.5.6} . 0{\cdot}5^0$	$= 1.0{\cdot}5^6$	$1.0{\cdot}5^6 .6 = 6.0{\cdot}5^6$

To find $E(X)$ we add the terms in $P_X(x).x$, by definition. We notice that n and p occur in all but the $x = 0$ row of $P_X(x).x$, and may be factored out of the sum, giving:

$$np\ [q^{n-1} + p^1 (n-1) q^{n-2} + p^2 \frac{(n-1)(n-2)}{1.2} q^{n-3} + p^3 \frac{(n-1)(n-2)(n-3)}{1.2.3} q^{n-4} + p^4 \frac{(n-1)(n-2)(n-3)(n-4)}{1.2.3.4} .q^{n-5} + p^{n-1}]$$

From column 3 you can see that we have taken out each n, then we have taken out a p from each term, reducing the exponent by 1 in each term, thus p^4 became p^3 and so on.
The terms in [] result from multiplying $(p+q)$ by itself $n-1$ times, i.e. $(p+q)^{n-1}$. As $q = 1-p$, $p+q = p + 1 - p = 1$, and 1 to any power is still 1. Thus $np\ [1]^{n-1}$ is $np(1)$, is np.

1.11 The expected value of a binomial random variable, showing the algebra and the arithmetic of its calculation.

The sum of these probabilities is 1. We can conceive the expected value as the sum of the product of the value of R and the probability of that value. The sum is taken over all values that R can take. In the shop example we have the values given in fig. 1.11.

Definition. Let R be a discrete random variable with possible values $r_1, r_2, \ldots, r_n$. Let $p(r_i) = P(R = r_i)$, $i = 0, 1, 2, \ldots, n$. Then the *expected value* of R is denoted by $E(R)$ and is defined as

$$E(R) = \sum_{i=0}^{n} r_i . p(r_i)$$

This value is also called the *mean value* of R, the symbol, μ, is used for the population mean, and the symbol $\bar{x}$ or $\bar{r}$ is used for the mean of some actual values.

In the above case the expected value equals 3 and is the same as our intuition led us to suppose. The mean is seen to be the central value of R in this particular case, and in general when the probability distribution is symmetric the mean is the central value. The mean is referred to as a measure of location, and as a measure of central tendency. The mean and the median are the two most common measures of statistical location.

1.8.2.2 Expected value of a binomial random variable

When the probability function involved is the binomial distribution the expected value is

$$E(X) = \sum_{i=0}^{n} x_i p(x_i) \qquad (1.8.3)$$

$$\text{with } p(x_i) = P(X = x_i) = P_X(x_i) = \binom{n}{i} p^i q^{n-i}$$

$$\text{Then } \sum_{i=1}^{n} p(x_i) . x_i = \sum_{i=0}^{n} \binom{n}{i} p^i q^{n-i} . x_i$$

$$= \sum_{i=0}^{n} \frac{n!}{x_i!\,(n-x_i)!} . p^i . q^{n-i} . x_i$$

We wish to show that the expected value, $E(X)$, of a binomial random variable is np. A table helps to make this clear (see fig. 1.11). When $x_i = 0$ there is a zero as the first product term. The remaining terms are given in fig. 1.11 for the general and the specific numerical case. From this figure it can be seen that the $x_i = 1, 2, \ldots, n$ terms are the binomial expansion of $(p+q)^{n-1}$ and since $p+q = 1$, this expansion sums to 1. Thus the sum of $P_X(x_i)\, x_i$ is then $np\,(p+q)^{n-1} = np$ proving for the general case that the mean of the binomial distribution is what we expected intuitively.

$E(X)$ is a number, parameter, associated with a theoretical probability distribution. The importance of such parameters lies in their ability to characterize these theoretical probability distributions completely. The mean and the variance (see below) are used to characterize a great many distributions and this accounts for their importance. Often the parameters of a particular distribution can be used to convert the unknown exact probabilities to the known probabilities of another distribution such as the normal distribution. Generally we don't know $E(X)$ as we do not have the whole population of values available. All we have is some observations, $x_1, x_2, \ldots, x_n$. We are familiar with the idea of an arithmetic average or mean value of a set of observations. The mean is denoted by $\bar{x}$ and defined as

$$\bar{x} = \frac{1}{n} \sum_{i=1}^{n} x_i$$

This mean gives an idea of the value of the unknown expected value $E(X)$ of the probability distribution. As the number of observations increases the mean becomes more and more similar to $E(X)$.

1.8.2.3 Variance of a random variable

Another important expected value is known as the variance. It is written as $Var(X)$ or $V(X)$ and defined as

$$Var(X) = E[X - E(X)]^2 \tag{1.8.4}$$

This expression shows how the variance is related to the expected value, $E(X)$. Alternative, but equivalent, expressions are often met. They can be derived by expanding the definition.

$$= E[X - E(X)]^2 = E\{X^2 - 2XE(X) + [E(X)]^2\}$$

The expected value of $E(X)$ is $E(X)$, because $E(X)$ is a constant like the 2.

$$= E(X^2) - 2[E(X)E(X)] + [E(X)]^2$$

$$= E(X^2) - 2[E(X)]^2 + [E(X)]^2$$

$$= E(X^2) - [E(X)]^2$$

Again for a set of observations, $x_1, x_2, \ldots, x_n$ there is an empirical calculation rule for the variance and it is defined as

$$Var(X) = \frac{1}{n} \sum_i (x_i - \bar{x})^2$$

which also has the alternative form

$$Var(X) = \frac{1}{n} \sum_i x_i^2 - (\bar{x})^2$$

The alternative form is useful in computation because it avoids calculating $(x_i - \bar{x})^2$ for each value of the random variable X. The variance of the binomial distribution is $np(1-p)$. The variance is a measure of the variability of the observed values about their mean (see p.241).

Example 1

The Meteorological Office classifies the type of sky visible in terms of the degree of cloudiness. It is usual to distinguish 11 categories, 0, 1, 2, . . ., 10, where 0 means no cloud, and 10 means no sky visible. Suppose such a categorization is made at a weather station over a given period. Let X be the random variable, taking the values 0, 1, . . ., 10. The frequency distribution of cloudiness is typically U-shaped. Suppose that $P_X(x)$ is

$$p_0 = p_{10} = 0{\cdot}20$$
$$p_1 = p_2 = p_8 = p_9 = 0{\cdot}10$$
$$p_3 = p_4 = p_5 = p_6 = p_7 = 0.04$$

Then
$$\begin{aligned} E(X) &= 0(0{\cdot}20) + 1(0{\cdot}10) + 2(0{\cdot}10) + 3(0{\cdot}04) + 4(0{\cdot}04) \\ &\quad + 5(0{\cdot}04) + 6(0{\cdot}04) + 7(0{\cdot}04) + 8(0{\cdot}10) \\ &\quad + 9(0{\cdot}10) + 10(0{\cdot}20) \\ &= 5{\cdot}0 \end{aligned}$$

And
$$\begin{aligned} E(X^2) &= 0^2(0{\cdot}20) + 1^2(0{\cdot}10) + 2^2(0{\cdot}10) + 3^2(0{\cdot}04) + \\ &\quad 4^2(0{\cdot}04) + 5^2(0{\cdot}04) + 6^2(0{\cdot}04) + 7^2(0{\cdot}04) + \\ &\quad 8^2(0{\cdot}10) + 9^2(0{\cdot}10) + 10^2(0{\cdot}20) \\ &= 40{\cdot}4 \end{aligned}$$

Consequently $\quad Var(X) = E(X^2) - [E(X)]^2 = 40{\cdot}4 - 25{\cdot}0 = 15{\cdot}4$

Example 2

Suppose we have a set of observations from a random sample of 10 parliamentary constituencies in 1966. Let our random variable, X, be the percentage vote polled for the Labour Party.

$$X: 27{\cdot}9,\ 35{\cdot}4,\ 37{\cdot}6,\ 44{\cdot}4,\ 44{\cdot}4,\ 45{\cdot}1,\ 50{\cdot}5,\ 55{\cdot}2,\ 59{\cdot}5,\ 65{\cdot}9$$

Then
$$\begin{aligned} \bar{x} &= \frac{1}{n}\,\Sigma\, x_i \\ &= \frac{1}{10}\,.\,(27{\cdot}9 + 35{\cdot}4 + \ \ldots\ + 65{\cdot}9) \\ &= \frac{1}{10}\,.\,465{\cdot}9 \\ &= 46{\cdot}59 \end{aligned}$$

And
$$\begin{aligned} Var(X) &= \frac{1}{n}\,\Sigma\,(x_i - \bar{x})^2 \\ &= \frac{1}{10}\,.\,[(27{\cdot}9 - 46{\cdot}59)^2 + (35{\cdot}4 - 46{\cdot}59)^2 + \ldots + (65{\cdot}9 - 46{\cdot}59)^2] \\ &= \frac{1}{10}\,.\,1196{\cdot}129 \\ &= 119{\cdot}6129 \end{aligned}$$

The standard deviation is $\sqrt{Var(X)} = 10{\cdot}937$.

Equivalently
$$\begin{aligned} Var(X) &= \frac{1}{n}\,\Sigma\, x_i^2 - (\bar{x})^2 \\ &= \frac{1}{10}\,22902{\cdot}41 - 46{\cdot}59^2 \\ &= 119{\cdot}6129 \end{aligned}$$

The population mean is $\mu = 49{\cdot}3$, the population standard deviation is $\sigma = 16{\cdot}152$.

The parameters $E(X)$ and $Var(X)$ measure particular characteristics of the distribution of one-dimensional random variables. When we have a two-dimensional random variable, (X, Y), we can calculate these expected values for each variable separately, but it is often of interest to determine whether there is some sort of relationship between the two random variables. The notion of relatedness can be interpreted in a variety of ways. Two parameters which are used very often as measures of association are defined in the following way.

Covariance
$$\begin{aligned} Cov(X, Y) &= E\{[X - E(X)]\,[Y - E(Y)]\} \\ &= E(X, Y) - E(X)E(Y) \qquad (1.8.5) \end{aligned}$$

Correlation
$$\begin{aligned} Corr(X, Y) &= \rho\,(X, Y) \\ &= \frac{Cov\,(X, Y)}{\sqrt{V(X)\,V(Y)}} \qquad (1.8.6) \end{aligned}$$

The correlation coefficient, $\rho\,(X,Y)$, is discussed in article 3.3.2. It is interesting to note how the simple idea of an expected value can be elaborated to cover four such important measures, which are so widely used in statistical inference.

1.8.3 Population and sample

These terms have been used freely and it is now time to try to define them more precisely and illustrate the ways in which they are often used in geography. The term *population* denotes a collection of all the separate objects under consideration. These objects are identified by at least one attribute that they have in common. The objects in any population have properties other than the particular one or ones being investigated, and so the same objects can be included in more than one population and their inclusion depends on the attributes being studied. The population may be farms of a given size, homes of undergraduate students, or visitors to national parks. In these cases the number of objects in the population is either too great for us to consider each one or it is not practical to do so, and yet we may wish to make some inference about the population on the basis of a portion of the population.

The idea of a sample is easy to grasp. A *sample* is just a portion of the population. In most cases a sample involves selection and the selection is made deliberately. The use of a selection procedure which prevents a predetermined outcome is a requirement common to most statistical inference. Such procedures are called random procedures and the collection of objects that arises from such a procedure is called a *random sample.* It is worthwhile emphasizing that objects are neither random nor not-random, as any of the objects in a population could be one of the objects in a random sample and in an ideal random procedure each object has the same likelihood of selection as any other object. The word random refers to the procedure. Randomness is not a property of an object nor is it a quantity. The randomness of the procedure is required to ensure that the sample of objects is selected without sufficient knowledge to predetermine the outcome. In this sense we shall say a random sample is not biased to a particular outcome or a particular range of values.

This requirement of lack of bias in a sample arises because one of the aims of statistical inference is to provide an estimate of some characteristic of the population on the basis of the evidence in the sample. The more closely the sample represents the population in miniature, the more accurate the estimate will be. If the selection of objects was restricted to one part of the range of values taken by the attribute in the population, the sample would not represent the full variation of the objects with respect to that attribute. Any estimated characteristic would be biased. The use of a random procedure reduces the chance of so unrepresentative a selection occurring and, secondly, it permits an estimate of the reliability of the estimated characteristic or parameter value to be made.

In geography, adapting these notions of sampling often appears difficult or impossible because it is not clear what the underlying population is that is supposedly being sampled. Consider first the total annual run-off at a particular site in some year. Define this as the random variable X_1. Then we could define for consecutive years the random variables $X_2, X_3, \ldots, X_n$, such that $(X_1, X_2, \ldots, X_n)$ could be considered as a sample of size n from the population of all possible yearly run-offs at that site. Secondly, consider the length of time a factory, defined in terms of a particular production sequence, survives in production. Let us study n such factories and measure their survival times as our random variable $T_1, T_2, \ldots, T_n$. Then $(T_1, T_2, \ldots, T_n)$ can be considered as a random sample from the population of all possible survival times of factories defined precisely in terms of that production sequence. Thirdly, for n factories in a particular year, consider a two-dimensional random variable (location, size) for each factory. Define X_1 as the least distance from a source of raw material and define Y_1 as the capacity to produce the product. Then $[(X_1, Y_1), \ldots (X_n, Y_n)]$ can be seen as a random sample from the population of all possible distance, capacity measures of the random variable (X, Y).

To cover these circumstances which arise frequently we can use the following definition.

Definition. Let X be a random variable with a particular probability distribution. Let $X_1, \ldots, X_n$ be n independent random variables each with the same distribution as X. Then $(X_1, X_2, \ldots, X_n)$ is called a random sample from the random variable X.

In this definition the population becomes a population of measurements and the random sample from a random variable X occurs by taking n measurements on X under essentially similar conditions. By essentially similar we mean that whatever differences there are, they are not considered to affect the outcome systematically. This emphasizes the need to ensure that environmental conditions are homogeneous, and that measurement procedures are comparable. For instance if the random variable was monthly run-off, we could record n months as $(X_1, \ldots, X_n)$ but we should ensure that we chose the same month from each annual record. The definition is extended readily to the many-dimensional case as the number of attributes is increased.

Finally in this section we shall show more exactly than we have been able to do so far, the links between probability distributions, random variables and hypotheses.

1.9 Hypothesis testing

Geographers often use maps to make assertions in the form of statements such as

1. Towns locate at lowest bridging points.
2. Papermills are market oriented.
3. An increase in run-off is associated with an increase in dissolved solids.
4. Understanding statistical inference helps geographers analyse maps.
5. Pollution increases downstream.

Our aim is to decide whether such statements are acceptable. We may decide without any evidence, we may base our decision on evidence which may not be relevant to the statement or we may make a decision to accept or reject the statement in terms of a statistical hypothesis.

Definition. A statistical hypothesis is an assertion about the probability function of one or more random variables or it is a statement about some characteristic of the population of measurements from which a representative sample has been taken.

The procedure for making such an assertion is quite straightforward; the stages may well be understood more readily if a particular example is used to fix the problem and the procedure.

Example. A river authority records pollution levels at 12 sites. These sites were selected randomly, but once chosen were used for all recording purposes. Legislation to control the disposal of effluent into the river was implemented at the beginning of a year and the question arose as to whether the legislation had been effective in reducing the level of pollution. A two-dimensional random variable $[(X_1 Y_1), \ldots, (X_n, Y_n)]$ was defined for the n sites such that X_i was the mean level of pollution at the ith site in the year immediately before legislation, and Y_i was the mean level of pollution at the ith site in the year immediately after legislation. These values are given below.

	Pollution mean level		
Site number	*Before legislation* X	*After legislation* Y	*Sign of* $(X-Y)$
1	5·6	5·6	0
2	7·1	6·3	+
3	6·4	6·7	−
4	5·8	5·3	+
5	4·9	4·0	+

Site number	*Before legislation* X	*After legislation* Y	*Sign of* $(X-Y)$
6	4·7	5·2	−
7	5·0	4·9	+
8	4·9	5·2	−
9	3·6	3·3	+
10	5·4	4·8	+
11	4·7	3·2	+
12	3·1	2·4	+

1.12 Mean pollution levels in a stream before and after legislation.

The figures show that at 1 site there was no difference, at 8 sites the figure was lower after legislation and at 3 sites it was higher. These values were used as evidence in support of the assertion that legislation had been effective in reducing pollution levels. The question we wish to resolve is in what sense can we say that figures such as these do give us sufficient reason to believe that the level of pollution has been reduced. Questions of this sort are resolved by

a) expressing the statements as hypotheses
b) applying a statistical test to these hypotheses
c) stating a rule for deciding whether to accept or to reject a particular hypothesis on the basis of the statistical test
d) submitting the evidence—the measurements—to these procedures.

a) *Hypothesis formulation.* The question we *wish* to resolve is whether these figures give us *sufficient* reason to believe that the level of pollution has been reduced. If we decide that the figures do not support this belief then we are saying that there is not sufficient reason to suppose that these figures show a difference in level at these sites. This statement of *no difference* is the hypothesis we are testing. The hypothesis to be tested is called the *null hypothesis* and it is denoted by H_0. If we decide that the figures do not give sufficient support to the null hypothesis then we reject it. Rejecting the null hypothesis is equivalent to accepting the *alternative hypothesis*, denoted by H_1. In other words we are saying we believe H_1 is likely to be true. If we decide that there is not sufficient evidence to accept the alternative hypothesis, we are saying that there is not sufficient evidence to reject the null hypothesis. We are not saying that the null hypothesis is true, but that it has not been shown to be false. This leaves the decision open to amendment if we have more evidence or if we use measurements that contain more information that can be exploited by a different test statistic.

These implications of acceptance and rejection of the statements we call hypotheses show how we designate one statement as the null hypothesis and the other as the alternative hypothesis. When we want to determine if a statement is likely to be false we put it as the null hypothesis. When we want to determine if a statement is likely to be true we put it as the alternative hypothesis. In the pollution example we want to decide whether it is true that the pollution levels have been reduced and so that becomes our alternative hypothesis. It is the alternative to the null hypothesis of no difference in the levels.

b) *Statistical testing.* To test a statistical hypothesis is to devise a rule which enables us to decide whether a null hypothesis should be rejected or not. This decision is based on the value of a test statistic and this is a function of the measurements we have made of the variables. There are many test statistics with particular names and designations, but we can denote the general concept of a test statistic by T.

A test statistic gives values to all points in the sample space and, as these numbers are used to decide whether to accept or reject H_0, we use them to measure the acceptability of the hypothesis. These numbers are most convenient if they are arranged in an ordered sequence of acceptability of the hypothesis. Thus increasing values of T might correspond to an increasingly strong indication that H_0 should be rejected. The values of T could be arranged in the opposite sense such that small values of T would indicate the decision to reject H_0. In either case when the decision to reject H_0 is made by using one end of the ordered set of values of T, the decision is said to be based on a *one-tailed test.* When the values of T are arranged in order and both large and small values are used as a basis for rejecting H_0, the decision is said to be based on a *two-tailed test.* Thus T generates an ordered sequence of values which can be used to provide:

i) a value of T, say T^*, that is arbitrarily taken to distinguish the decision *accept* H_1, from the decision *accept* H_0;

ii) a measure of the discrepancy between T^* and the value of T, say T', that is actually observed.

c) *Critical region or rejection region.* The test statistic assigns numbers to all points in the sample space. The critical region is the subset of those numbers which leads to the decision to reject the null hypothesis. The numbers corresponding to the remaining points in the sample space are referred to as the *acceptance region* for T. The boundary between the acceptance and rejection region is an arbitrarily chosen value of T. It is conventional to choose the bound of the critical region so that those values of T which are at least as extreme as the bounding value are the most extreme 5 per cent, or some other small percentage of the total set of values of T. It is worthwhile emphasizing that a decision to accept or reject a hypothesis is made *on the basis* of the evidence yielded by the test statistic. The decision is neither certainly correct nor incorrect. Incorrect decisions can be made in two ways and have special names.

A type I error occurs if the null hypothesis is rejected when it is true.

A type II error occurs if a false null hypothesis is accepted.

The names are less important than an awareness that an inference can always be incorrect in one of these two ways.

There is a probability of making each type of error. The probability of making a type I error is usually denoted by the symbol α. This is referred to as the *significance level.* This term is also used to denote the preselected arbitrary bound for the rejection region.

If H_0 is true the greatest probability of rejecting H_0 is α, consequently the probability of accepting H_0 is at least $1-\alpha$.

If H_0 is false the decision to accept H_0 is said to have probability β. The probability of rejecting H_0 is $1 - \beta$, and this is said to measure the power of the statistical test. The power of a test varies with the discrepancy between the H_0 and the true H_0, the size of α, the sample size and the number of random variables.

The true situation	*The decision*	
	Accept H_0	*Reject H_0*
H_0 true	Decision correct probability $= 1 - \alpha$	Type I error probability $= \alpha$
H_0 false	Type II error probability $= \beta$	Decision correct probability $= 1 - \beta$

1.13 Hypothesis testing and type I, type II errors.

The critical region contains those values of T that are at least as large as some particular value. It is conventional to express this region as containing the 100 α per cent largest or smallest values of T. Again by convention α is taken to be, 5, 2·5 or 1. *Under the assumption of the null hypothesis a value as extreme as or more extreme than the value of T corresponding to such an α occurs 100 α per cent of the time.* By making α sufficiently small, the decision to accept H_0 can always be made; equally by making α sufficiently large, H_0 can be rejected. The balance between these two is arbitrarily decided. To a considerable extent it depends on whether the cost of accepting a false H_0 is higher than that of rejecting a true H_0. The choice has been formalized by the standard α levels which represent widely held opinions on what risks are acceptable and unacceptable. The choice remains. The important point is that the choice of α does not affect the likelihood of the assertion, and the decision about the assertion is referred to that likelihood. In this book it is often convenient to use an α: 0·05 as the nominal significance level for both one-tailed and two-tailed tests. In appropriate cases comments are made about the implications of the α level. The standard procedure for expressing a significance level is α: 0·05 which denotes the fraction of the values of T that lead to rejection. The total probability of all values is 1·00, thus 0·05 means 0·05/1·00 or 5 per cent. Similarly α: 0·025 means 2·5 per cent and 0·01 means 1 per cent; consequently they are sometimes referred to as the 5 per cent $2\frac{1}{2}$ per cent and 1 per cent significance levels. They imply that $1 - \alpha$ is 0·95, 0·975, 0·99 respectively; $1 - \alpha$ is the acceptance region.

In discrete distributions this nominal significance level may not be able to be attained exactly by any paticular value of T. If we set α as the value of T, say T_α, which defines the largest 5 per cent of all values of T, and we set this level of α before we analyse our observed measurements, then when T is a set of discrete values we may find that there is no value of T that has exactly 5 per cent of the remaining values of T at least as large as it. Instead we find two values of T, say T_i and T_{i+1}, such that T_i is less than T_α, and T_{i+1} is greater than T_α. As such discrete probability distributions are usually the bases of distribution-free and non-parametric statistical tests it is useful to give both the nominal α for rejecting H_0 and the exact significance level of T_{i+1}.

Example (continued from previous example, using details in fig. 1.12)
Our observations form natural pairs of measurements taken at different times at the same sites along the river. As the (X, Y) values are matched or paired, we consider that a comparison of the paired values could provide the basis for the test statistic. We might argue that if there is no difference in pollution level then the probability that an X value will be *larger* than its paired Y value will be the *same* as the probability that it is *smaller* than its paired Y value. Similarly if there has been a general reduction in pollution level then the probability that an X value is *larger* than its corresponding Y value will be *greater* than the probability that an X value is *smaller* than its corresponding Y value. The (X, Y) values can be reduced to one sequence if we characterize these relations by writing, as in fig. 1.12

(i) $(X_i > Y_i)$ as a +
(ii) $(X_i < Y_i)$ as a −
(iii) $(X_i = Y_i)$ as a 0

We wish to determine whether the statement is true that the level of pollution has been reduced. This becomes the H_1. Accepting that statement means that we reject the statements

a) that there is no difference
b) that there has been an increase in pollution.

Thus we could rewrite our null hypothesis as

$$H_0 : P(X > Y) \leqslant P(X < Y)$$

which is composed of

a) $P(X > Y) = P(X < Y)$
b) $P(X > Y) < P(X < Y)$

Now (a) is equivalent to $P(+) = P(-)$
and (b) is equivalent to $P(+) < P(-)$
We can express the alternative hypothesis as

$$H_1 : P(X > Y) > P(X < Y)$$

which is equivalent to $\quad P(+) > P(-)$

For the moment consider that part of H_0 which says that $P(+) = P(-)$. If this is true then we should expect some discrepancies in the number of plusses and minuses; however, we should not expect so many discrepancies that made the differences either all positive or all negative. If we did find 12 plusses we should be reluctant to accept H_0. The test statistic we shall use gives the probability distribution of plusses and minuses under the assumption that there has been no change and that the outcome at any one site is unaffected by the outcome at any other site. That is to say that if the difference at one site is positive, the difference at any other site is equally likely to be negative as positive.

We define our test statistic, T, as the number of plusses, t. We ignore tied values; consequently there are 12 points in the sample space corresponding to $t = 0, 1, 2, \ldots, 11$ plusses.

From the earlier discussion in this chapter we recognise that under H_0, T is a binomial random variable with $p = (1 - p) = \frac{1}{2}$, $n = 11$ and t taking values from $0, 1, 2, \ldots, n$. The distribution of T is given by

$$P(T = t) = \binom{n}{t} p^t (1 - p)^{n-t}$$

The probability distribution is tabulated, fig. 1.14, and the graph corresponding to those probabilities is presented too (fig. 1.15).

	t	$\binom{n}{t}$	$\binom{n}{t} p^t (1-p)^{n-t}$		
	0	1	0·0004883		
	1	11	0·0053713		
	2	55	0·0268565		
	3	165	0·0805695		
	4	330	0·1611390		
	5	462	0·2255946		
	6	462	0·2255946		
	7	330	0·1611390	0·8867144	
T_i	8	165	0·0805695		Values of (T=t) in the rejection region
T_α: 0·05	. . .	. . .	. . .	. . .	
T_{i+1}	9	55	0·0268565		
	10	11	0·0053713		
	11	1	0·0004883	0·1132856	
				1·0	

1.14 Values of T and a rejection region.

From the table we see from the sequence of values of the test statistic ordered as a sequence from $t = 0$ to $t = 11$ that

$$P(t = 8) = 0{\cdot}0805695 : \textit{point probability}$$

$$\text{and that } P(t \geqslant 8) = 0{\cdot}1132856 : \textit{cumulative probability}$$

$$\text{and thus } P(t < 8) = 1 - 0{\cdot}1132856 = 0{\cdot}8867144$$

The H_1 indicates that our critical region is defined on one tail. Suppose that we have not preselected a boundary value for T, then we can see that if we reject H_0 on this evidence we are using a rejection region that contains a little over 11 per cent of all the possible outcomes under that null hypothesis. That is to say that if there were really no difference in pollution level and designating the members of each pair of values as X or as Y quite arbitrarily then a result giving 8 or more plusses would occur 11 per cent of the time. The exact significance level is 0·1132856. At a conventional α of 0·05 the H_0 would be accepted on the basis that there was not sufficient evidence to reject it. This is equivalent to saying that there is not sufficient evidence to accept H_1. When we use the sign test we can

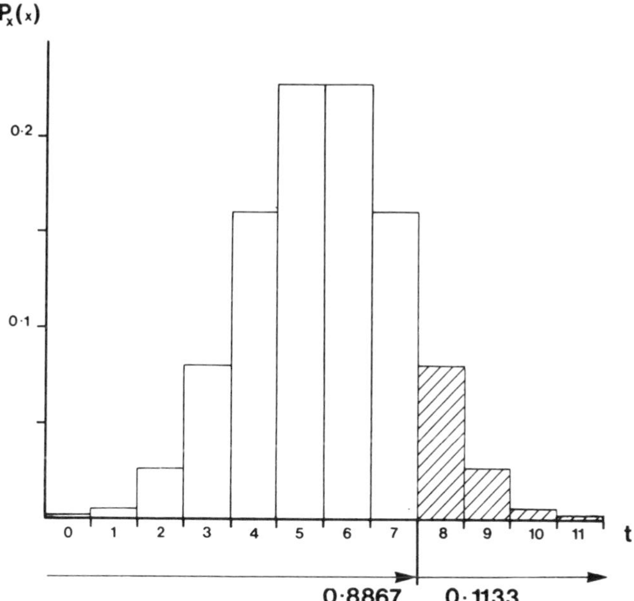

1.15 Probability distribution and a rejection region.

reach the same decision without the calculation by using table 3b in appendix 2. In this table a one-tailed α : 0·05 is equivalent to a two-tailed α : 0·10 because the probability distribution is symmetric and it shows the probability of the *less* frequently occurring sign. In the example this means we want to know whether 3 or fewer minus signs is in the one-tailed rejection region of α : 0·05. We enter the table with $n = 11$ and in the column α : 0·10 we see that 2 is the greatest number of minus signs that lies in this region. We accept H_0, but we don't know the exact probability. Exact probabilities identical to the ones we calculated could be obtained from tables like table 3a but which are more extensive.

This decision based on this test statistic about the assertion made from the observations is a personal opinion. If H_0 is true, accepting H_1 on the basis of this evidence is equivalent to accepting the risk of 0·1132856 of making a type I error. The cost of this decision could be considerable in terms of environmental deterioration. Similarly, if H_0 is false, failing to reject it would involve more expense in effluent treatment by users and in monitoring and litigation by the river authority and the community. Judgement is exercised on the assertion at this stage of the study, and it is based on quite explicit rules and results. This seems to many people to be a sounder basis for opinion and a better way to establish experience than by allowing decisions to be made at arbitrary points in the sifting of evidence.

If our H_0 had been $P(+) = P(-)$

and H_1 had been $P(+) \neq P(-)$ such that we were prepared to find that pollution had increased *or* decreased, then we should have used a two-tailed test.

Suppose we took a nominal $\alpha : 0{\cdot}05$, then this corresponds in the two-tailed test to values of $t \leqslant 2$ and of $t' \geqslant (n-t) = 9$. The exact significance level for $t \leqslant 2$, $t' \geqslant 9$ is 0·0654322.

The sign test is one of many tests based on the distribution of a binomial random variable. There is no need for $p = (1-p)$. All statistical tests are based on at least one probability distribution, when the H_1 is composite then more than one distribution is involved.

Many of the tests discussed in this book do not have tabulated values of exact probabilities for large n, often because the calculations involved are very great and because the particular, exact probability distribution becomes increasingly similar to another probability distribution which is more easily calculated and for which extensive tables do exist. The most usual approximating distributions are

a) the normal distribution
b) the chi-square distribution (pronounced kie).

The way in which one probability distribution approximates another probability distribution is illustrated in the next section. In the first case we see how the *binomial* is approximated by the *Poisson*: they are both discrete distributions but the Poisson is easier to calculate. In subsequence cases we use the more usual *continuous* distributions of the *normal* and *chi-square*.

The whole procedure of stating hypotheses, choosing a test statistic, building up the probability distribution of the values of that test statistic and seeing what a particular decision implies is discussed throughout the text. Different details are elaborated as appropriate. Increasing use of statistical shorthand notation means that the statements become increasingly concise. The elaborations should be completed by the student by reference to these early chapters. Statistical test is linked to map situation and the common assertions from such maps.

Exercises

1.1 Compute: 3!, 4!, 5!, 8!

1.2 Compute: $\dfrac{8!}{5!}, \dfrac{8!}{3!}$

1.3 Compute: $\binom{7}{2}, \binom{8}{5}, \binom{6}{6}, \binom{5}{0}, \binom{11}{2}, \binom{11}{9}$

1.4 $\dfrac{n!}{(n-1)!} = ?, \quad \dfrac{n!}{(n-2)!} = ?$

1.5 Show that $n!/n = (n-1)!$

1.6 Verify algebraically $\binom{n}{r} = \binom{n}{n-r}$

1.7 Verify algebraically $\binom{n}{r} = \binom{n-1}{r-1} + \binom{n-1}{r}$

1.8 Substitute $k = 6, r = 3$ in

$$\binom{k}{r} + \binom{k}{r-1} = \binom{k+1}{r}$$

Can you show this is true in general?

1.9 Check $\binom{5}{0}+\binom{5}{1}+\binom{5}{2}+\binom{5}{3}+\binom{5}{4}+\binom{5}{5} = 2^5$

Can you see that this generalizes symbolically to

$$\binom{n}{0}+ \ldots + \binom{n}{n} = (1+1)^n = 2^n$$

1.10 Given $E = 4$, $C = 3$, $H = 3$, how many distinguishably different arrangements are there?

1.11 In how many ways can a group of 2 men, 2 women and 3 children be chosen from a group of 5 men, 7 women and 8 children?

1.12 Seven different shop types occupy 7 adjacent sites. How many different arrangements are there?

1.13 Evaluate $\sum_{i=1}^{3} \binom{5}{i}$

1.14 Evaluate $\sum_{i=0}^{3} \binom{5}{i}$

1.15 Evaluate $\sum_{i=2}^{4} \binom{6}{i}\left(\frac{1}{2}\right)^2$

1.16 Evaluate $\sum_{i=2}^{4} \binom{6}{i}\left(\frac{1}{3}\right)\left(\frac{2}{3}\right)^{6-i}$

1.17 Compute and plot the binomial probability distribution for

a) $n = 6, p = \frac{1}{4}$
b) $n = 6, p = \frac{3}{4}$
c) calculate $E(X)$, $Var(X)$ for each of these.

1.18 For $n = 5, p = q$, calculate $E(X)$. Is this a possible outcome? Comment, but recall $E(X)$ refers to a probability distribution.

1.19 Given 4 distinct maps and 4 distinct assertions such that each map has just 1 correct assertion. Let X be the number of correct matchings.

1. Determine the probability distribution, $P_X(x)$ of $X = 0, 1, 2, 3, 4$.
2. Construct $P_X(x)$ and the *cdf* $F_X(x)$.

Hint: Say correct order is 1, 2, 3, 4 then calculate the number of possible arrangements and count for X.

1.20 Verify $\dfrac{(N-n+1)!}{(N-n)!} = \dfrac{(N-n+1)}{1}$

Compute for $N = 10$, $n = 4$.

1.21 Simplify p^n/p^{n-1}.

1.22 Verify $N\left(\frac{1}{N}\right)^k = \dfrac{1}{N^{k-1}}$ for $N = 6$, $k = 3$
$N = 2$, $k = 7$.

1.23 After a course in map interpretation students are given some maps and asked to match map with assertion. What are the appropriate H_0, H_1? For exercise 19 what α would you choose and what X?

1.24 What is the appropriate H_1 for the following hypotheses:
1. H_0: The distance of a resort from large urban populations does not affect the popularity of the resort.
2. H_0: The density of vegetation cover does not affect run-off.
3. H_0: Towns are located randomly.

Can you restate H_0, H_1 to imply one-tailed tests?

1.25 What is the appropriate H_0 for the following hypotheses:
1. Mean annual water consumption is increasing.
2. Steel rolling mills are market-oriented.
3. Banks are clustered in the same part of a city.

What definitions need specifying carefully?

1.26 If you record days as wet or dry, with a dry day defined as $\leqslant 0{\cdot}01''$ moisture, what measurement scale are you using? What sort of statistical test would you use?

1.27 If you asked people how accessible they thought towns were from a centre, what measuring scale would be implied?

1.28 How can perception be measured on a continuous quantity scale?

1.29 a) Do the two scales on fig. (i) provide the same ordering relationship for the 63 towns?

b) Do you consider these two scales are equivalent scales for measuring locational relations?

(i)

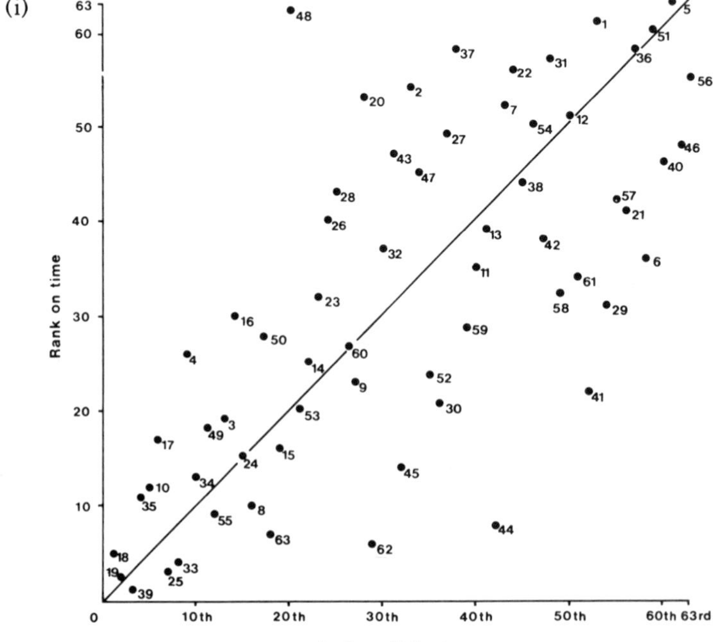

c) Does the map (ii) provide more order information than the graph?
d) What inversions are seen on the graph (iii)?

(ii)

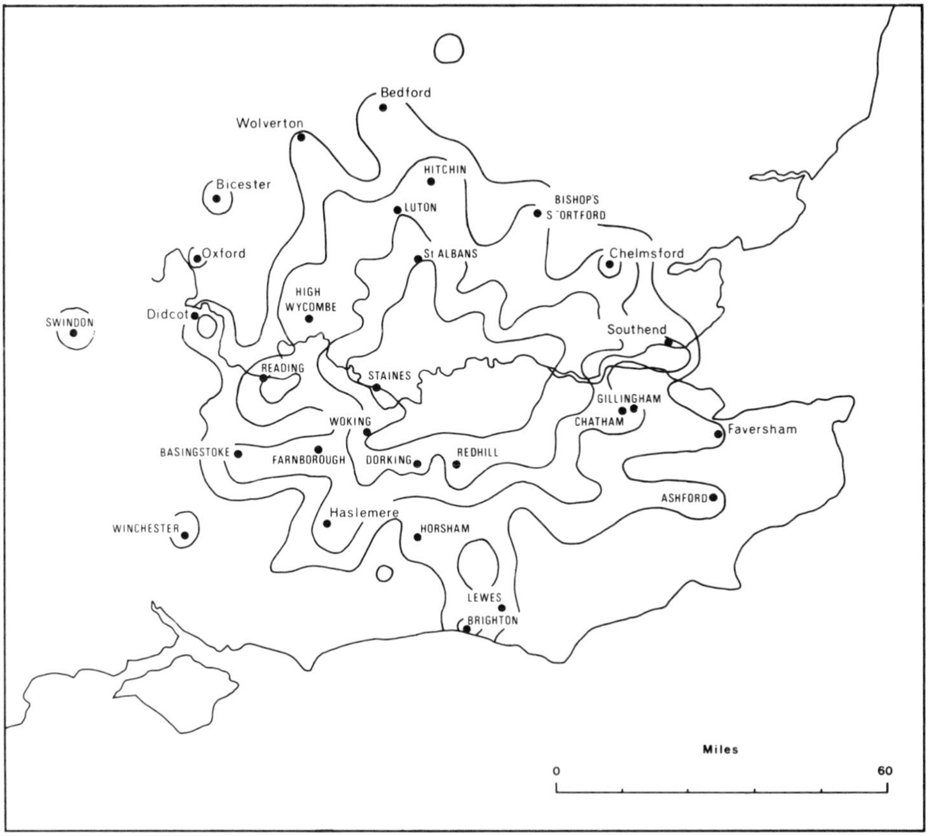

Bedford
Wolverton
HITCHIN
Bicester
BISHOP'S S'ORTFORD
LUTON
Oxford
St ALBANS
Chelmsford
HIGH WYCOMBE
SWINDON
Didcot
Southend
READING
STAINES
GILLINGHAM
CHATHAM
WOKING
Faversham
BASINGSTOKE
FARNBOROUGH
DORKING
REDHILL
ASHFORD
Haslemere
WINCHESTER
HORSHAM
LEWES
BRIGHTON
Miles
0
60

(iii)

MINUTES
TIME
0
10
20
30
40
50
60
70
80
DISTANCE
MILES

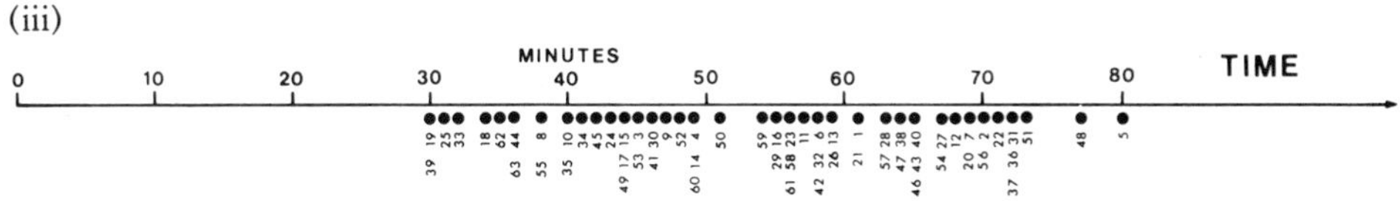

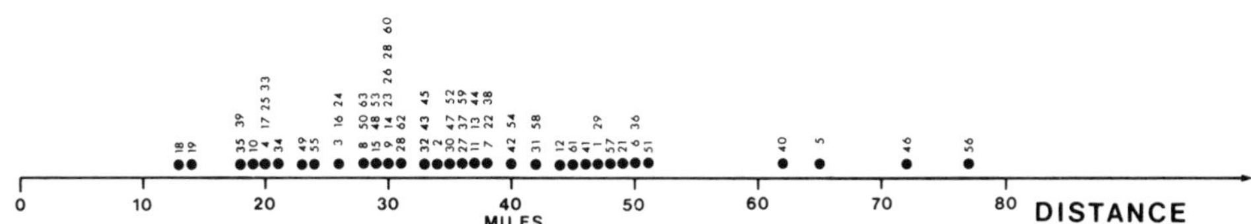

1.30 A linear scale can be constructed such that the first object, 4 units away from a second object, is not twice as far from that second object as a third object 2 units from it. We are not accustomed to scales such as this one because we generally prefer additive scales. To construct a linearly non-additive scale we can use a diagonal scale such that

$$\begin{aligned} 1 \text{ dinch} = 1 \text{ diagonal inch} &= \sqrt{1} \text{ inch} \\ 2 \text{ dinches} &= \sqrt{2} \text{ inches} \\ n \text{ dinches} &= \sqrt{n} \text{ inches.} \end{aligned}$$

Such a scale and an illustration of its graphical construction are shown below.

On this scale objects' distances from the origin are simply the sum of their coordinate values.

a) How far from the origin are objects b, c, d?

b) In geography we are finding that because an object is, say, twice as far from a fixed object, say a town, as another object, it does not imply that it has twice or half the interaction with the fixed object. For interest plot the locational relations of some of your own data on an additive scale and on a dinch scale.

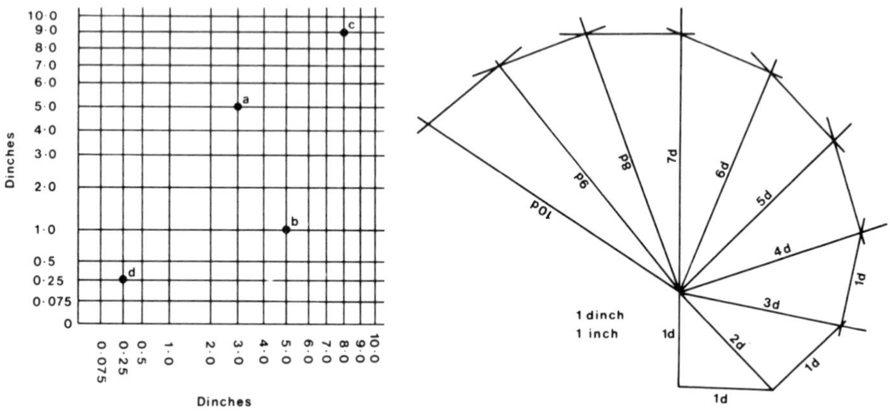

a (3.5d) 8d from origin n dinches √n inches
b (5.1d) 6d from origin
c (8.9d) 17d from origin
d (·25.·25d) ·5d from origin

1.31 On the figure below, in what sense do any of the 9 objects have the same location? If the location class is a coalfield, say, with costs increasing monotonically in all directions, would this affect your answer?

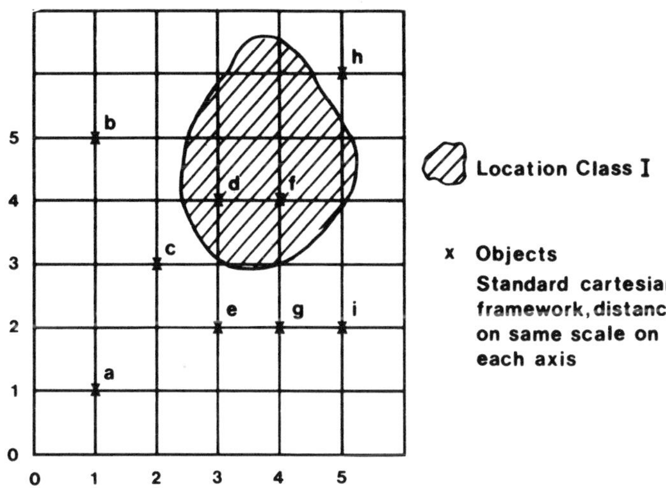

1.32 We are trying to test the general idea that students with experience in statistical inference get higher marks in geography examinations than students without such experience. A first-year class of undergraduates without previous experience in statistics is split into two groups such that each student is allocated randomly to one group.

a) Would you put this assertion as H_0 or H_1?
b) Provide H_1 or H_0.
c) Put H_0 and H_1 symbolically as on p.41.
d) Is a one- or two-tailed test implied?
e) Which type of error would you try to minimize?

Section 2/Analysing point symbol maps

Analysing point symbol maps

2.1 The construction of point symbol maps

Mapping is the procedure of representing elements (objects and their attributes) at the earth scale by elements (symbols) at the map scale. The symbols used are frequently put into one of four classes: point, line, area and volume symbols. These terms are accepted by usage as a basis for classifying map symbolism. While the terms *point* and *line* are accurate designations of the symbols used and a consistent basis for distinguishing the resulting maps, the use of the terms *area* and *volume* symbolism cannot be justified on either basis. Area symbolism usually refers to discrete changes in the density of shading corresponding to changes in the intensity of the property being represented for contiguous but distinct arbitrary area units. Volume symbolism usually refers to contour lines representing equal but arbitrary levels of intensity in some continuously variable property. The conventions of map construction mentioned in the previous section refer to the procedures for using such symbolism and a number of texts discuss map construction in terms of those procedures (Monkhouse and Wilkinson 1964; Robinson 1953). As we are concerned with procedures for map inference we shall only discuss those features of map construction which may affect propositions, measurement scales and the test statistics used.

Point symbols are the most straightforward means of representing a finite set of objects in terms of their locational relationships. The basic point mapping procedure is to relate a set of objects on the earth's surface to an origin in terms of a standard distance scale and to match this by a set of point symbols on a map whose distance measure is some monotonic function of the standard distance measure, with one-to-one correspondence as the principle.

This principle is one of a number of possible relationships. The three most important possibilities are understood readily from the following diagrams.

In the first case for each map element (a symbol) there is at least one earth element (an object). This sort of relationship is called a *surjective* mapping and is sometimes referred to as an *onto* mapping. The word mapping is used here more generally than in geography and is not restricted to the cartographic notion of a map. Surjective mappings arise regularly in point symbol maps.

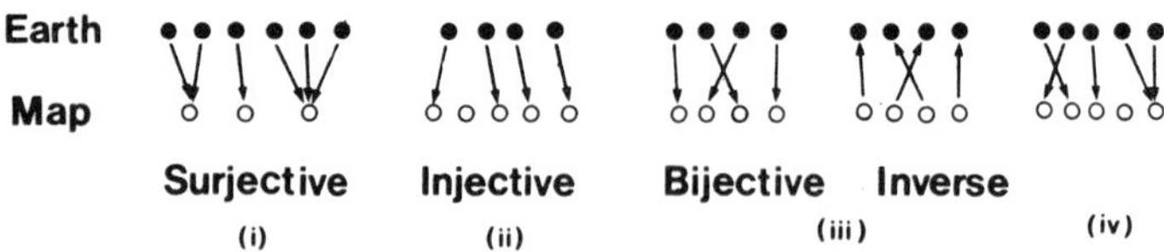

2.1 Mapping relationships.

In the second case the distinct elements (objects) at earth scale are mapped as distinct elements (symbols) at map scale. Such a relationship is called an *injective* mapping and is sometimes referred to as a one-one mapping. Its importance to us is in the idea of the third mapping shown above which is known as a *bijective* mapping. A mapping is bijective if it is both surjective and injective. The injective mapping ensures that each object at earth scale has a distinct representation on the map and the surjective mapping ensures that each map symbol represents at least one earth object. Bijective mappings are referred to sometimes as *one-one correspondence* mappings. This unique association between the elements in the two sets means that we can define the inverse mapping, which is also bijective, and so move freely from one set of elements to the other. These and other relationships between the two sets of elements occur in cartography. The strongest relationship is bijective and it seems reasonable to assume that the principle of point symbol mapping is to provide a bijective map.

2.1.1 Bijective mappings

One implication of point symbol mapping being a one-one matching procedure for discrete elements (objects, symbols) is that this mapping is a counting procedure as well as a procedure for measuring location. Thus the map contains elements that correspond to the counting set. There is choice of scale transformation and choice of symbol size, but these choices are quite distinct from the mapping procedure which is a matching procedure. If the choice of scale or of symbol size is constrained such that one-one correspondence is precluded, then the geographer can either alter location correspondence or omit objects. The scale transformation involved often makes some loss of information inevitable, and this loss can either be in number or in location. There seems to be tacit acceptance in geography that as some locational imprecision is frequently necessary, countability is paramount. Perhaps this decision in favour of number against location is at first a surprising decision for geographers to make. However, if the map is to be used for an assertion that involves location class or an ordinal measure of location the decision is justified readily. If the assertion uses continuous distance measures then the inaccuracy can be avoided by suitable alterations in the relationship between the distance measures of the map and the earth. Such alterations may make the presentation of a standard map too unwieldy to be worthwhile, but the information can be retained in a computer for analysis.

2.1.2 Surjective mappings

When the relationship between earth and map distance scales precludes bijective mapping the traditional response is to provide a surjective mapping. The standard procedure is to use point symbols whose size is graded so that a symbol of x times the *area* of the unit symbol represents x objects. This expedient is worthwhile because a numerical ratio of $1:x$ is kept by a symbol-area ratio of $1:\sqrt{x}$. If visual discrimination between symbols of adjacent sizes is possible, enumeration is retained. In practice, this device involves increasing the diameter of the symbol sufficiently to include the greatest number of objects needed to be shown in the unit distance on the map.

Although circles, triangles and squares are all used as point symbols, the most common symbol is the circle, presumably because it is so easily drawn by hand, it can be expanded readily and it is the most compact plane figure. However one disadvantage is the unused space between adjacent circles. In order to combine enumerability with a symbol density that corresponds to the object density the square is a more satisfactory symbol. The system of graphical rational patterns, *GRP*, introduced by Bachi (Bachi 1968) is based on the unit square in a conventional arrangement of symbols, whose pattern, like that of dominoes, enables correct numerical assignments to be made without the need to count individual symbols or to compare diameters. This system is a recent stage in formalizing the arbitrariness of point symbol mapping. The construction of maps by computers together with automatic plotting systems removes the main impediment to the widespread adoption of this excellent system which is especially appropriate to surjective mapping. An example illustrates the problems of such maps.

Example

The aim is to produce a point symbol map of the factories in Humberside for 1965. The relationship of distance scales from earth to map puts the unit distance on earth scale as 1 kilometre. The number of factories in a square kilometre varies from 0 to 100. One factory is matched by 1 symbol in the *centre* of the unit kilometre square, and the regular arrangement of symbols up to 9 is shown diagrammatically. Ten factories are shown as a square in the centre of the unit kilometre and the pattern for tens imitates that for units up to the dense unit square representing 100 factories. The proportion of black to white is that of the number of objects shown to the maximum that could be shown in the unit square. The density of symbols is related precisely to the number of factories although this does not ensure that the impression of the correct number is gained by the average viewer.

In such maps the size of the unit symbol is determined by the number of objects and the length discrimination that can be kept under the particular scale transformation. A bijective mapping is replaced by a surjective mapping, and a continuous distance measure by an ordinal measure or by a location category. The hypotheses formulated from such maps are judged on the basis of test statistics that acknowledge these changes in measurement.

The remainder of this section is developed by considering the three categories of proposition in turn. First we shall consider propositions of randomness,

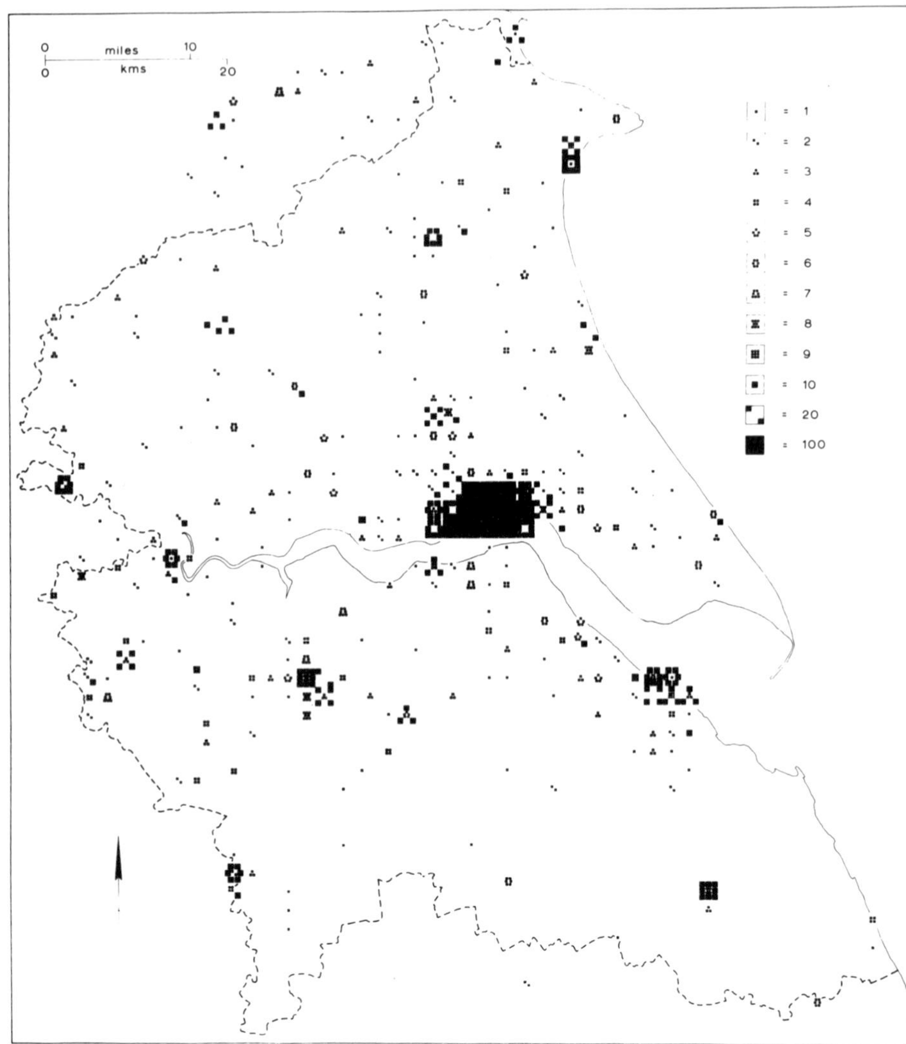

2.2 Industries of Humberside using graphical rational pattern symbols.

propositions of independence and measures of association next and propositions of identity of distribution or equality of location last. There is a close relationship between point symbolism and measurement scale, between elaboration of symbolism to proportional and divided symbol and number of attributes, and between these and the appropriate hypotheses with their test statistics. I must stress that affinity not necessity links the tests and the maps. The tests are introduced in an order that appeals to me and which I have found convenient. Many of the tests can be used with other types of map: for example the tests introduced in section 3 on line symbols are appropriate to maps in this section just as the converse often applies. The relevance of particular tests often depends on the experience of the user and his ingenuity in stating the hypothesis.

2.2 Propositions of randomness

Property: location class.
Measurement scale: nominal, frequencies.
Statistical procedure: probability distributions for binomial, multinomial, hypergeometric, Poisson random variables.

We shall be confined in this chapter to considering some of the inductive procedures that are appropriate for judging the acceptability of a statement which asserts that a set of objects as depicted by a particular point symbol map is a realization of a random procedure. The inductive procedures are all probability distributions of a random variable that is assumed to satisfy certain conditions. These conditions are discussed for each probability distribution separately. In all the cases considered the random variable is the frequency of occurrence of objects by location class. In the general case the point symbol map is a bijective mapping of the objects, although a surjective mapping is satisfactory if the location classes are defined in terms of the unit squares used in constructing the map. These inductive procedures have been developed in an attempt to characterize this concept of randomness. If this is not the feature of the map that is of interest then there may be good reason to use measures that characterize some other attribute such as density or pattern.

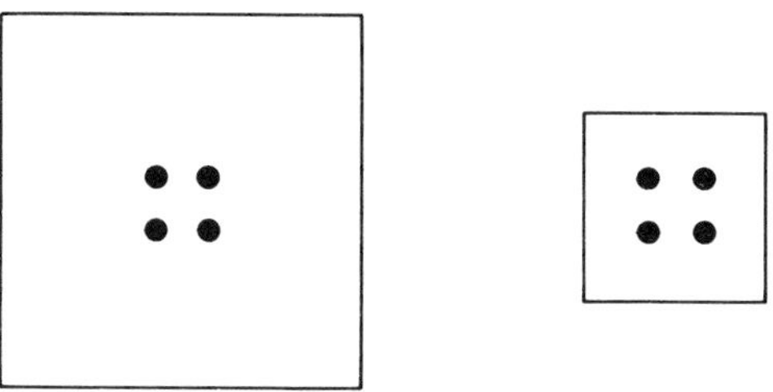

2.3 Changes in density.

Density is expressed as the ratio of the number of symbols to the area of the map. Variations in either of these affect the value of the ratio. A change in the zero ordinates alters the density as illustrated in fig. 2.3. This difficulty has been discussed thoroughly by botanists and by statisticians working on botanical problems. In particular the mean and variance of the density vary with the size of areal unit used, and when a fixed size of areal unit is used many processes can yield the same densities in that area (Greig-Smith 1964; Holgate 1972; Pielou 1969).

The same number of objects in two maps of the same area have the same density, but the characteristic of most interest would be given the name *pattern*, rather than

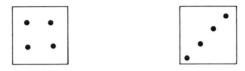

2.4 Changes in pattern.

density as in fig. 2.4. Pattern is a difficult characteristic to measure. As a simple illustration consider the following diagrams in fig. 2.5, where a regular *domino* pattern has been subjected to various modifications. In diagrams 2 to 5, the standard pattern, S, is recognizable: in diagram 2 the pattern has been transformed monotonically on one axis, in diagrams 3 and 4 this pattern has been *translated* and *rotated*, while diagram 5 is the *image* of diagram 2. The last three diagrams introduce *stochastic variability* or *random variation* into the relative locations of the five symbols. Diagrams 2, 3, 4 and 5 can all be related precisely to diagram 1, which is the standard pattern. This standard can be compared with diagrams 6, 7 and 8, and on the basis of a *goodness-of-fit* test (article 3.5.2) some decision could be made about the appropriateness of this standard pattern or of some other standard pattern as a description of the actual pattern. As the number of symbols or objects varies from one situation to another it is not possible to use such standard patterns, and we must adopt a different approach and ask a similar but less specific question of a map.

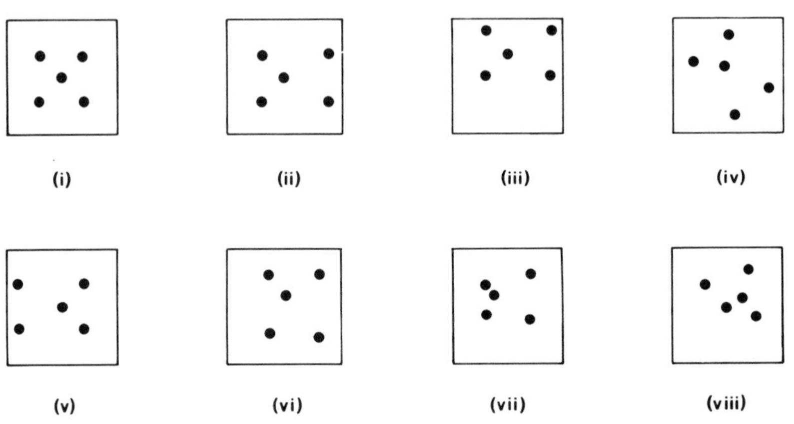

2.5 Characterizing pattern.

This question is whether a distribution as depicted by a set of point symbols can be considered to be a random selection from a particular probability distribution. We cannot have one visual standard which can be compared graphically with an actual map to determine whether in this sense it is random or not, as there are infinitely many arrangements of symbols that could constitute a random distribution. As a great many maps may have random arrangements of symbols and yet not look alike, our standard for comparison is a probability standard and our assertion that a particular arrangement is a realization of a random procedure is put in terms of a probability distribution, or a probability density function if the random variable is continuous. We shall consider some of the probability functions that can be used in map description. We shall discuss their assumptions and perhaps be surprised to find that two-dimensional position and orientation are not entailed by the basic formulations used by geographers in assertions that a point symbol map is or is not a random arrangement. These assumptions are important because they constrain the meaning of the term 'random'. We shall describe map situations that are equivalent to these assumptions, and then we shall be doing no more than asserting that a particular probability function is taken to be the appropriate probability function for describing the incidence of symbols on a map, and, by extension, of the objects they represent. The expected values under this probability function can then be compared to some actual values and a judgement made as to the acceptability of that closeness. Thus random is used to refer to a simple probability function over the map space. Non-random will imply that the particular probability function leads us to reject the null hypothesis that the expected distribution under the assumptions of H_0 is a sufficiently good representation of the actual distribution. This use of random is different from allowing stochastic variability around a determined or fixed arrangement of symbols. Some reconciliation of the two ideas is given at the end of this part (article 2.2.4.2).

When we speak of a map being a realization of a random procedure we are not implying that there is something that allocates factories or people to sites in some region. We are suggesting that we use some simple model to establish what sort of distribution would arise if such a random procedure were used experimentally. This then provides us with some tangible sense of random without supposing that human behaviour is manipulated by some procedure in the same way in which we operate the experimental procedure. Nor are we saying that judgement and decision are irrelevant to behaviour. This is emphasized later (article 2.2.4.2) when the Poisson distribution function is used as the model to describe the arrangement of the homes of undergraduate students at Birkbeck.

There is no prototype of random. Random variables can take on values in accordance with a variety of constraints. We shall now derive the basic probability distributions that are implicit in assertions of randomness in point symbol maps.

2.2.1 Randomness and a binomial random variable

Suppose we have a map of a region, R, with area, A, which is composed of k regions each of area, a. There are N symbols to be allocated randomly in this map and we wish to ascertain what is the probability distribution that n objects occur in a region of interest, RI, of area a. If these symbols are to be described as a random arrangement they must be a realization of a random procedure. One such procedure has the simple assumptions that:

a) each region of area a has the same chance of receiving each symbol, i.e. equal probability;
b) if a region of area a receives one symbol then this does not affect the probability that it will receive any other symbol, i.e. the probabilities of the outcomes of each allocation are independent.

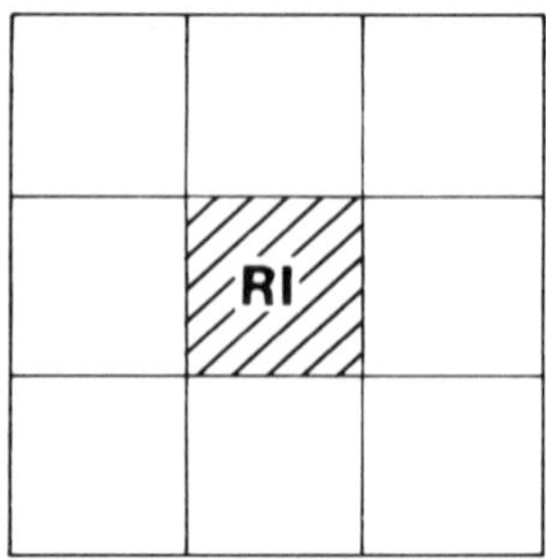

9 possible RI's each with equal probability $\frac{1}{9}$
of receiving one or more of N = 10 objects.
The particular RI has p = $\frac{1}{9}$
The remaining region has probability q = 1 − p = $\frac{8}{9}$
The likelihood of any one RI receiving n = 0, 1, 2 .... N
particular objects is given as

$$p^n q^{N-n}$$

The likelihood of any RI receiving any n = 0, 1, 2 .... N
objects is given as

$$\binom{N}{n} p^n q^{N-n}$$

2.6(a) Region of interest and a binomial random variable.

These are the assumptions for a binomial random variable.

If there are k such regions of area, a, the probability that any specified symbol will be allocated to a particular RI is $\frac{1}{k}$. This is illustrated by a simple map, fig. 2.6 (a), with $k = 9$ small regions each with probability $p = \frac{1}{9}$ of receiving any *specified* symbol and of $q = 1 - p = \frac{8}{9}$ of not receiving that specified symbol. For $N = 10$ the likelihood that any one of those regions, say the RI, will receive $n = 0, 1, 2, \ldots, N$ specified symbols is given by

$$p^n \, q^{N-n}$$

As the symbols are considered to be indistinguishable we are not concerned with any particular symbol and consequently any n of the N symbols can be received by an RI. Consequently the appropriate probability function for determining the likelihood that n of the N symbols will occur in an RI is given by the binomial function as:

$$b\,(n; N, p) = \binom{N}{n} p^n \, q^{N-n} \tag{2.2.1}$$

n	$\binom{N}{n}$	p^n	q^{N-n}	$\binom{N}{n}p^n q^{N-n}$
0	1	1·0	0·307946	0·30795
1	10	0·1111	0·346439	0·38493
2	45	0·01235	0·38974	0·21651
3	120	0·001372	0·438462	0·07219
4	210	0·0001524	0·4932702	0·01579
5	252	0·00001694	0·554929	0·00237
6	210	$0{\cdot}0^5\,1882$	0·624295	0·00025
7	120	$0{\cdot}0^6\,2091$	0·70233	0·00002
8	45	$0{\cdot}0^7\,232$	0·79012	$0{\cdot}0^6 82$
9	10	$0{\cdot}0^8\,26$	0·889	$0{\cdot}0^7 23$
10	1	$0{\cdot}0^9\,3$	1·0	$0{\cdot}0^9 3$

2.6(b) Probability distribution of the event in fig. 2.6(a).

The probabilities corresponding to $n = 0, 1, \ldots, N$ are given in fig. 2.6(b). Under the assumptions of a binomial random variable, the probability of $n \geqslant 4$ symbols in *RI* is 0·01842. With a significance level at $\alpha : 0{\cdot}05$ a distribution of symbols in the map such that at least 4 were in *RI* would be treated as non-random. There are 1024 $= 2^N$ points in the sample space. The random variable is the frequency of symbols, and by implication the frequency of objects in *RI*. The event is $N \geqslant n$ and the probability of that event is the basis for accepting that under the assumptions of the binomial probability function the arrangement of symbols is random or that it is non-random. All the events are possible; improbable events are used for rejecting the H_0 of randomness in favour of H_1 of non-randomness under these assumptions.

2.2.1.1 *The homogeneity and the contiguity of the small regions*

The k regions do not have to be the same size: their similarity was convenient for deciding the value of p but it was not essential to the argument. We can consider any *RI* of arbitrary area a such that a is non-zero and less than the total area A, such that the *RI* has probability $p = \frac{a}{A}$ of receiving any symbol and a probability of $\frac{(A-a)}{A} = q$ of its not occurring in *RI*. In the case just considered this gives $p = \frac{1}{9}$, $q = \frac{8}{9}$ as before, but in fig. 2.7 we have $p = \frac{8}{20}$, $q = \frac{12}{20}$, corresponding, say, to the administrative limits of a town. In such a case we might well be interested in the

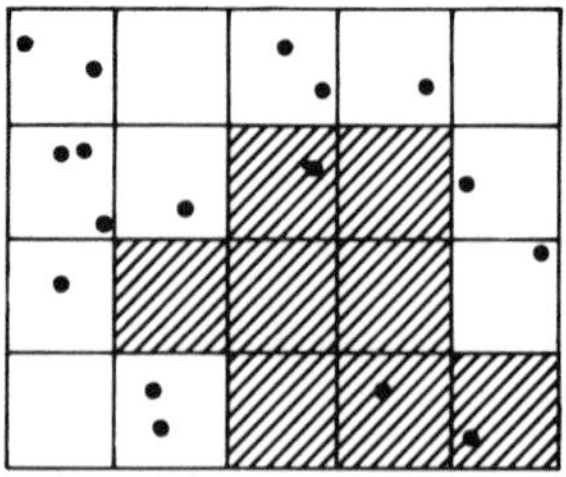

2.7(a) Region of interest of arbitrary size.

[hatched square] $= RI = p = \frac{8}{20} = 0{\cdot}4$

$q = \frac{12}{20} = 0{\cdot}6$

n	$\binom{N}{n}$	p^n	q^{N-n}	$\binom{N}{n} p^n q^{N-n}$
0	1	$0{\cdot}4^0 = 1{\cdot}0$	$0{\cdot}6^{17} = 0{\cdot}00016927$	0·000169
1	17	$0{\cdot}4^1 = 0{\cdot}4$	$0{\cdot}6^{16} = 0{\cdot}00028211$	0·001918
2	136	$0{\cdot}4^2 = 0{\cdot}16$	$0{\cdot}6^{15} = 0{\cdot}00047019$	0·010231
3	680	$0{\cdot}4^3 = 0{\cdot}064$	$0{\cdot}6^{14} = 0{\cdot}00078364$	0·034104
				Cump = 0·046422

2.7(b) Probability distribution of the event in fig. 2.7(a)

probability that of N objects, n of them would occur in the town under the assumption that this frequency was a binomial random variable. The frequency observed is 3 of the 17 objects in the RI which is 0·4 of the area of the total region (fig. 2.7). The point probability of the event $n = 3$ is

$$\binom{17}{3} 0{\cdot}4^3 \,.\, 0{\cdot}6^{14} = 0{\cdot}034104$$

and of the event $n \leqslant 3$ is

$$\sum_{n=0}^{3} \binom{17}{n} 0{\cdot}4^n \,.\, 0{\cdot}6^{17-n}$$

$$= \binom{17}{0} 0{\cdot}4^0 \,.\, 0{\cdot}6^{17} + \binom{17}{1} 0{\cdot}4^1 \,.\, 0{\cdot}6^{16} + \binom{17}{2} 0{\cdot}4^2 \,.\, 0{\cdot}6^{15} + \binom{17}{3} 0{\cdot}4^3 \,.\, 0{\cdot}6^{14}$$

$$= 0{\cdot}046422.$$

The event $n \leqslant 3$ lies in the rejection region defined by α: 0·05; no n corresponds to this bound, for $n \leqslant 4 \simeq \alpha$: 0·126.

For convenience, the *RI* in fig. 2.7 was composed of 8 of the 20 equally sized small areas of the total region, but a continuous boundary could have been used without affecting the assumptions of the frequency being a binomial random variable. These assumptions of equal and independent probabilities of outcome do entail an implicit condition that the *RI* and the remainder of the map are sufficiently large to make those assumptions reasonable. When this implicit condition is not met a different probability function is required (article 2.2.3), because the conditions are those defining a hypergeometric random variable.

An *RI* can consist of non-contiguous component areas. Consider a region, R, with k regions a_i, such that $\sum_{i=1}^{k} a_i = A$, is the total area of the region. Define a region of interest, *RI*, as $\sum_{i=1}^{m} a_i = a$, with $0 < m < k$, is the area of the region of interest (fig. 2.8). The probability of a symbol occurring in a_i is

$$p_i = a_i/A,$$

and

$$\sum_{i=1}^{k} p_i = \sum_{i=1}^{k} a_i/A = A/A = 1$$

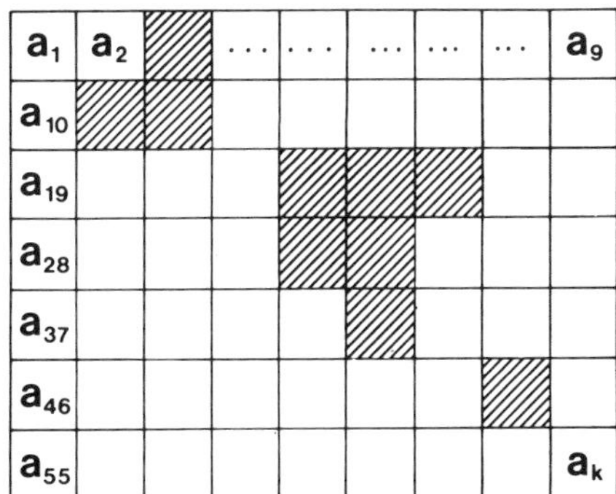

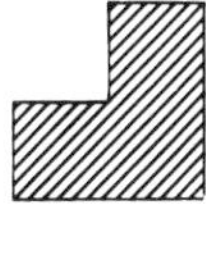

Region of Interest; RI

$$RI \leftarrow a_3 + a_{11} + a_{12} + a_{23} \cdots\cdots a_{53}$$

$$= P_3 + P_{11} + P_{12} \cdots\cdots\cdots\cdots P_{53}$$

$$= \frac{a_3}{A} + \frac{a_{11}}{A} + \cdots\cdots + \frac{a_i}{A} + \cdots\cdots + \frac{a_k}{A}$$

$$= \frac{\Sigma a\,RI}{A}$$

2.8 Unconnected region of interest.

Then the probability of a symbol occurring in RI is

$$p_{RI} = \sum_{i=1}^{m} a_i/A$$
$$= a/A$$

and in $R - RI$ the probability is $1 - p_{RI}$ such that there is no restriction that the a_i be contiguous.

The binomial model is appropriate so long as we are concerned with just two classes of occurrence; in RI, not in RI. By putting $p_i = a_i/A$, and thus $p_{RI} = a/A$, and $p_{R-RI} = (A-a)/A$, we must ensure that a and $(A-a)$ are sufficiently large to be able to receive N objects, because we are assuming that the occurrence of any one object in these two location classes does not affect the probability of the occurrence of any other object in the two location classes.

The other parameter, N, of the binomial probability function may be composed of $N_1, N_2, \ldots, N_j$ symbols without affecting the validity of the model for calculating the probability that n of the symbols will occur in an RI, because the binomial refers to the dichotomy around a, $(A-a)$ for RI, $R-RI$, and *not* to the number of classes of the N_j. Suppose $N = N_1 + N_2$; say the distinction refers to two different years, or to two different kinds of object, then if we wish to determine the probability that n *specified* objects *without* regard to whether they are from N_1 or N_2 will be found in an RI with probability $p = a/A$ then the probability function is

$$b(n; N_1 + N_2, p) = p^n q^{N_1+N_2-n}$$

The distinction into attribute class is not involved in this derivation as the important distinction is whether the objects are in the RI or in $R-RI$, and it is not required that n contains n_1 of the N_1 objects and n_2 of the N_2 objects. Thus for any n objects from $N = \Sigma N_j$ objects we have:

$$b(n; N_1 + N_2, p) = \binom{N_1 + N_2}{n} p^n q^{N_1+N_2-n} \qquad (2.2.2)$$

There is no restriction to two classes of N, so long as the composition of n is unspecified.

2.2.2 Randomness and a multinomial random variable

This restriction to one RI makes the binomial model inappropriate to those problems that involve a number of location classes corresponding, say, to travel-time zones, cost increments, soil type or land use. For such multi-dimensional discrete random variables the binomial can be generalized to the multinomial distribution.

Consider a region, R, of area, A, partitioned into k mutually exclusive RI, designated as $RI_1, RI_2, \ldots, RI_k$, such that the area of RI_1 is a_1, of RI_2 is a_2, and of RI_k is a_k. Again the a_i need not be composed of contiguous units. Suppose that N objects occur in the region, R, with n_i objects in RI_i, such that $\sum_{i=1}^{k} n_i = N$.

Let $p_i = P(a_i) = a_i/A$, and suppose that p_i remains constant for each of the N objects. Again we have $\sum_{i=1}^{k} p_i = 1$. We define the random variables, X_i, to be the number of times that RI_i receives an object, with $i = 1, 2, \ldots, k$. Clearly $\sum_{i=1}^{k} X_i = N$. Thus we wish to know what the probability is that $X_1 = n_1, X_2 = n_2, \ldots, X_k = n_k$. The argument is a straightforward extension of the argument used already for establishing the binomial probabilities when the random variable had just two cases X_1 corresponding to the number of objects, n, received by RI, and X_2 corresponding to the number of objects, $N-n$, received by $R-RI$.

The probability that n_i *specified* objects occur in an RI is given by the product of their respective probabilities, $p_1^{n_1}, p_2^{n_2}, \ldots, p_k^{n_k} = \Pi p_i^{n_i}$* just as in the binomial case we had $p^n q^{N-n}$ when $p = p_1$, $q = (1-p_1) = p_k$ and $n = n_1$, $N-n = n_k$. Again, as the objects are considered to be indistinguishable, the probability that any n_i objects can occur in an RI is given by the *multinomial* coefficient, which tells us how many ways there are of arranging N objects with n_1 of one kind, n_2 of a second kind, to n_k of the kth kind. This is written and calculated as,

$$\begin{bmatrix} N \\ n_i \end{bmatrix} = \frac{N!}{n_1!\, n_2! \ldots n_k!} \qquad i = 1, \ldots, k.$$

combining the arrangements with the probabilities we have

$$\begin{aligned} P(X_1 = n_1, X_2 = n_2, \ldots, X_k = n_k) &= \begin{bmatrix} N \\ n_i \end{bmatrix} \Pi p_i^{n_i} \\ &= \frac{N!}{n_1!\, n_2!, \ldots n_k!} \cdot p_1^{n_1} p_2^{n_2} \ldots p_k^{n_k} \qquad (2.2.3) \end{aligned}$$

It is plain that if $k = 2$ we have

$$\begin{aligned} P(X_1 = n_1, X_2 = n_2) &= \frac{N!}{n_1!\, n_2!} \cdot p_1^{n_1} p_2^{n_2} \\ &= \binom{N}{n} p^n q^{N-n} \end{aligned}$$

Example

Consider the situation shown in fig. 2.9. There are 3 RI of equal area thus $p_1 = p_2 = p_3 = \frac{1}{3}$, and 9 objects occur such that $n_1 = 6$, $n_2 = 3$, $n_3 = 0$. The point probability of getting any 6 objects in RI_1, any 3 objects in RI_2 and no objects in RI_3 is given as

$$\begin{aligned} P(X_1 = n_1, X_2 = n_2, X_3 = n_3) &= \begin{bmatrix} N \\ n_i \end{bmatrix} \Pi p_i^{n_i} \qquad i = 1, 2, 3 \\ &= \frac{N!}{n_1!\, n_2!\, n_3!} \cdot p_1^{n_1} p_2^{n_2} p_3^{n_3} \end{aligned}$$

* The Greek letter capital pi, Π, is used to denote a product of probabilities.

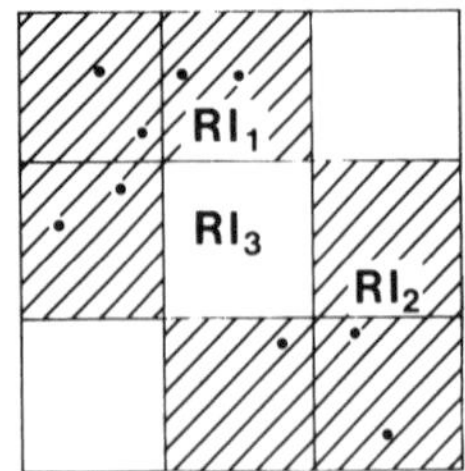

$n_1 = 6;\ \ P_1 = 1/3$

$n_2 = 3;\ \ P_2 = 1/3$

$n_3 = 0;\ \ P_3 = 1/3$

$$P\left[(n_i, RI_i)\right] = P(6,3,0) = \frac{9!}{6!3!0!}\left(\frac{1}{3}\right)^6\left(\frac{1}{3}\right)^3\left(\frac{1}{3}\right)^0$$

$$= \frac{9.8.7.}{3.2}\cdot\left(\frac{1}{3}\right)^9$$

$$= 0{\cdot}004268$$

2.9 Region of interest and a multinomial random variable.

$$= \frac{9!}{6!\ 3!\ 0!}\cdot\left(\frac{1}{3}\right)^6\cdot\left(\frac{1}{3}\right)^3\cdot\left(\frac{1}{3}\right)^0$$

$$= \frac{9.8.7.}{3.2.1.1}\left(\frac{1}{3}\right)^9$$

$$= 0{\cdot}004268$$

In general, any inference about such likelihoods is affected by the definition of the event of interest. If the event specifies n_1 in RI_1, n_2 in RI_2, . . ., n_k in RI_k, then equation 2.2.3 is appropriate. However in such a case the implication is that each arrangement of the n_1 is different. In the example the event is defined so that (6, 3, 0) is different from (6, 0, 3), (0, 3, 6), (0, 6, 3), (3, 6, 0), (3, 0, 6), but each of these permutations has the same probability. Consequently the probability of those 3 values of n_i arising indiscriminately in the RI_i is the sum of the 6 permutations. In this case when the RI_i are arbitrary labels for the location classes, such that any arrangement of the n_i is allowed in the RI_i then equation 2.2.3 has to be modified. The modification consists of summing the probabilities of all the distinct permutations of the n_i. These distinct permutations are the combinations of the n_i. Thus if the 9 objects were allotted such that $n_1 = 4$, $n_2 = 4$, $n_3 = 1$ then only 3 of the permutations result in distinct arrangements; that is $3!/2!\ 1! = 3$. The full table of probabilities is given for this example to consolidate the idea.

(n_1, n_2, n_3)	$\frac{n}{n_j}$	$\begin{bmatrix} N \\ n_i \end{bmatrix}$	$\Pi p_i^{n_i}$	*Point probability*	*Cumulative probability*	X^2
(3, 3, 3)	$\frac{3!}{3!}$	$\frac{9!}{3!3!3!}$	$\cdot\frac{1^3}{3}\cdot\frac{1^3}{3}\cdot\frac{1^3}{3}$	0·085353	0·085353	0
(4, 3, 2)	$\frac{3!}{0!}$	$\frac{9!}{4!3!2!}$	$\cdot\frac{1^3}{3}\cdot\frac{1^3}{3}\cdot\frac{1^3}{3}$	0·384088	0·469441	0·66
(4, 4, 1)	$\frac{3!}{2!}$	$\frac{9!}{4!4!1!}$	$\cdot\frac{1^3}{3}\cdot\frac{1^3}{3}\cdot\frac{1^3}{3}$	0·096022	0·565463	2·00
(5, 2, 2)	$\frac{3!}{2!}$	$\frac{9!}{5!2!2!}$	$\cdot\frac{1^3}{3}\cdot\frac{1^3}{3}\cdot\frac{1^3}{3}$	0·115226	0·680689	2·00

(n_1, n_2, n_3)	$\frac{n}{n_j}$	$\begin{bmatrix}N\\n_i\end{bmatrix}$	$\Pi p_i^{n_i}$	*Point probability*	*Cumulative probability*	X^2
(5, 3, 1)	$\frac{3!}{0!}$	$\frac{9!}{5!3!1!}$	$\cdot\frac{1^3}{3}\cdot\frac{1^3}{3}\cdot\frac{1^3}{3}$	0·153635	0·834324	2·66
(5, 4, 0)	$\frac{3!}{0!}$	$\frac{9!}{5!4!0!}$	$\cdot\frac{1^3}{3}\cdot\frac{1^3}{3}\cdot\frac{1^3}{3}$	0·038409	0·872733	4·66
(6, 2, 1)	$\frac{3!}{0!}$	$\frac{9!}{6!2!1!}$	$\cdot\frac{1^3}{3}\cdot\frac{1^3}{3}\cdot\frac{1^3}{3}$	0·076818	0·949551	4·66
(6, 3, 0)	$\frac{3!}{0!}$	$\frac{9!}{6!3!0!}$	$\cdot\frac{1^3}{3}\cdot\frac{1^3}{3}\cdot\frac{1^3}{3}$	0·025606	0·975157	6·00
(7, 1, 1)	$\frac{3!}{2!}$	$\frac{9!}{7!1!1!}$	$\cdot\frac{1^3}{3}\cdot\frac{1^3}{3}\cdot\frac{1^3}{3}$	0·010974	0·986131	8·00
(7, 2, 0)	$\frac{3!}{0!}$	$\frac{9!}{7!2!0!}$	$\cdot\frac{1^3}{3}\cdot\frac{1^3}{3}\cdot\frac{1^3}{3}$	0·010974	0·997105	8·66
(8, 1, 0)	$\frac{3!}{0!}$	$\frac{9!}{8!1!0!}$	$\cdot\frac{1^3}{3}\cdot\frac{1^3}{3}\cdot\frac{1^3}{3}$	0·002743	0·999848	12·66
(9, 0, 0)	$\frac{3!}{2!}$	$\frac{9!}{9!0!0!}$	$\cdot\frac{1^3}{3}\cdot\frac{1^3}{3}\cdot\frac{1^3}{3}$	0·000152	1·000000	18·00

$\frac{n}{n_j}$ gives the number of distinguishable arrangements among arbitrarily labelled *RI*, when n is the number of *RI* and n_j is the number of *RI* with different frequencies of objects. X^2 is the value to be referred to χ^2 distribution; tabulated here for convenience and discussed in article 2.3.2.

2.10 Exact probabilities and χ^2 approximations for the event in fig. 2.9.

The column of cumulative probabilities shows the implications of the event $P(X_1 \leqslant n_1, X_2 \leqslant n_2, X_3 \leqslant n_3)$.

2.2.3 Randomness and a hypergeometric random variable

In using the binomial distribution as our model we assumed that the two mutually exclusive location classes *RI* and *R-RI* were large enough to receive as many as *N* objects. Indeed this is implicit in the assumptions of equal and independent probabilities. We shall now consider the whole region, *R*, to be composed of small areas, which we can think of as cells or sites, designated as a_i such that each a_i is just large enough to take one object.

The region has *A* cells, a_i, altogether, with *a* of them in *RI* and $A-a$ of them in *R—RI*. There are *N* objects to be distributed, but each time an object is placed, one of these *A* cells is also allotted, so that *N* objects require *N* of the *A* cells. If these *N* objects are allocated randomly to the a_i regardless of whether the a_i are in *RI* or *R—RI*, it is clear that the ratio of available sites in *RI* and in *R—RI* varies with each allocation. Let us define our random variable, *X*, to be the number of occupied a_i in *RI*. Suppose $X = n$, then we wish to ascertain the probability that n of the N objects will occupy n of the a cells in RI, at the same time as $N-n$ of the objects occupy $N-n$ of the $A-a$ cells in $R-RI$. Fig. 2.11 clarifies this arrangement.

	Objects in		
a_i *cells*	*RI*	$R - RI$	*Total cells*
Occupied	n	$N - n$	N
Empty	$a - n$	$A - a - N + n$	$A - N$
Total cells	a	$A - a$	A

2.11 Region of interest and a hypergeometric random variable.

The n objects can be put into the a sites in $\binom{a}{n}$ ways, while, for each of these ways, the $N-n$ objects remaining can be put into the $A-a$ sites outside RI in $\binom{A-a}{N-n}$ ways. Thus the number of different ways of allocating N objects such that n are in the a cells of RI and $N-n$ are in the $A-a$ cells outside RI is given by the product.

$$\binom{a}{n}\binom{A-a}{N-n}$$

But there are $\binom{A}{N}$ ways of allocating N objects among A sites in the region, thus the probability that exactly n of the objects will be in the RI is

$$P(X = n) = \frac{\binom{a}{n}\binom{A-a}{N-n}}{\binom{A}{N}} \qquad (2.2.4)$$

A discrete random variable having the probability distribution given by equation 2.2.4 is said to have a *hypergeometric distribution.* It is different from the binomial because it takes account of the change in N and A available after each allocation. This is a most important discrete random variable and is the basis of a number of tests that have applications to propositions arising from a number of map types. A more extensive treatment of the distribution is deferred until article 2.3.1 when it provides a convenient introduction to an important probability distribution of a continuous random variable, known as the chi-square distribution. One point should be made now, namely that the binomial coefficient $\binom{m}{n} = 0$ for $n > m$ if n and m are non-negative. We obviously cannot use more than a sites or A sites, but in equation 2.2.4 a probability of zero is assigned to that event.

Remark
It is worth emphasizing that in the case of the binomial, the multinomial and the hypergeometric distributions we have just discussed, and in the case of the Poisson distribution which follows, the position of the cells, a_i, or of sets of these, the RI_i, within the map framework *does not affect* the probability function. The probability of each object's outcome in or not in an RI has been constant. We could have considered the a_i as a string of cells just as correctly; indeed all we have done is to put sections of such a string above one another (fig. 2.12). The order in the string or string sections is equally able to be rearranged. These are the conditions of spatial homogeneity that are assumed.

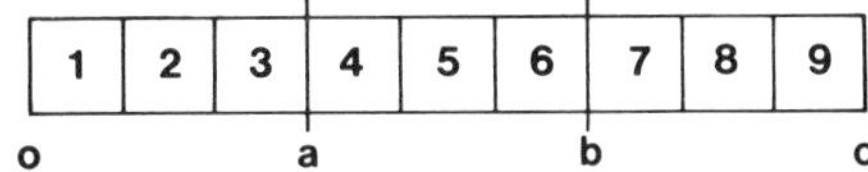

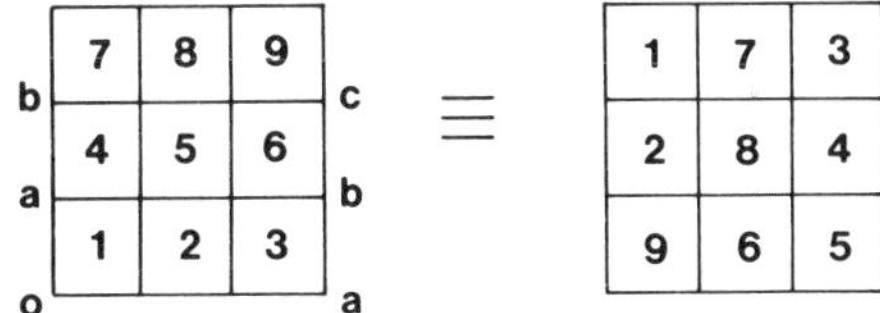

2.12 Arbitrariness of position and random variables considered in this article.

2.2.4 Randomness and the Poisson distribution

Accounts of randomness in point patterns include references to the Poisson distribution rather more often than to the distributions that we have met. Any intention of understanding what is implied by the ideas of randomness depends on an understanding of how all these important distributions are related to point patterns and to each other. The binomial model is a basic model and one to which the Poisson provides an approximation. The usefulness of an approximating distribution arises because it is rare to find an empirical case in which the proportion of the area of the RI is comparatively large when N is comparatively small. The difficulties of evaluating $b\,(n;N,p)$ rapidly become considerable as N increases. The binomial model remains appropriate conceptually for considering the nature of randomness, but the Poisson distribution provides a very convenient method of evaluating the probabilities. It also stands as a distinct distribution with a distinct cartographic analogue. We shall first understand how it arises as an approximation to the binomial. The detailed explanation of the relationship of the two distributions shows what is meant by one distribution providing an approximation to another, and your familiarity with the binomial will help to give you a feeling for the Poisson and for the mechanics of its calculation which many students find strange initially.

2.2.4.1 The Poisson distribution as an approximation to the binomial distribution
The expected number of occurrences in binomial trials is given by Np. In the sort of situation geographers meet, N is large and p is small because the RI is small compared to the total region represented by the map. The probability of an object occurring in such an RI is small, but as N is large the expected number of occurrences is moderate. By the assumptions of the binomial model, N and p are constant, consequently their product Np is constant in any particular situation. When this product is used in the Poisson probability function it is traditionally denoted by the Greek letter, lambda, λ. The expression for the binomial $b(n;N,p)$ becomes, by substitution, that of the Poisson $P(n;\lambda)$. To calculate binomial probabilities for various values of n we used the two parameters N and p according to the rule

$$b(n;N,p) = \binom{N}{n} p^n q^{N-n} \qquad (2.2.5)$$

We now need the corresponding rule for $P(n;\lambda)$ in terms of its parameter λ. This rule for the number of occurrences, n, of a random variable, X, in an RI of area a is given by

$$P(n;\lambda) = \frac{\lambda^n}{n!} \cdot e^{-\lambda} \qquad (2.2.6)$$

This expression clearly needs some explanation, first because it is not apparent how equation 2.2.5 is related to equation 2.2.6, and secondly because the *irrational* constant, e, appears in this text for the first time. We need five steps.

Step 1
An irrational number, like $\sqrt{2}$ or π, is an endless, non-repeating decimal and can be used in mathematical expressions in the same way that other *real* numbers are used.* The value of e to five decimal places is 2·71828. It is important because it arises often in mathematical expressions and it is the base of the *natural logarithm.* The standard notation for a logarithm is

$$\log_a a^x = x \qquad (2.2.7)$$

where the *base* of the logarithm is whatever real number a stands for. The a could be, say, 2, or 10 in which latter case they are called *common* logarithms. When e is the base they are called natural logarithms. The exponent to which the base is raised, x, is called the logarithm of the number a^x.

Example 1

$$\log_2 2^3 = \log_2 8 = 3$$

So that we say 3 is the logarithm of 8 to the base 2.

Example 2

$$\log_{10} 8 = \log_{10} 10^{\cdot 9031} = 0{\cdot}9031$$

So that we say 0·9031 is the common logarithm of 8 (i.e. to the base 10).

* $\frac{1}{4} = 0{\cdot}25000\ldots$; $2{\cdot}00\ldots$; $\frac{1}{7} = 0{\cdot}142857142857\ldots$ are real numbers but are not irrational.

Example 3

$$\log_e 8 = \log_e \; e^{2\cdot 0794} = 2\cdot 0794$$

So that we say 2·0794 is the natural logarithm of 8 (i.e. to the base e which is approximately 2·71828). Thus example 3 is equivalent to

$$\log_{2\cdot 71828} \; 8 = \log_{2\cdot 71828} \; 2\cdot 71828^{2\cdot 0794} = 2\cdot 0794$$

In general, if $\log_a N = x$ then $N = a^x$. Thus

1. if we know $\log_2 N = 3$ then we know $N = 2^3 = 8$
2. if we know $\log_{10} N = 0\cdot 9031$ then we know $N = 10^{0\cdot 9031} = 8$
3. if we know $\log_e N = 2\cdot 0794$ then we know $N = e^{2\cdot 0794} = 8$

In algebraic terms we write $\log_e e^x = x$. It is customary to abbreviate $\log_e$ to ln, so we find ln N and know this means $\log_e N$. If we have $\ln N = x$ then we know $N = e^x$. A useful theorem of logarithms is

$$\log_a (x^y) = y \log_a x \tag{2.2.8}$$

This can be seen by putting

	$u = \log_a (x^y)$	
so that	$a^u = x^y$	from 2.2.7
then put	$v = \log_a x$	
so that	$a^v = x$	from 2.2.7
consequently	$(a^v)^y = a^u$	by substitution
giving	$yv = u$	

Then by substitution we have

$$y \log_a x = \log_a (x^y) \qquad \text{as required.}$$

For $\ln (p)^n$ we write $n.\ln (p)$.

Step 2

The values of e^x are calculated by summing the first n terms of an infinite series of terms known as *Taylor's Series* for e^x. This is written as

$$e^x = \sum_{k=0}^{\infty} \frac{x^k}{k!} = \frac{x^0}{0!} + \frac{x^1}{1!} + \frac{x^2}{2!} + \frac{x^3}{3!} + \ldots + \frac{x^n}{n!} + \ldots \tag{2.2.9}$$

We can use this series to evaluate e itself. As $e^1 = e$,* we put

$$e = e^1 = \sum_{k=0}^{\infty} \frac{1^k}{k!} = \frac{1^0}{0!} + \frac{1^1}{1!} + \frac{1^2}{2!} + \frac{1^3}{3!} + \frac{1^4}{4!} + \frac{1^5}{5!} + \ldots$$

$$= 1 + 1 + 0\cdot 5 + 0\cdot 16667 + 0\cdot 04166 + 0\cdot 00833$$

so with $k = 5$ $\qquad e^1 = 2\cdot 71666$

* Any number raised to the power 1 is itself.

The increment diminishes rapidly because the denominator is a factorial: $e \simeq 2.71828$ when $k = 8$ and is used often as an acceptable working approximation to e. Exponents can be negative as well as positive. We can find e^{-x} as well as e^{x}. In general, $n^{-x} = \frac{1}{n^x}$. To evaluate e^{-x} we use the Taylor's Series expansion with $-x$ instead of x giving the same terms as before for $x = 1$, but added and subtracted alternately.

A great many important algebraic expressions are evaluated in a similar way by using a series that converges to a value for that expression. We need to know one more such expression before we can show the relation between the binomial and the Poisson distribution functions and that is the expression for

$$\ln(1-p) = -\sum_{k=1}^{\infty} \frac{p^k}{k} = -p - \frac{p^2}{2} - \frac{p^3}{3} - \frac{p^4}{4} \ldots \qquad (2.2.10)$$

Step 3

You will recall that the problem is to show why equation 2.2.5 can be approximated by equation 2.2.6. With $q = (1-p)$, if we put $n = 0$ in equation 2.2.5 we get

$$b(0; N, p) = \binom{N}{0} p^0 (1-p)^{N-0}$$

$$= (1-p)^N$$

Thus

$$\ln b(0; N, p) = \ln(1-p)^N$$

$$= N \ln(1-p)$$

$$= -N \sum_{k=1}^{\infty} \frac{p^k}{k} \quad \text{from equation 2.2.10}$$

Now if we put $\lambda = Np$, and $p = \frac{\lambda}{N}$. Recalling that $\frac{p^k}{k} = \frac{p^k}{1} \cdot \frac{1}{k}$ we can write

$$\ln b(0; N, p) = -N \sum_{k=1}^{\infty} \left(\frac{\lambda}{N}\right)^k \cdot \frac{1}{k}$$

$$= \sum_{k=1}^{\infty} -\frac{\lambda^k}{kN^{k-1}} \quad \text{because } N \cdot \left(\frac{1}{N}\right)^k = \frac{1}{N^{k-1}}$$

$$= -\lambda - \frac{\lambda^2}{2N} - \frac{\lambda^3}{3N^2} - \ldots$$

When N is large, terms after the first one have a rapidly diminishing effect on the sum, so for this reason we write

$$\ln b(0; N, p) \simeq -\lambda^*$$

* $\simeq$ means approximately equal.

and $$b(0;N,p) \simeq e^{-\lambda}$$

As we are using $\lambda = Np$ we can write

$$b(0;N,p) = p(0;\lambda)$$

Nearly there.

Step 4

Now there remains the problem of terms other than $n = 0$. We can calculate the probabilities for $n = 1, 2, \ldots N$ occurrences by a *recursive formula*, that is a formula that uses the previous result as the basis for calculating the next value. To do this we need to find whether there is a consistent relationship between two adjacent terms in the binomial: that is between

$$b(n;N,p) = \binom{N}{n} p^n (1-p)^{N-n} \qquad (2.2.11)$$

and $$b(n-1;N,p) = \binom{N}{n-1} p^{n-1} (1-p)^{N-n+1}$$

$$[N-(n-1)] \qquad (2.2.12)$$

To do this we put these equations in the form of a ratio with equation 2.2.11 as the numerator, and expand the terms to give*

$$\frac{b(n;N,p)}{b(n-1;N,p)} = \frac{\dfrac{N!}{n!(N-n)!} \cdot p^n (1-p)^{N-n}}{\dfrac{N!}{(n-1)!(N-n+1)!} p^{n-1} (1-p)^{N-n+1}}$$

$$= \frac{N!}{n!(N-n)!} \cdot \frac{(n-1)!(N-n+1)!}{N!} \cdot \frac{p^n (1-p)^{N-n}}{p^{n-1} (1-p)^{N-n+1}}$$

$$= \frac{(N-n+1)}{n} \cdot \frac{p}{(1-p)}$$

$$= \frac{pN-p(n-1)}{n(1-p)} \qquad (2.2.13)$$

As $pN = Np$, we can replace pN by λ to give

$$\frac{P(n;\lambda)}{P(n-1;\lambda)} = \frac{\lambda-p(n-1)}{n(1-p)} \qquad (2.2.14)$$

As p approaches zero, that is when the area of a is very small compared to the area A, then $p(n-1)$ approaches zero, while $(1-p)$ approaches 1. Thus, $\lambda - p(n-1)$ is dominated by λ and $n(1-p)$ tends towards n. The ratio on the left hand side of equation 2.2.14 can be expressed as

$$\frac{P(n;\lambda)}{P(n-1;\lambda)} \simeq \frac{\lambda}{n} \qquad (2.2.15)$$

so $$P(n;\lambda) \simeq P(n-1;\lambda) \cdot \frac{\lambda}{n}$$

which is the recursive formula we wanted.

* **Recall exercises 4, 5, 20 and 21 in section 1.**

Step 5

We can now substitute our first term $P(0;\lambda) = e^{-\lambda}$ and calculate for $n = 1, 2, \ldots, k$, by

$$P(1;\lambda) \simeq \frac{\lambda}{1} . e^{-\lambda}$$

$$P(2;\lambda) \simeq \left(\frac{\lambda}{1} \cdot e^{-\lambda}\right) \cdot \frac{\lambda}{2} = \frac{\lambda^2}{1\cdot2} \cdot e^{-\lambda}$$

$$P(3;\lambda) \simeq \left(\frac{\lambda^2}{1\cdot2} \cdot e^{-\lambda}\right) \frac{\lambda}{3} = \frac{\lambda^3}{1.2.3} \cdot e^{-\lambda}$$

and in general

$$p(n;\lambda) \simeq \left(\frac{\lambda^{n-1}}{(n-1)!} \cdot e^{-\lambda}\right) \cdot \frac{\lambda}{n} = \frac{\lambda^n}{n!} \cdot e^{-\lambda} \qquad (2.2.16)$$

This, then, is the derivation of equation 2.2.6 and it is the general Poisson probability function. Have a well-deserved rest.

Example

Working through the arithmetic procedures of calculating the probabilities for n of a binomial and a Poisson distribution function will confirm your appreciation of their mechanics as well as provide an illustration of the approximation. Putting $N = 50$, $a/A = p = 0{\cdot}01$, $\lambda = Np = 0{\cdot}5$, set $n = 0, 1, 2, 3, 4$, gives the results tabulated below (fig. 2.13).

n	$b(n;50,0{\cdot}01)$	$P(n;0{\cdot}5)$
0	0·605006	0·606531
1	0·305559	0·303265
2	0·075618	0·075816
3	0·012221	0·012636
4	0·001450	0·001580

$$b(0;50,0{\cdot}01) = \binom{50}{0} 0{\cdot}01^0 . 0{\cdot}99^{50} \qquad P(0;0{\cdot}5) = e^{-0\cdot5}$$

$$= 1.1.0.99^{50}$$

$$b(1;50,0{\cdot}01) = \binom{50}{1} 0{\cdot}01^1 . 0{\cdot}99^{49} \qquad P(1;0{\cdot}5) = \frac{0{\cdot}5}{1} . e^{-0\cdot5}$$

$$= 50.0.01.0.611117 \qquad = 0{\cdot}5.0{\cdot}606531$$

$$b(2;50,0{\cdot}01) = \binom{50}{2} 0{\cdot}01^2 . 0{\cdot}99^{48} \qquad P(2;0{\cdot}5) = \frac{(0{\cdot}5)^2}{1{\cdot}2} . e^{-0\cdot5}$$

$$= 1225.0{\cdot}0001.0{\cdot}617290 \qquad = \frac{0{\cdot}25}{1{\cdot}2} . 0{\cdot}606531$$

and so on.

2.13(a) Comparison of exact binomial probabilities and their Poisson approximation.

Computers and many electronic calculating machines calculate values for e automatically. When there is no access to such machines tables of common logarithms can be used in the following way, generally with diminished accuracy.

$$e^{-\lambda} = -\lambda \log_{10} e \qquad \log_{10} e \simeq 0{\cdot}43429$$

$$\text{putting } \lambda = 0{\cdot}5 = -0{\cdot}5\,(0{\cdot}43429)$$
$$= -0{\cdot}217145$$

To convert this to a real number it needs referring to an *antilogarithm*, i.e. we put

$$\text{antilog}_{10} \quad = 1 - 0{\cdot}217145$$
$$= \bar{1}{\cdot}78285$$
$$\simeq 0{\cdot}6065$$

From a table of logarithms we find that $\left.\begin{array}{l}\bar{1}{\cdot}7825 = 0{\cdot}606\\ \bar{1}{\cdot}7832 = 0{\cdot}607\end{array}\right\} \Rightarrow 0{\cdot}6065$

2.13(b) Calculating $e^{-\lambda}$ using logarithms.

2.2.4.2 The Poisson process

As $\dfrac{\lambda^n}{n!}$ is a Taylor Series of the same form as equation 2.2.9 for e^{λ}, the expression

$$\sum_{n=0}^{\infty} P(n;\lambda) = e^{-\lambda} e^{\lambda} = e^0 = 1 \qquad (2.2.17)$$

when summed over all values of n. Consequently the Poisson distribution is a probability function in its own right, and is used for calculating the probability of getting exactly n occurrences in a space a. These occurrences can be seen as events along a line, such as time or distance, as events in a plane, such as a map, or events in higher dimensional space. The assumptions of the model are that the conditions of the experiment remain constant over the space, and that non-overlapping intervals of the line, or regions of interest in the plane are independent. That is that the outcome in one *RI* does not affect the outcome in any other *RI*. If this *RI* is of arbitrary size, a, then the probability of finding exactly n objects in that *RI* of area a is

$$P(n;\lambda_a) = e^{-\lambda_a} \frac{(\lambda_a)^n}{n!}$$

where λ_a is just given the subscript a to indicate that it refers to an arbitrary space of interest. The probability of *no occurrence* is $e^{-\lambda_a}$ and of *one or more occurrences* is $1 - e^{-\lambda_a}$.

The value of λ is a measure of the density of occurrences along the line or in the plane: *that is, it is a measure of the expected number of occurrences per unit of measurement.* The higher the density, the smaller the chance of finding no point. A map of a region of area A may be divided into RI_i of area a. There are N such RI_i, such that N_0 is the number of areas with no objects, N_1 the number with one object, to N_n the number of areas with n objects. The total number of objects is then found from the sum of products, $0(N_0) + 1(N_1) + 2(N_2) +, \ldots + n(N_n) = S$. The total number of objects divided by the total number of areas is the mean number per area, a, and is denoted as λ_a. Thus $S/N = \lambda_a$. The average refers to a, but it is independent of the shape of the areas. This then is the procedure for using the Poisson probability function as the model for testing the assertion that the incidence of objects in a region is random. The random variable, X, is the number of objects in the a, when X assumes the possible values $0, 1, 2, \ldots, n, \ldots$ in

$$P(X = n) = P(n; \lambda_a) = e^{-\lambda_a} \frac{(\lambda_a)^n}{n!} \tag{2.2.18}$$

Example

A map of Greater London showing the location of the homes of Birkbeck geography undergraduate students is given as fig. 2.14. Depending on your preconceptions this may encourage a variety of assertions about the distribution. We might suppose that the yield is homogeneous within the region. If this is the case then the frequency of domicile within small areas a may be taken as a Poisson random variable, and the Poisson distribution function is the appropriate model for calculating the probability that there will be N_0 areas which contain homes of no students at Birkbeck, N_1 areas with homes of one student to N_n areas with homes of n students. Common sense may seem to encourage the belief that certain parts of Greater London should have a higher proportion of the young teachers or the other professional people who constitute the majority of Birkbeck students, and that as a result there will be clustering. This belief may be reinforced by the recognition that house prices and land use are unlikely to be homogeneous in the region and these may also foster clustering.

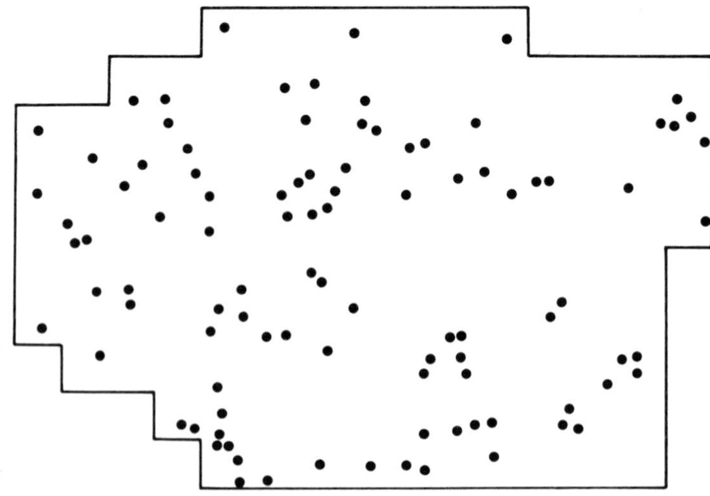

2.14 Distribution of homes of Birkbeck undergraduates.

The map is divided into 126 areas, a, where a is 4 square miles. There are 59 areas with no student homes, 43 with 1, 17 with 2, 6 with 3, and 1 with 4 student homes. Thus there are 0 (59) + 1 (43) + 2 (17) + 3 (6) + 4 (1) = 99 students in the sample. The average number of students per area a is 99/126 = 0·785714 which is λ_a. Under Poisson expectation we find the following probabilities:

$$P(X=0) = e^{-\lambda_a} = 0{\cdot}455794$$

$$P(X=1) = \frac{\lambda_a^1}{1!} e^{-\lambda_a} = 0{\cdot}358266$$

$$P(X=2) = \frac{\lambda_a^2}{2!} e^{-\lambda_a} = 0{\cdot}140692$$

$$P(X=3) = \frac{\lambda_a^3}{3!} e^{-\lambda_a} = 0{\cdot}036848$$

$$P(X=4) = \frac{\lambda_a^4}{4!} e^{-\lambda_a} = 0{\cdot}007238$$

In order to see whether this model provides an adequate description of the actual figures we must multiply each probability by N. The full result is tabulated (fig. 2.15).

1	2	3	4	5	6
n	N_n	$n(N_n)$	$P(n;\lambda)$	$N.P(n;\lambda)$	X^2
0	59	0	0·455794	57·43	0·043
1	43	43	0·358266	45·14	0·101
2	17	34	0·140695	17·73	0·030
3	6	18	0·036848	4·64	0·399
4	1	4	0·007238	0·91	0·003
$n \geqslant 5$	0	0	$1-\sum_{n=0}^{4} P(n;\lambda)$	126—125·85	
Total	126	99			0·576

$$\lambda = S/N_n = 99/126 = 0{\cdot}785714$$
$$X_A = 99; A = \text{number of unit areas} = 126$$
$$\text{then } \lambda = E(X_A)/A = 99/126$$

2.15 Probability distribution of the event in fig. 2.14 as a Poisson random variable.

It is plain from a comparison of columns 2 and 5 that the Poisson model provides expected frequencies that are remarkably similar to the actual frequencies. The two sets of figures are sufficiently similar for us to accept that the Poisson process is an adequate statement of the distribution of students homes in the region. We conclude that the evidence does not contradict the belief that Greater

London is homogeneous in its yield of Birkbeck geography undergraduates, and that their frequency is a random variable with a Poisson distribution with parameter λ_a. Birkbeck attracts students from much further afield than Greater London and there is no reason to suppose the pattern is the same throughout the southeast, nor that it is true, for, say, postgraduates within Greater London. There is no suggestion that any individual undergraduate makes anything but a well-reasoned case for his decision to come to Birkbeck, but just that the whole of Greater London appears to be homogeneous for such decisions.

The Poisson distribution is applicable to a wide variety of empirical phenomena, but as with any mathematical model there are underlying assumptions that must be recognized and met. To meet them exactly makes the results of the model fully acceptable. Often the conditions are not met certainly, and some models are more acceptable than others because their conditions seem more appropriate to the actual situation. Let us examine informally the assumptions of the Poisson distribution function as they apply to maps and the regions they represent. The background to this discussion is a region, R, over which objects occur. We define X_a as the number of objects in a small cell of that region of size a. This discrete random variable can take on values 0, 1, 2, ... Informally the assumptions are

1. The number of objects occurring in non-overlapping cells are independent random variables.
2. The distribution of the number of objects occurring in any area depends only on the size of the area and not on its position with respect to other areas.
3. a) If the area is sufficiently small the probability of getting exactly one occurrence is proportional only to the size of that area.
   b) The probability of getting two or more occurrences in a sufficiently small area is negligible.

These assumptions interpreted in this areal context mean that the number of objects occurring in non-overlapping cells of the region represent independent random variables, and that the probability of more than one object occurring in a very small area of the region is negligible. In the region considered, the density, λ, of objects is assumed constant over the whole region. That means that in an area of size A units there would be on the average λA objects. Thus if X_A represents the number of occurrences of objects within a specified region of area A, then $E(X_A) = \lambda A$, and then $\lambda = [E(X_A)]/A$ is the expected density of objects. When these assumptions are met we say we have a *Poisson process.*

This example is attractive because it involves people making decisions in a space that appears to vary considerably. My original choice of example to illustrate the use of the Poisson model concerned the occurrence of meteoroid craters in the USA as revealed by satellite photographs. For earth or lunar craters the incidence is expected to be Poisson because the flux of meteoroids is uniform in space with only slight gravitational modification. The success of a model applied to inanimate phenomena does not mean that an equally successful application can be made with animate behaviour even if there are gross similarities, because man's decision-space is often inhomogeneous.

Imagine that we have a completely random process on the plane, in which the occurrence of objects is described by the homogeneous Poisson model. Let us transform the plane by measuring from one point on a different scale. The homogeneous Poisson process is then converted into a non-homogeneous Poisson process. Consequently the numbers of objects occurring in non-overlapping areas, a, of the transformed plane are still independent, but have distributions which now depend on position and orientation of the area in relation to the origin, as well as to the area's size as before. So long as the function relating the original to the transformed plane is known, no difficultly arises and the inverse transformation will restore the original process.

By assuming that equality of area is given by distance scale in the region and in the map, and referring frequency of occurrence in such areas to the homogeneous Poisson process, we may well be missing some important generalizations on the nature of locational relations in human geography. The attempts to devise scales that assign numerals to events such that those numerals conform recognizably more closely with man's response to physical distance is a recognition of the inhomogeneity of man's decision-space. So long as the transformation is one-one the task is comparatively straightforward. More complex situations are all too easy to envisage and to support on other considerations. Patterns of houses, towns and factories may be a compound of many scales, such as cost, time and aesthetics, and these are not necessarily simple scales: composite scales may be compounded by having multiple origins. Thus the failure of a simple, homogeneous Poisson process may not mean rejecting the process as the suitable model, but of ensuring that the space is homogeneous for the appropriate measuring scales. Perhaps in Greater London, distance, cost and time are comparable scales, whereas beyond that region different transformations are needed (fig. 2.16).

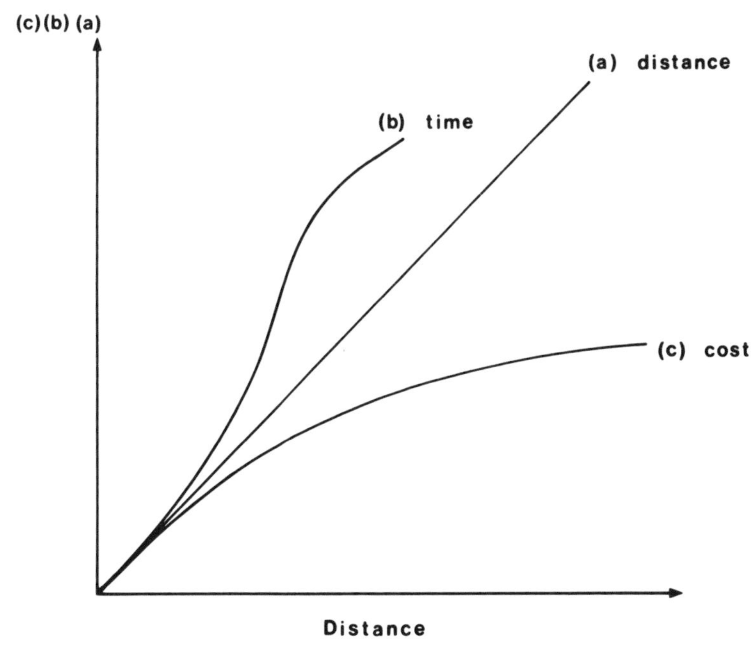

2.16 Schematic comparison of different scales.

Another approach is to extend the process. For example, we could envisage a Poisson process generating events such as the occurrence of a factory, and these events then becoming sources of a secondary Poisson process which generated similar events locally, that is within some *RI* or over some time interval. This sort of compound Poisson process could operate within *RI* that were homogeneous with respect to costs or to profitability of the industry. Such regions could be defined by some boundary condition and the Poisson process giving rise to the initial event could be uniform within that region. Each such outcome of this process could lead to a second such process, also Poisson, for subsequent factories within that region. The result would yield a clustered distribution (Fisher 1972; Daley and Vere-Jones 1972). The distribution of factories under some such compound point process provides a model for the papermaking industry in England and Wales at selected periods and within regions bounded by certain costs. The evolution of one distribution over time can also be accommodated by an extension of point processes as indicated by Gani (1972). The application of such processes in these elaborate ways to geomorphological problems is implicit in the work of Marcus (1972). Surely the more thorough exploration of what transformations of distance give appropriate measurement scales for response space, together with the use of compound point processes for events in such space, will form a profitable and exciting basis of the development of geography.

All such developments arise from the simple notion of a probability function of a random variable. I hope this first part of section 2 will provide the reader with a useful start to understanding those parts of the recent geographic literature that make frequent reference to Poisson processes. It is my experience that very few undergraduate students grasp the implications of using this probability function as the basis for an assertion of randomness. The binomial and the Poisson distributions are related to other important functions that are referred to by geographers. These are very briefly discussed in the exercises. Nearest-neighbour statistics are discussed by many geographers; unfortunately their full understanding involves considerations that are beyond the intended scope of this book. The use of descriptive statistical measures of mean or median centres in the plane is straightforward and does not involve assumptions of population form unless inferences are made about that population. Once such inferences are made and their appropriate test statistics invoked the concepts involved are again beyond the deliberate restriction of this book.

We shall now treat the situations that arise when objects are distinguished with respect to a number of attributes and a number of location classes.

2.3 Propositions of independence (1)

Properties: location class, one other attribute
Measurement scale: nominal, categories, on both attributes
Statistical procedures: hypergeometric random variable (Fisher's Exact test), chi-square approximation

It is natural to progress from maps used to show point symbols distinguished by location class, to maps showing one other property. Such maps are common and involve no real change in symbolism apart from making the symbols distinguishable into at least two classes, by shading, shape or size. Such maps contain symbols that carry information on two attributes of the same set of objects. A typical requirement is for objects to be put simultaneously into one of two possible classes for each of two attributes, as for example success or failure, location class 1 or location class 2. These location classes might be urban, non-urban; glaciated, periglacial; treated, non-treated. The possibilities are numerous and arise in all aspects of geography. In such circumstances one plain requirement is that the geographer make some statement about the dependence of the two classifications. If there is some association between the two classifications, then knowing that an object belongs to a particular location class is equivalent to increasing the information about its survival. Conversely the statement that two classifications are independent is the same as asserting that knowing an object belongs to one dichotomy does not tell us anything about its membership of the other dichotomy.

2.3.1 Fisher's exact test

Consider that we have N similar objects from two location classes with n objects from location class 1 and $N-n$ objects from location class 2. These N objects are also put into one of the two mutually exclusive categories, say survived or failed, such that there are R survivors and $N-R$ failures. We assume that the N objects are from a common, homogeneous population of survivors and failures which have been allocated arbitrarily to the two location classes, such that n of the N objects were put in location class 1 and the remaining $N-n$ were put into location class 2. The number of objects in both dichotomies, that is four categories, is fixed : these are the marginal totals or marginal frequencies. If the assumption of a homogeneous population is correct then the differences in cell frequencies are due entirely to chance. If the objects in location class 1 and location class 2 are not identical in terms of survival, then the combined sample is not homogeneous and the objects in each location class will be biased towards survival or failure.

Consequently we imagine a pool of N objects, classified on two dichotomies with fixed marginal totals, n, $N-n$, R, $N-R$. Then we suppose that these N objects are allocated randomly to one of the four possible conjunctions of the categories. We wish to determine the probability distribution of these cell values with the fixed marginal totals. Let the random variable, X, be the number of occurrences in one cell, say the conjunction of location class 1 and survival. Let r be any possible value of X. Then we want to know the probability function that gives us the distribution values of r. That is $P(X = r)$. All this can be put diagrammatically (fig. 2.17).

The diagram overleaf shows that we need only calculate the probability for any one cell, because, with fixed marginal totals, knowing one cell enables all the other cell frequencies to be calculated. The diagram also reveals the similarity of this situation to that of the previous part (article 2.2.3) when the appropriate distribution was that of a hypergeometric random variable.

We wish to determine the probability of getting r survivors in a sample of size n when there are R survivors altogether in N objects altogether. The number of ways

	Attribute		
	Survived	*Failed*	*Total*
Location class 1	r	$n-r$	n
Location class 2	$R-r$	$N-R-n+r$	$N-n$
Total	R	$N-R$	N

2.17 Notation for Fisher's exact test, 2 × 2 case.

of getting r things from R is simply the number of ways of combining R objects taking r each time. This is $\binom{R}{r}$. For each such combination there remain $\binom{N-R}{n-r}$ ways of taking $n-r$ non-survivors in location class 1 from the total of $N-R$ non-survivors. Thus, for a sample of size n containing r survivors and $n-r$ non-survivors there are exactly $\binom{R}{r}\binom{N-R}{n-r}$ different ways of yielding such a sample. However if we took no account of the number of survivors we could get $\binom{N}{n}$ different samples of size n from N objects altogether. Thus only $\binom{R}{r}\binom{N-R}{n-r}$ of the $\binom{N}{n}$ possible ways of getting n of the N objects in location class 1 will contain exactly the r survivors required. Thus the point probability that location class 1 has a particular number, r, of survivors is given by

$$P(X=r) = \frac{\binom{R}{r}\binom{N-R}{n-r}}{\binom{N}{n}} \tag{2.3.1}$$

$$= \frac{\dfrac{R!}{r!\,(R-r)!}\;\dfrac{(N-R)!}{(n-r)!\,[(N-R)-(n-r)]!}}{\dfrac{N!}{n!\,(N-n)!}}$$

$$= \frac{R!\,(N-R)!\,n!\,(N-n)!}{r!\,(R-r)!\,(n-r)!\,(N-R-n+r)!\,N!}$$

This is the probability distribution of a hypergeometric random variable. It arose from considering the probability of getting r objects in a sample of size n. It is plain that the derivation is equivalent to the probability of getting r objects in a sample of size R. The equivalent statement is:

$$P(X=r) = \frac{\binom{n}{r}\binom{N-n}{R-r}}{\binom{N}{R}} \tag{2.3.2}$$

The probability distribution refers to the entire 2 X 2 table with its fixed marginal frequencies arising from dichotomizing objects on each of two properties. Note that in this probability function the sum of the upper symbols in the numerator gives the upper symbol in the denominator and the sum of the lower symbols in the numerator gives its lower symbol in the denominator.

Example
Figs. 2.18 and 2.19 depict a common situation. We may suppose that these figures summarize all the known information on the survival of factories in the region.

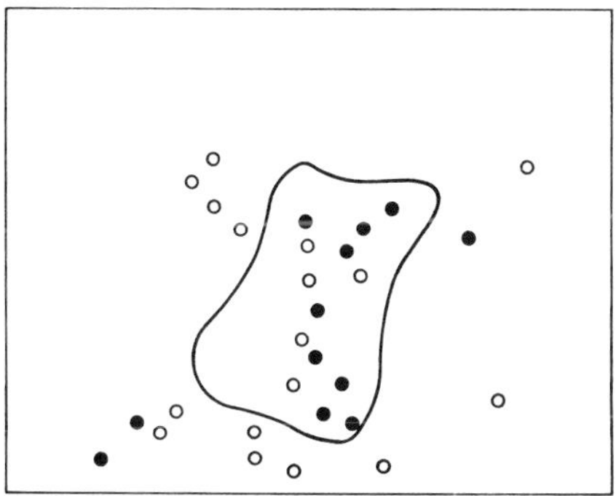

2.18 Objects and attributes as a hypergeometric random variable.

	Attribute		
	Survived	*Failed*	
Urban	9	5	14
Non-urban	3	12	15
	12	17	29

2.19 Summary table for fig. 2.18.

We wish to test the assertion that urban factories are more likely to survive than non-urban factories. As this is the statement whose truth we wish to establish we put it as our alternative hypothesis. This is the statement we accept if we reject the null hypothesis that the proportion of factories in the non-urban location class is the same or is less than the proportion in the urban location class. Define the rejection region at a nominal $\alpha : \leqslant 0{\cdot}05$. We wish to calculate the probability that $r \geqslant 9$ with the given marginal totals. The point probability is given by substituting the numbers in fig. 2.19 in the equation 2.3.1 as

$$P(X=9) = \binom{12}{9}\binom{17}{5}\Big/\binom{29}{14}$$

$$= \frac{12!\ 17!\ 14!\ 15!}{9!\ 3!\ 5!\ 12!\ 29!}$$

$$= 0{\cdot}017553$$

The cumulative probability is given by summing the above equation over the values $r = 10, 11, 12 = R$. The cell frequencies and the point probabilities corresponding to these values of r are tabulated in fig. 2.20.

a	b	c	d	P	X^2	X_y^2
12	2	0	15	0·0000017535	21·448	18·541
11	3	1	14	0·000105	15·43485	12·61287
10	4	2	13	0·00202	10·0755	7·8229
9	5	3	12	0·01755	5·855	4·177
8	6	4	11	0·078986	2·7727	1·6587
7	7	5	10	0·198595	0·8292	0·2845
6	8	6	9	0·289618	0·02437	0·04891
5	9	7	8	0·248244	0·3581	0·04891
4	10	8	7	0·124122	1·8304	0·9519
3	11	9	6	0·035105	4·4414	2·9936
2	12	10	5	0·005265	8·1909	6·1738
1	13	11	4	0·001105	13·0791	10·4927
0	14	12	3	0·0000015532	19·1059	15·9502

r a	n − r b
R − r c	(N − R) − (n − r) d

$N = a + b + c + d$

$P = f(x) = \binom{R}{r}\binom{N-R}{n-r} / \binom{N}{n}$

$X^2 = N(ad - bc)^2/(a+b)(c+d)(a+c)(b+d)$

$X_y^2 = X^2$ with Yates correction

$= N(|ad - bc| - 0{\cdot}5N)^2/(a+b)(c+d)(a+c)(b+d)$

2.20 Summary tables of all possible 2 × 2 values for fig. 2.18 showing their exact probabilities and their X^2 approximation.

The exact cumulative probability is 0·019685 which lies in the rejection region, consequently our decision is to reject H_0 and accept H_1. The rejection region could have been bounded by an $\alpha : \leqslant 0{\cdot}02$ and have led to the same decision.

Let us examine the implications of this decision by tabulating all the possible cell frequencies from these fixed marginal totals together with their point probabilities (fig. 2.20 above). The decision to reject H_0 is made because if the 29 factories were from a common population of survivor-failure measurements, the allocation of these factories at random to cells with these marginal totals would result in that particular set of four values about 2 times in 100. This is sufficiently unlikely to make it more reasonable to suppose that the population is not homogeneous. The dissimilarity is taken to be that objects in location class 1 have a higher, but unspecified, proportion of survivors.

The distribution of exact probabilities is asymmetric because the marginal frequencies are unequal (fig. 2.21). This becomes an important consideration if H_0 is undirected and requires a two-tailed test. If we wish to know whether we can assert justifiably that the likelihood of survival is different in one location class, then the null hypothesis is stated as

H_0 : there is no difference in survival by location class

H_1 : there is a difference in survival

That is, H_1 states that the classes are not independent, but it does not state how they are dependent. They could be dependent in that a disproportionate number of survivors are urban or that a disproportionate number of survivors are non-urban. When both possibilities are accepted, a two-tailed test is required. If we specify one of the alternatives then we reduce the part of the probability distribution to which the decision to reject is referred; this restriction increases the discrimination of any test because it confines attention to a reduced range of the test statistic.

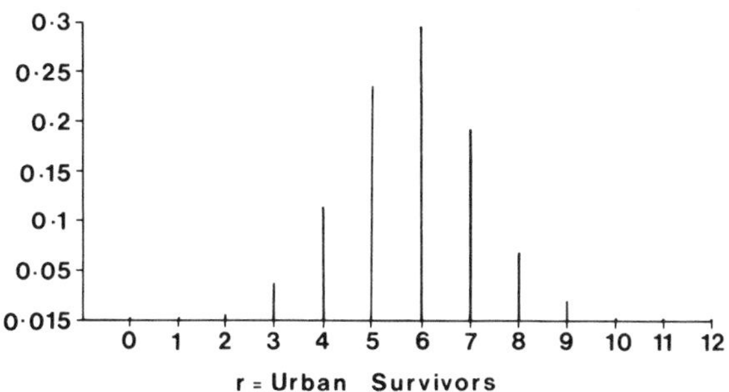

2.21 Probability distribution for fig. 2.20.

In the case where the column or row marginal frequencies are equal, the two-tailed test is double that of the one-tailed test, because the distribution is symmetric. When it is asymmetric, we have to consider the set of frequencies that have a probability as extreme in the opposite direction as in the calculated direction. The need for this can be seen from the previous example and the cell frequencies in fig. 2.20 give tangible expression to the problem. We need to provide a formula that enables us to do this in the general case without the need to calculate all the tables and their values.

In essence we have argued that we expect the proportion of survivors in location class 1, that is r/n, and the proportion of survivors in location class 2, that is $(R-r)/(N-n)$, to be in the ratio $Rn/N : R(N-n)/N$. We have defined the one-tailed rejection region for the H_0 of independent population proportions of survivors to be those values of r that cause r/n to be greater than $(R-r)/(N-n)$ by a particular amount corresponding to the significance level. Any value of r which falls in the rejection region is sufficiently large, that is extreme in one direction, to lead to one-tailed rejection. Let this actual value of r used in the test be denoted by t, then

$$t = \frac{r}{n} - \frac{R-r}{N-n} \tag{2.3.3}$$

In the case of a two-tailed test we want those values of r that cause r/n to be either greater than or less than $(R-r)/(N-n)$ to the same extent. Call this unknown value of r, x, and, because it falls in the opposite one-tailed rejection region, define $-t$ as

$$-t = \frac{x}{n} - \frac{R-x}{N-n} \tag{2.3.4}$$

Thus we can equate the right-hand sides of t, $-t$, and solve for x.

$$\frac{x}{n} - \frac{R-x}{N-n} = -\left(\frac{r}{n} - \frac{R-r}{N-n}\right)$$

$$\frac{x}{n} - \left(\frac{R}{N-n} - \frac{x}{N-n}\right) = -\frac{r}{n} + \frac{R-r}{N-n}$$

$$\frac{x}{n} + \frac{x}{N-n} = -\frac{r}{n} + \frac{R-r}{N-n} + \frac{R}{N-n}$$

$$\frac{(N-n)x + nx}{n(N-n)} = \frac{-(N-n)r + n(R-r) + nR}{n(N-n)}$$

$$\frac{Nx - nx + nx}{n(N-n)} = \frac{-Nr + nr + nR - nr + nR}{n(N-n)}$$

$$\frac{Nx}{n(N-n)} = \frac{-Nr + 2nR}{n(N-n)}$$

$$Nx = -Nr + 2nR$$

$$x = \frac{-Nr + 2nR}{N} \tag{2.3.5}$$

When the row marginal frequencies are equal $2n = N$, and when the column marginal frequencies are equal $2R = N$. Consequently $x = (R-r)$ and $x = (n-r)$ respectively and this implies that rows are interchanged and columns are interchanged.

Example
For $N = 29$, $r = 9$, $n = 14$, $R = 12$ we substitute in equation 2.3.5 and get x as

$$x = \frac{-29\,(9) + 2\,(14)\,(12)}{29}$$

$$= 75/29$$

$$= 2{\cdot}586$$

Again the problem arises of how to interpret fractional values for discrete random variables : rounding down to 2 means that the lower-tail rejection region is smaller, and rounding up to 3 means that the lower-tail rejection region is larger, than the upper-tail rejection region.

$$P(r \geqslant 9) = 0{\cdot}019685$$

$$P(r \leqslant 2) = 0{\cdot}006372$$

$$P(r \leqslant 3) = 0{\cdot}041477$$

Thus $$P(r \geqslant 9) + P(r \leqslant 2) = 0{\cdot}026057$$

and $$P(r \geqslant 9) + P(r \leqslant 3) = 0{\cdot}061162$$

If H_0 had required a two-tailed test, both results should be stated, particularly as an $\alpha : 0{\cdot}05$ is in general nominal only for discrete probability distributions. This result emphasizes that considerable advantage often accrues from giving careful thought to the statement of H_1 and the restriction of the rejection region to one tail.

This example is used to illustrate the procedure and its rationale. In general an inference from such data would take account of other considerations. In the case of factories over some time period as implied by survival we should be concerned with the way urban and non-urban location classes are defined : for example the definition could be in terms of an administrative boundary or in terms of a cost surface. Comparability would have to be attempted. In the case of survival of a plant species on for example slag tips, one of which was raw slag and the other burnt slag, the exposure, slope and so on would influence the assertion.

Tables

Calculating these exact probabilities is not so great a chore with modern electronic desk calculators or computers. Indeed the successive calculations provided in the text should be a useful learning exercise in familiarization with the way the probabilities change as cell frequencies change. Once done, however, the tedious exercise need not be repeated as tables exist of exact probabilities for the hypergeometric function for all cases of expected cell frequencies not great enough to warrant an approximating distribution. In the present example we use table 4 in appendix 2. We enter table 4 with $A = 15$, $B = 14$ and $a = 12$. Then we see that $b \leqslant 5$ has a probability of 0·020 which, for three decimal places, is exactly the same as our calculated value 0·019685. On a one-tailed test at $\alpha : 0{\cdot}05$ we should accept H_1. For a two-tailed test we consider the table

3	11
9	6

, with $A = 15$, $B = 14$, $a = 9$ and $b = 3$. In the table we see for A, B, a that a $b \leqslant 3$ has a probability of 0·041, which compares to our calculated value of 0·041477.

The binomial provides a good approximation if N is large as then the ratio of n/N tends to be small and the probability of r tends to be practically constant.

Recall that the hypergeometric distribution was introduced in article 2.2.3 to take into consideration the fact that $P = R/N$ was not able to be constant. The usual approximating distribution is the χ^2 distribution which is considered next.

2.3.2 Chi-square and 2 X 2 contingency tables

Chi-square is usually recommended as the probability distribution for the test statistic calculated to determine whether cell frequencies in a cross-classification are independent. Arrays of frequencies with r rows and c columns are known generally as contingency tables. When there are two rows and two columns such arrays are referred to as 2 X 2 tables or fourfold tables. We shall use the previous data as the basis for illustrating the logic of the approach as well as the arithmetic procedures involved in the calculation of the test statistic. We shall denote this test statistic by X^2 and it is referred to the distribution of a chi-square random variable which is denoted by the Greek letter chi, χ^2. We shall compare χ^2 probabilities with the exact probabilities.

The general question we are asking is the same as in article 2.3.1 : are the two classifications independent? To accept that they are independent is equivalent to accepting that the proportion, P_1, of factories that go out of business in location class 1, 5/14, is essentially the same as the proportion, P_2, of factories that go out of business in location class 2, 12/15. This entails that these proportions are equal and essentially the same as the overall proportion of non-survivors, namely 17/29. These proportions P_1 and P_2 are not identical and so we have to decide whether the difference between the actual proportions, 12/15, 5/14, and the expected proportion, 17/29, is sufficiently slight to have arisen by chance or whether it is great enough to suggest a real difference in survival propensity between the two location classes.

The overall proportion of failures is 17/29 = 0·586207. For this to be the proportion of urban failures, that is location class 1, we need a value such that

$$\frac{\text{(urban, failure)}}{14} = \frac{17}{29}$$

thus

$$\text{(urban, failure)} = \frac{14.17}{29} = 8{\cdot}2069$$

For non-urban failures the corresponding value is

$$\frac{\text{(non-urban, failure)}}{15} = \frac{17}{29}$$

thus

$$\text{(non-urban, failure)} = \frac{15.17}{29} = 8{\cdot}7931$$

Similar calculations give the results for urban survivors, and non-urban survivors. The four values given in fig. 2.22 are the *expected values* under the null hypothesis of independence in the two classifications.

	Attribute		
	Survived	*Failed*	*Total*
Location class 1 Urban	5·7931	8·2069	14
Location class 2 Non-urban	6·2069	8·7931	15
Total	12	17	29

2.22 Expected values under H_0 for fig. 2.19.

We have set the proportion of urban failures, 8·2069/14, and of non-urban failures, 8·7931/15, to be equal to the overall proportion of failures, 17/29 = 0·586207. If these expected values are substantially different from the observed frequencies we shall take this as evidence that survival and failure are related to location class.

If we compare these expected values (fig. 2.22) with the frequencies in each of the 13 possible 2 X 2 contingency tables for these fixed marginal totals, we note that:

a) they are most similar to

(i) $\begin{array}{c|c} 6 & 8 \\ \hline 6 & 9 \end{array}$ (ii) $\begin{array}{c|c} 5 & 9 \\ \hline 7 & 8 \end{array}$ (iii) $\begin{array}{c|c} 7 & 7 \\ \hline 5 & 10 \end{array}$

b) these tables have the highest *exact* probabilities

c) the tables with the lowest *exact* probabilities are the most different from the expected values, namely

(i) $\begin{array}{c|c} 0 & 14 \\ \hline 12 & 3 \end{array}$ (ii) $\begin{array}{c|c} 12 & 2 \\ \hline 0 & 15 \end{array}$

This decrease in likelihood with an increase in departure from the frequencies expected on the assumption of independence indicates one way of estimating a probability without calculating exact probabilities, and that is to compute a test statistic based on the size of this discrepancy. Given a general 2 X 2 contingency table with observed frequencies, O_{ij}, and computed expected frequencies, E_{ij}, the test statistic, X^2, is calculated from

$$X^2 = \sum_{i=1}^{2} \sum_{j=1}^{2} \frac{(O_{ij} - E_{ij})^2}{E_{ij}} \tag{2.3.6}$$

and the value of X^2 is referred to tables of a χ^2 random variable (see table 2 in appendix 2). An informal justification for this notes, first, that the exact probabilities are those of a hypergeometric random variable. Secondly, that the binomial approximation to the hypergeometric distribution is good when N is sufficiently large to leave the probability of allocation to any cell unaffected. Thirdly, if this procedure of allocating the N objects to the cells is repeated a very large number of times, then as this number of repetitions becomes very large so the distribution of the test statistic, i.e. the values yielding the proportions, gets closer and closer to that of a χ^2 random variable : χ^2 is the asymptotic distribution of the test statistic. Frequencies resulting from nominal measurement are distributed approximately as chi-square when they are processed in equation 2.3.6 or an equivalent formula.

This formula is evaluated as shown in fig. 2.23.

O_{ij}		E_{ij}	$(O_{ij}-E_{ij})$	$(O_{ij}-E_{ij})^2$	$(O_{ij}-E_{ij})^2/E_{ij}$
O_{11}	9	5·7931	3·2069	10·2842	1·7753
O_{12}	5	8·2069	−3·2069	10·2842	1·2531
O_{21}	3	6·2069	−3·2069	10·2842	1·6569
O_{22}	12	8·7931	3·2069	10·2842	1·1696
	29	29·0			$X^2 = 5{\cdot}8548$

All entries rounded from ten-figure results.

2.23 Details of X^2 calculation for fig. 2.19.

The sum of the values in column 5 gives the value of X^2 that is used to provide an estimate of the probability of getting differences from the expected values as large as those observed. The value of X^2 is referred to standard tables showing values of χ^2 at standard significance levels; usually at α: 0·05, 0·025 and 0·01, with given numbers of *degrees of freedom*. In this case there is only 1 degree of freedom because once one of the four cell values has been determined the remaining three follow automatically from the fixed marginal totals. In contingency tables the degrees of freedom value is calculated by $(r-1)(c-1)$ which for $r=c=2$ gives 1 degree of freedom. A value of 5·8548 lies somewhere between χ^2 values of 5·024 and 6·635 at α levels of 0·025 and 0·010 respectively : by interpolation this gives a probability of approximately 0·01774 for $\chi^2 = 5{\cdot}8548$. For $H_0 : P_1 = P_2$ and $H_1 : P_1 \neq P_2$ at α : 0·05 with χ_1^2 we accept H_1.

Actual values of r must be integers. The use of a chi-square continuous curve as an estimation of the discrete probabilities of these actual values is liable to error. The approximation is likely to be poorest when the sample size, N, is small, or when the expected frequencies in particular cells are small. Many writers consider an overall sample size of less than 40 to be small and an expected frequency to be small if it is less than 5. To make X^2 more sensitive in the sense of making a continuous function a closer approximation to a discrete distribution a *correction for continuity* was suggested by Yates in 1934. This correction involves subtracting 0·5 from each positive difference of $(O_{ij} - E_{ij})$ and adding 0·5 to each negative difference before squaring.

Calculations of X^2 are facilitated by using the observed frequencies directly and avoiding the intermediate calculation of expected frequencies.

$$X^2 = \frac{N(O_{11}O_{22} - O_{12}O_{21})^2}{(O_{11} + O_{12})(O_{21} + O_{22})(O_{11} + O_{21})(O_{12} + O_{22})} \qquad (2.3.7)$$

With Yates' correction this becomes

$$X^2_{\text{Yates}} = \frac{N(|O_{11}O_{22} - O_{12}O_{21}| - 0{\cdot}5\,N)^2}{(O_{11} + O_{12})(O_{21} + O_{22})(O_{11} + O_{21})(O_{12} + O_{22})} \tag{2.3.8}$$

When $|a - b|$ means the numerical value of the difference regardless of sign : i.e. the difference is taken as positive.

Substituting in the above equations we get:

$$X^2 = \frac{29\,[(9)(12) - (5)(3)]^2}{14.15.12.17}$$

$$= 5{\cdot}8548$$

$$X^2_{\text{Yates}} = \frac{29\,[|(9)(12) - (5)(3)| - 0{\cdot}5\,(29)]^2}{14.15.12.17}$$

$$= 4{\cdot}1715$$

To use X^2 as a test statistic in the one-tailed case the following procedure is followed. The hypotheses are; $H_0 : P_1 \leqslant P_2$ and $H_1 : P_1 > P_2$ at α: 0·05. If P_1 is less than or equal to P_2 we accept H_0 without further calculation. If P_1 is greater than P_2 we need to decide whether it is large enough to accept H_1. We base this decision on half of the probability represented by the particular value of X^2. In the above case, $X^2 = 5{\cdot}8548$ which has an interpolated probability of 0·01774, consequently the one-tailed probability is $0{\cdot}5(0{\cdot}01774) = 0{\cdot}00887$ and we reject H_0 in favour of H_1. The probability distribution and the *cdf* of the exact probabilities corresponding to fig. 2.20 are shown as graphs (figs. 2.24, 2.25), with the horizontal axis scaled as χ^2 values. As α : 0·05 is equivalent to the 0·95 quantile of the *cdf*, the one- and two-tailed probabilities refer to this sort of graph.

Exact probabilities should be used for small N and small expected frequencies; chi-square becomes increasingly appropriate as N increases. Its extension to larger tables is taken up after a number of *measures of association* in 2 X 2 contingency tables have been discussed.

2.4 Measures of association

Contingency tables arise in many cases when a proposition of independence is made for a point symbol map. If the null hypothesis of independence is accepted, the result is taken to mean that there is no association between the variables as contained in the particular categorization and the test statistic used. If the null hypothesis is rejected, this is taken to mean that there is some dependence between the row and column classifications. The alternative hypothesis indicates whether dependence is in one direction or in both directions and the improbability of independence is contained in the significance level. A number of measures exist which attempt to characterize the degree of dependence between the classifications. No single measure has been devised which is entirely satisfactory. Such a measure may be elusive because the idea of dependence or of association is imprecise and different definitions of the idea result in different measures. A discussion of the better-known measures illustrates the variety of characterizations of the notion of

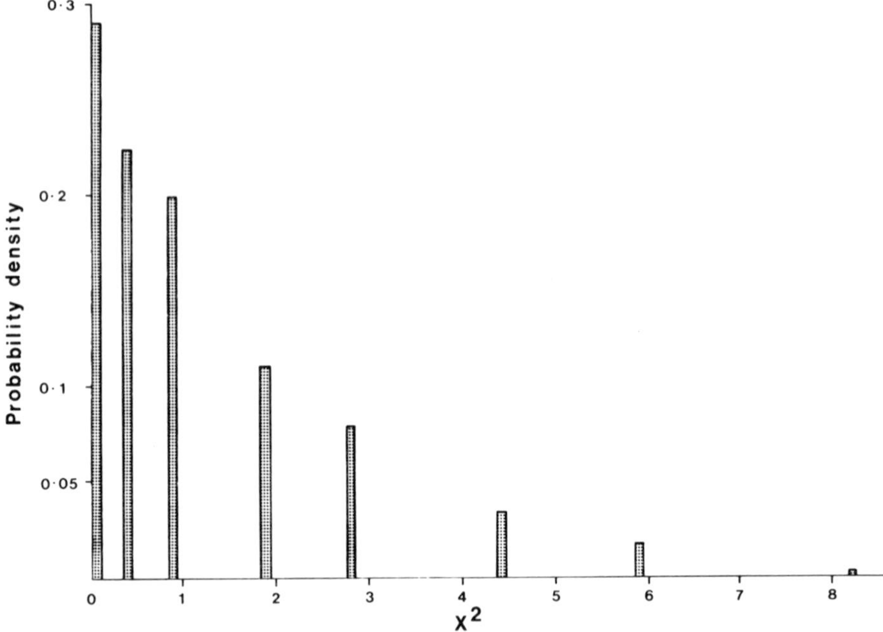

2.24 Comparison of exact probabilities and X^2 values for fig. 2.19.

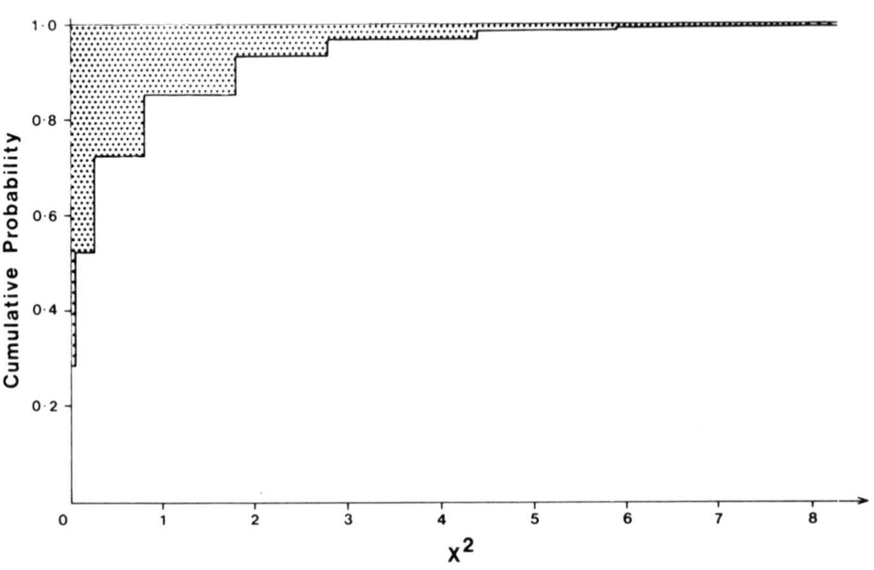

2.25 Comparison of *cdf* and X^2 values for fig. 2.19.

association in 2 × 2 tables which have resulted from trying to capture the concept in one value. Some turn out to be different forms of the same equation, others are important because they are special cases of correlation coefficients or because they generalize to higher-dimensional situations.

Chi-square provides a test of independence such that increasing values of the statistic imply decreasing likelihood of independence. We may suppose that there is some largest value for chi-square for tables with particular degrees of freedom which could be used to confine a measure of association, based on a ratio of actual chi-square to maximum chi-square, to a standard range of, say, (0, 1) or (−1, +1). The largest value is attained when only one cell in each row and each column has a non-zero entry, and when these entries are equal. In general the largest value of chi-square is given by $N\,[(r-1)$ or $(c-1)]$ for whichever of r or c is the smaller.

For each of the measures, the letter D denotes the coefficient used to show dependence and a subscript designates the author. There is no accepted designation for most of these coefficients: they are usually referred to as so-and-so's coefficient. Each measure is applied to the same 2 × 2 table taken from fig. 2.19 of the previous article which is repeated for convenience.

	Survived	*Failed*	
Urban	9	5	14
Non-urban	3	12	15
Total	12	17	$29 = N$

2.26 Summary table for fig. 2.18, repeated.

2.4.1 Contingency coefficients

The first three contingency coefficients are equivalent and in any particular case there is no gain in calculating more than one of them. However they all occur in publications and it may be useful to see their form and identity.

2.4.1.1 Cramer's coefficient D_C

Cramer's coefficient (Goodman and Kruskal 1954) is the quotient of the actual X^2 to the largest X^2 as

$$D_C = \frac{X^2}{N(r-1)} \qquad \text{For } r \leqslant c \qquad (2.4.1)$$

Thus in a 2 × 2 table where $r = c = 2$ it becomes

$$D_C = \frac{X^2}{N}$$

In fig. 2.26 we have

$$D_C = 5{\cdot}8548/29(2-1)$$
$$= 0{\cdot}201890$$

As the marginal frequencies are not equal in this 2 X 2 table the largest value the quotient can reach is 21·9328/29 = 0·7563. This coefficient of association lies between 0 and 1, but can only attain the value 1 when the row and column totals are equal and can only attain the value 0 when the row and column frequencies are equal and even.

2.4.1.2 *Tschuprow's coefficient* D_T

This coefficient is related to Cramer's measure (Goodman and Kruskal 1954, 1959) and is given as

$$D_T = \sqrt{\frac{X^2}{N\sqrt{(r-1)(c-1)}}} \qquad (2.4.2)$$

which for 2.26 gives

$$D_T = \sqrt{\frac{5{\cdot}8548}{29\sqrt{(1)(1)}}}$$
$$= 0{\cdot}44932$$

And for the most extreme table we have $D_T = 0{\cdot}85999$. Both measures were introduced to standardize X^2 so that it would fall between 0 and 1 providing a measure of association such that it is 0 when the row and column classifications are independent and 1 when there is complete association.

2.4.1.3 *Pearson's contingency coefficient* D_P

This is defined as

$$D_P = \sqrt{\frac{X^2}{X^2 + N}} \qquad (2.4.3)$$

which for 2.26 gives a value

$$D_P = \sqrt{\frac{5{\cdot}8548}{(5{\cdot}8548 + 29)}}$$
$$= 0{\cdot}40985$$

For the maximum value of the test statistic in this case, $D_P = 0{\cdot}65204$. When the classifications tend to independence X^2 and D_P are both small. As the degree of dependence increases so X^2 and D_P increase and tend to 1. It is clear that $X^2 + N$ must always be greater than X^2 so that D_P can never reach 1. Indeed the maximum value for X^2 is $N(r-1)$ so that the maximum value of D_P can be expressed as

$$D_{P\,\max} = \sqrt{\frac{N(r-1)}{N(r-1) + N}}$$
$$= \sqrt{\frac{N(r-1)}{Nr-N+N}}$$
$$= \sqrt{\frac{(r-1)}{r}}$$

From this it can be seen that the value of D_P is related to the degrees of freedom. We infer, then, that two separate contingency tables on similar data are comparable in terms of this statistic only if the minimum value of rows or columns is the same.

These three coefficients are applicable to X^2 for contingency tables of r, c greater than 2, however in the case of such arrays other considerations, which are dealt with in the next part of this section, diminish their usefulness. As the statistics are simple functions of X^2, a test of significance in terms of χ^2 is equivalent to a test of significance on the coefficients.

2.4.1.4 *Kendall's phi coefficient* D_K

The *phi* coefficient arises as the special case of Kendall's *tau*, τ, coefficient or rank correlation when applied to a fourfold contingency table (Gibbons 1971, p.269). Both *phi* and *tau* are special cases of the general coefficient of correlation known as Pearson's product moment coefficient of correlation; these are discussed in more detail later (article 3.3.2). The *phi* coefficient is defined by

$$D_K = \frac{O_{11}O_{22} - O_{21}O_{12}}{\sqrt{(O_{11}+O_{12})(O_{21}+O_{22})(O_{12}+O_{22})(O_{11}+O_{21})}} \tag{2.4.4}$$

Giving for fig. 2.26

$$D_K = \frac{108 - 15}{\sqrt{14\,.\,15\,.\,17\,.\,12}}$$
$$= 0{\cdot}44932$$

and for the most extreme of the 2 X 2 tables

$$D_K = \frac{180 - 0}{\sqrt{14\,.\,15\,.\,17\,.\,12}}$$
$$= 0{\cdot}86966$$

For the most extreme table in the opposite direction

$$D_K = \frac{0 - 168}{\sqrt{14\,.\,15\,.\,17\,.\,12}}$$
$$= -0{\cdot}81168$$

This measure of association preserves direction and ranges from -1 when O_{11} *and* O_{22} are zero, to $+1$ when O_{12} *and* O_{21} are zero: that is when unlike cells are empty.

The definition of D_K in equation 2.4.4 is very similar to the definition of X^2 as in equation 2.3.7. As Cramer's coefficient, which is equivalent in 2 X 2 tables to a second coefficient due to Pearson, is based on $\frac{1}{N}X^2$, then D_K and D_C are related as follows:

$$X^2 = \frac{N(O_{11}O_{22} - O_{12}O_{21})^2}{(O_{11}+O_{12})(O_{21}+O_{22})(O_{12}+O_{22})(O_{11}+O_{21})}$$

thus $$D_C = \frac{1}{N}X^2$$

$$= \frac{N(O_{11}O_{22}-O_{12}O_{21})^2}{N(O_{11}+O_{12})(O_{21}+O_{22})(O_{12}+O_{22})(O_{11}+O_{21})}$$

and $$D_K = \frac{(O_{11}O_{22}-O_{12}O_{21})}{\sqrt{(O_{11}+O_{12})(O_{21}+O_{22})(O_{12}+O_{22})(O_{11}+O_{21})}}$$

$$= \pm \quad D_C = \pm\sqrt{\frac{X^2}{N}}$$

Now $D_C = 0{\cdot}201890$ and $\sqrt{D_C} = 0{\cdot}4493$ as expected.

The sign is retained accordingly as $(O_{11}O_{22}) \lessgtr (O_{12}O_{21})$.
As $D_K = \pm\sqrt{\frac{X^2}{N}}$, then $\pm\sqrt{X^2} = D_K\sqrt{N}$, which is approximately normally distributed, this coefficient can provide a one-tailed test of association.
Put: H_0 : there is no positive association between the classifications (e.g. urban, survival)

H_1 : there is positive association between the classifications.

Reject if $D_K\sqrt{N} >$ than some specified α, say $\alpha : 0{\cdot}05$. Now an $\alpha : 0{\cdot}05$ is the $(1-\alpha) = (1-0{\cdot}05) = 0{\cdot}95$ quantile of a probability distribution, which in this case is the standard normal distribution. Reference to tables of the standard normal distribution shows that the 0·95 quantile has a value of 1·6449. Now $D_K\sqrt{N} = 0{\cdot}4493\sqrt{29} = 2{\cdot}4196$ which is $> 1{\cdot}6449$ and thus we reject H_0 in favour of H_1. Of course the word *positive* in H_0 and H_1 could be replaced by *negative*.

These are the most useful and important of the statistics used to measure the degree of dependence in contingency tables. They are all some function of chi-square, and a test of significance based on X^2 is equivalent to a test based on $f(D)$ where $f(D)$ is the particular functional relationship of the measure of dependence and X^2.

2.4.2 Cross-product ratio

Although we may conclude that the measures of dependence, D, are significant when X^2 is significant, we still do not have ameasure of that dependence which is comparable between all contingency tables. This variability is illustrated clearly in the table opposite, fig. 2.27.

Now if we are correct in supposing that there is some relationship between survival and location class it is reasonable to expect that relationship to be independent of the numbers of objects in the sample. In table (iii) in fig. 2.27 the number of survivors has been multiplied by 2 and the number of failures by 5. In table (iv) the number of factories in the urban class has been increased by 4 times and in the rural class 2 times. X^2 has increased as sample size has increased because the size of X^2 for a given departure from the frequencies expected under independence is approximately proportional to sample size. As the amount of data increases, proportionately small deviations are multiplied. This arises because

	Survived	*Failed*	*Total*	*Survived*	*Failed*	*Total*	*Survived*	*Failed*	*Total*	*Survived*	*Failed*	*Total*
Location class 1 Urban	0_{11}	0_{12}	r_1	9	5	14	18	25	43	36	20	56
Location class 2 Non-urban	0_{21}	0_{22}	r_2	3	12	15	6	60	66	6	24	30
Total	c_1	c_2	N	12	17	29	24	85	109	42	44	86
		(i)			(ii)			(iii)			(iv)	

	(i)	(ii)	(iii)	(iv)
	$X^2 = \dfrac{N(O_{11}O_{22}-O_{12}O_{21})^2}{r_1 r_2 c_1 c_2}$	$= 5{\cdot}8548$	$= 16{\cdot}2836$	$= 15{\cdot}3332$
	$D_C = \dfrac{X^2}{N}$	$= 0{\cdot}20189$	$= 0{\cdot}14939$	$= 0{\cdot}17829$
Cross-product ratio $= CPR$				
	$CPR = \dfrac{O_{11}O_{22}}{O_{12}O_{21}}$	$= 7{\cdot}2$	$= 7{\cdot}2$	$= 7{\cdot}2$

2.27 Comparison of X^2, D_C, CPR for fig. 2.26.

the quotient $(O_{ij}-E_{ij})^2/E_{ij}$ increases by a factor of k when the observed and expected values are increased by k; that is $(kO_{ij}-kE_{ij})^2/kE_{ij} = k^2\,(O_{ij}-E_{ij})^2/kE_{ij}$ $= \frac{k^2}{k} \cdot \frac{(O_{ij}-E_{ij})^2}{E_{ij}} = k.\ \frac{(O_{ij}-E_{ij})^2}{E_{ij}}$.

Yet, in fig. 2.27 the urban location class objects in table (iv) are no more successful as a class than they are in table (ii); there are just more objects. A similar situation occurs in table (iii). In such cases where there seems to be a basic association it is disconcerting to find that this property is not apparent in the X^2 – based measures of that association. This inconsistency does not prejudice the use of X^2 as a test of independence, but emphasizes the nebulous character of association. Let us imagine that there is a nucleus of association in table (ii) that is retained in all contingency tables in which the row and column frequencies are multiples of that basic table. Call all such tables the *equivalence class* of tables with the same degree of association. Then tables (iii) and (iv) are just two members of that equivalence class (Mosteller 1968).

An index of association that is invariant under such multiplication is the *cross-product ratio*, *CPR*, given as:

$$CPR = \frac{O_{11}O_{22}}{O_{12}O_{21}} = \frac{(r_1c_1O_{11})(r_2c_2O_{22})}{r_1c_2O_{12}\ \ r_2c_1O_{21}} \quad \text{for } r_i > 0,\ c_i > 0 \quad i = 1, 2\ \ldots \qquad (2.4.5)$$

In table (ii) $CPR = 108/15 = 7{\cdot}2$. The value of *CPR* ranges from 0 to ∞. A value of 1 corresponds to independence; indeed the column and row margainal totals are independent if and only if $CPR = 1$. A positive association with $CPR = 7{\cdot}2$ can be considered as equal in degree but opposite in direction to a negative association with $CPR = \frac{1}{7{\cdot}2}$. To convert the cross-product ratio into a more traditional measure of association it should be confined to the interval (−1, +1) and be symmetric about zero. Taking $\log_e CPR$ provides the symmetry and transforming ln *CPR* to give *CPR′* by

$$CPR' = \frac{2}{\pi} \arctan \ln CPR$$

maps it onto the interval (−1, +1) with 0 for independence. A discussion of the geometry of 2 × 2 tables and of the difference between *CPR′* and D_C is found in Fienberg and Gilbert (1970).

There is no general test of significance for *CPR*, but the idea of a basic association in a contingency table can be used to reveal more clearly the form of the association in the table. The idea of an equivalence class of contingency tables formed by column and row multiplications can be used to yield a standardized table which retains the same *CPR* as the original table, but which expresses the cell frequencies as numbers that sum to either unit row and column totals or to a unit grand total. These numbers show the probability of a particular row class conditional upon a given column class.

Any contingency table can be standardized by making the marginal totals uniform. This is accomplished by successive multiplication of the cell frequencies by the reciprocal of the marginal totals. The procedure is grasped readily from an example for the 2 × 2 table used already.

	Survived	*Failed*	*Total*	*Required*	*Survived*	*Failed*	*Total*
Location class 1	9	5	14	1	0·6429	0·3571	1·0
Location class 2	3	12	15	1	0·2	0·8	1·0
Actual total	12	17	29		0·8429	1·1571	
Required total	1	1					
	Table (i)				Table (ii)		

0·7627	0·3086	1·0713	0·7284	0·2716	1·0
0·2373	0·6914	0·9287	0·2713	0·7287	1·0
1·0	1·0		0·9997	1·0003	
Table (iii)			Table (iv)		

0·7285	0·2715	1·0	0·36425	0·13575	0·5
0·2715	0·7285	1·0	0·13575	0·36425	0·5
1·0	1·0		0·5	0·5	1·0
Table (v)			Table (vi)		

2.28 Steps in reducing a 2 × 2 table to standard form.

We wish to make table (i) in fig. 2.28 sum to unit margins. First we multiply the first row by 1/14 and the second row by 1/15 to give table (ii). This gives incorrect column totals and so we multiply the first column by 1/0·8429 and the second column by 1/1·1571 to give table (iii) in which the row totals are incorrect. This procedure is continued until the row and column sums are as near to the required sum as required. Five pairs of row, column multiplications were needed in this particular case to yield the required uniform margins. The last cycle is shown as tables (iv) and (v). The cross-product ratio of table (v) is (0·7285) (0·7285)/(0·2715)(0·2715) = 7·1998 which indicates that this particular table is one of the equivalence class tables for the original: indeed it is the standard table. By dividing the entries by 2 we derive table (vi), giving values which can be seen as conditional or transitional probabilities, representing in this case the distribution of factories given their survival category or their survival distribution given their location class. Such tables are useful for determining whether a number of contingency tables with different frequencies on the same classification categories are members of the same equivalence class. Comparisons for different towns, different countries or different sample sizes are facilitated: the standardization of international trade movements for countries before and after trade agreements can indicate whether the structural balance has been affected (Lewis 1974).

Chi-square is applicable to classifications with more than two categories. As an approximating distribution it is both appropriate and useful in many situations in maps where it is a common expedient to present a substantial amount of information into point symbols that are made proportional to some property or which are subdivided to correspond to a many-category classification. We shall consider the implications of such maps and the procedures for testing hypotheses of independence for such maps when the attributes are measured on nominal, ordinal and continuous scales.

2.5 Propositions of independence (2)

Properties: location class, one other attribute.
Measurement scale: nominal, elemental, multiple classes.
Statistical procedure: chi-square in $r \times c$ contingency tables.

The next change in the information increase in a map is to distinguish more than two classes for each of two attributes or variables. When we confine our attention to two attributes we shall assume that one of them is location and the other is not locational. If both attributes were not locational there would be no reason to present the objects cartographically. When there are more than two variables we shall see that propositions about two or more non-locational attributes arise quite naturally from maps, and analysis of these propositions is then a proper part of the map analysis.

This increase in information is accomplished cartographically by the simple expedient of either making the symbols proportional in size to the non-locational attribute or of subdividing the symbols in a distinctive way with each division corresponding to a class of the non-locational attribute. Both procedures can be combined.

It seems to be a slight change to go from using symbols of different sizes in order to maintain enumerability with a minimum reduction in locational relations, to using symbols whose size is proportional to some non-locational property of the objects. In my view it is a major change in emphasis. In the first case it is a device to contend with the effect of scale transformation on matching, in the second it is not a response to problems of matching but to the intention of showing simultaneously another property of the objects which, it is anticipated, will be related to the locational attribute. As a result location information is usually reduced deliberately. Geographers use divided and proportional symbols to indicate whether any regularities are evident in the data mapped. The construction of these symbols often requires that distance is replaced by position in a sequence or by location class.

A similar problem occurs with the attribute that is shown by the proportional symbol. The use of a continuously graded symbol implies that the property to be shown is measured on a more restricted scale than an ordinal scale and that this refinement is preserved. There is a convention in cartography that the relation between magnitude of property and area of symbol is preserved by finding the square root of all the magnitudes and using the square roots for the radiuses of the circles or the sides of the squares. A deliberate distortion is recommended by some cartographers who suggest that as the visual response to area is not linearly related to actual area the radiuses should be modified. One suggestion was for each magnitude, x, to take the value $x^{0\cdot57}$ rather than the customary $x^{0\cdot5}$, which is the square root. Thus, taking logarithms we write 0·57 $\log_a x$ instead of 0·5 $\log_a x$ (Robinson 1953). The apparent precision introduced by this procedure is not convincing if the purpose of the map is either to enable accurate measurements to be made from the symbols or to illustrate an assertion, because in the first case the accuracy comes from the data not the map, and in the second case an assertion is convincing in terms of the test statistic not the impression. When the non-locational attribute is used as an ordinal measure then the symbols need only clearly discriminate adjacent ranks (see diagram in section 2, exercise 31). When, as is often the case, the range of values is put into k classes, these *classes* need distinguishing. Of course the cartographer can easily superimpose a *class* distinction on continuously graded symbols, thus emphasizing the information actually *used* in the assertion.

The divided symbol, proportional or uniform, is liable to be abused in such a way that its accessible information rapidly diminishes as its ostensible information increases. However, the frequent use of both proportional and divided symbols testifies to the geographer's belief in their usefulness. Their usefulness is enhanced by an explicit statement of the restrictions imposed and by the information and measurement levels used in the proposition and its test statistic.

2.5.1 Chi-square and $r \times c$ contingency tables

Just as the maps are an immediate generalization of the previous situation, so the 2×2 contingency table can be extended to a contingency table with r rows and c columns. Such a table can be used to summarize the information from several samples, each sample corresponding to a separate location class, for which we want to test the hypothesis that the probabilities relating to the non-locational attribute do not differ from sample to sample. In other situations, the information may be treated as a single sample in which each object is classified uniquely into one of r distinct categories on one attribute and into one of c distinct classes on the other attribute. Both situations are treated in a similar way. The measurement need be nominal only, but, when at least one attribute can be ordered, a number of modifications are possible which increase the sensitivity of the test statistic. As it is often realistic either to treat location as ordered from one or more origins, or to exploit the symbols proportional to the non-locational property by collecting them into ordered categories, we shall consider a number of these extensions of chi-square.

The multinomial probability distribution is appropriate to most $r \times c$ contingency tables, but the calculation of the exact probabilities soon becomes unprofitable as the number of classes and the row and column totals increase. In the 2×2 case, the hypergeometric probability function was appropriate for small N. When N increased sufficiently so that the allocation of one object did not noticeably affect the ratios in the classes the binomial was an appropriate distribution. In each case the continuous probability density function of a χ^2 random variable was used as an acceptable approximation. In an entirely analogous way χ^2 will again be used to approximate the exact multinomial discrete probabilities. Incidentally, when the number of objects is comparatively small the hypergeometric is appropriate in its multivariate form. This involves so many factorials that its calculation is not recommended beyond, say, the 2×3 case, although in a particular situation the student experimenting with programming a computer might consider its use a doubly profitable exercise.

The general $r \times c$ table is arranged in the following format and notation (fig. 2.29).

	c_1	c_2	...	c	*Total*	
r_1	a_{11}	a_{12}	...	a_{1c}	n_1	$a_{1.}$
r_2	a_{21}	a_{22}	...	a_{2c}	n_2	$a_{2.}$
.	.	.		.	.	.
.	.	.		.	.	.
.	.	.		.	.	.
r	a_{r1}	a_{r2}	...	a_{rc}	n_r	$a_{r.}$
Total	c_1	c_2	...	c	N	
	$a_{.1}$	$a_{.2}$	...	$a_{.c}$		$a_{..}$

2.29 The general $r \times c$ contingency table with alternative notations for marginal totals.

The rows represent the categories of one attribute, say location class, or the separate samples, and the columns represent the categories of the second attribute. The number of objects in each of the rc cells is designated by a_{ij} where i refers to the row number and j to the column number : i goes from 1 to r, j goes from 1 to c. The row totals are shown as n_i or as $a_{i\cdot}$, with the dot meaning the values are summed over that attribute, in this case the j variable or column. The column totals are shown as c_j or $a_{.j}$. The row and column totals are often referred to as marginal totals. The sum of the row or column marginal totals is the grand total, N, or $a_{..}$ in dot notation.

Example

Three samples were taken by a random procedure, one for each of three land-use zones in east London, and each building was given a score on a variety of criteria; these scores were combined into a measure of decay at one of four levels. The result is summarized as a contingency table with $r = 3$, $c = 4$, giving 12 cells, a_{ij}, altogether.

Location class	*Decay class*				
	1	2	3	4	*Total*
1	69	30	12	9	120
2	47	32	5	3	87
3	57	19	17	10	103
Total	173	81	34	22	310

2.30 Contingency table showing location class and decay class of buildings in part of London.

Row totals reflect sample size. Column totals correspond to the number of buildings allotted to each decay category on the same criteria. The hypothesis is that the probability that a building will be allocated to a particular decay class is independent of the location class from which the sample was taken. In other words probabilities in the same column are equal to each other. The alternative is that this does not hold for at least two probabilities in the same column. There is no need to specify these probabilities because the null hypothesis states that the probability of an object being in the jth class is the same for all location class populations. It says nothing about what that probability is, nor about the column class.

The test statistic is calculated as a generalization of the chi-square statistic introduced for the 2 × 2 table. For each cell value, a_{ij}, we calculate an expected value, E_{ij}. This is the value that is expected if the null hypothesis is true, such that a column probability is the same for each sample. The column proportions are c_j/N and so for each sample with its total of n_i objects we expect that there will be $\frac{c_j}{N} \cdot n_i$ objects in the jth column category. These expected values have to be calculated for each ijth cell and so we have two summation signs. The test statistic, X^2, is written as

$$X^2 = \sum_{i=1}^{r} \sum_{j=1}^{c} \frac{(a_{ij} - E_{ij})^2}{E_{ij}} \text{ with } E_{ij} = \frac{c_j \cdot n_i}{N} \qquad (2.5.1)$$

This is referred to the χ^2 distribution with $(r-1)(c-1)$ degrees of freedom as an approximation to the exact distribution. To derive the exact probability distribution involves calculating all the possible contingency tables with the *same* row totals. If the row totals were also allowed to vary so long as their sum was N, 310 in this case, then there would be a great many more possible contingency tables and the exact probability distribution would be more difficult to enumerate. The least possible number of tables occurs when both row and column totals are fixed. That was the situation presented in article 2.3.2 for which we calculated the exact distribution and compared the chi-square approximation to it. Fortunately all three exact probability distributions are approximated satisfactorily by the *pdf* of a χ^2 random variable with $(r-1)(c-1)$ degrees of freedom, *df*. As the expected values increase in size for each cell, the exact distributions get closer and closer to the χ^2 *pdf* and X^2 provides an increasingly good approximation. The approximation is more likely to be poor when some of the expected values for a particular contingency table are very small. Cochran (1954) suggests the situation may not be as serious as other authors have thought, and he provides the well-known yardstick for contingency tables with more than 1 *df* that 'if relatively few expectations are less than 5 (say in 1 cell out of 5 or more, or 2 cells out of 10 or more), a minimum expectation of 1 is allowable in computing X^2'.

In the example we wish to test the null hypothesis that the distribution of decay scores is the same for buildings in each location class sample. We shall reject this hypothesis if our test statistic has a value greater than 12·59 which corresponds to a rejection region of α : 0·05 of the χ^2 distribution with $(r-1)(c-1) = (3-1)(4-1) = 6$*df*. (Appendix 2, table 2).

The expected value in cell a_{11} is given by

$$\frac{n_1 c_1}{N} = \frac{120.173}{310} = 66{\cdot}9677$$

and in cell a_{32} (row 3, column 2) by

$$\frac{n_3 c_2}{N} = \frac{103.81}{310} = 26{\cdot}9129$$

The complete table of expected values is shown as fig. 2.31.

Location class	*Decay class*				
	1	2	3	4	*Total*
1	66·9677	31·3548	13·1613	8·5161	120
2	48·5516	22·7323	9·5419	6·1742	87
3	57·4806	26·9129	11·2968	7·3097	103
Total	173	81	34	22	310

2.31 Expected values under H_0 for fig. 2.30.

The test statistic is calculated from equation 2.5.1 which for cell a_{11} is

$$\frac{(a_{11} - E_{11})^2}{E_{11}} = \frac{(69 - 66{\cdot}9677)^2}{66{\cdot}9677} = 0{\cdot}0617$$

and for cell a_{32} is

$$\frac{(a_{32} - E_{32})^2}{E_{32}} = \frac{(19 - 26{\cdot}9129)^2}{26{\cdot}9129} = 2{\cdot}3265$$

The complete table of individual cell contributions is given as fig. 2.32.

Location class	*Decay class*				
	1	2	3	4	
1	0·0617	0·0585	0·1025	0·0275	
2	0·0496	3·7783	2·1619	1·6319	
3	0·0040	2·3265	2·8793	0·9902	
					14·0719

2.32 X^2 contributions of each cell for fig. 2.30.

And $X^2 = 14{\cdot}0719$. Since this is greater than 12·59 we reject H_0 in favour of H_1, and conclude that the building decay scores are distributed differently in the location classes. This inference is the basis for further work which might well be

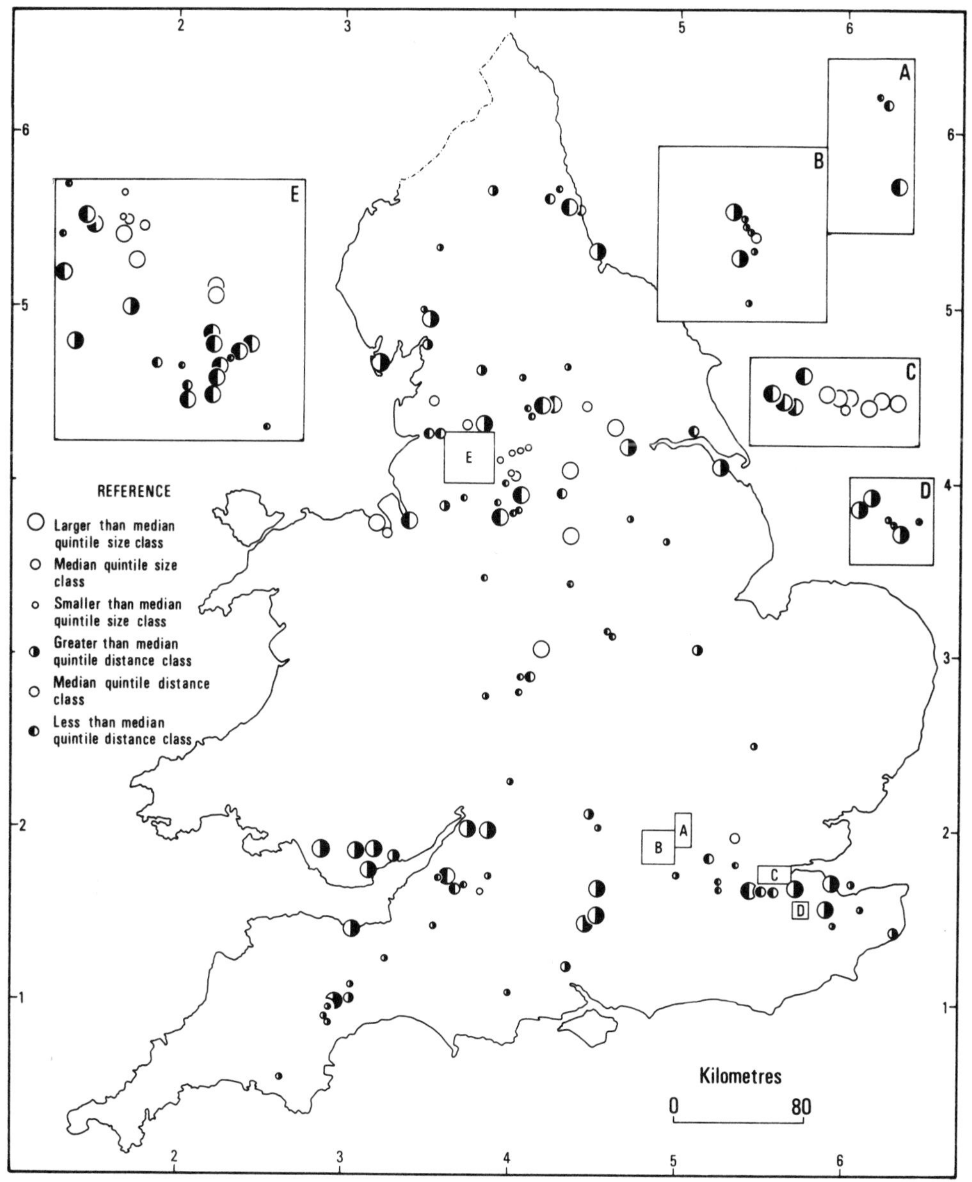

2.33 Capacity and distribution of papermaking mills in the United Kingdom, 1965.

influenced by a careful inspection of the actual and expected values. Notice that our inference is that the building decay *scores* are distributed differently, not that building decay is. That link depends on the acceptability of the measurement of building decay.

The same procedure is used when a sample of size N is taken and each element of that sample is then classified on two attributes simultaneously. Again we associate each element with the a_{ij}th cell, but emphasize that now the row as well as the column totals are variable. The null hypothesis is that an element's being classified in the ith row is independent of its being classified in the jth column. The alternative to independence is that for one or more rows and columns the allocations are not independent, thus we infer dependence within the table, but not for any particular row and column.

Example

Consider the distribution of the 164 papermills in England and Wales in 1965 (fig. 2.33). For each mill we define (X_i, Y_i) such that X_i is the distance from the nearest market able to absorb the quantity of paper it produces and Y_i as its capacity to produce paper, measured in size of papermaking machinery. The $(X_i Y_i)$ are seen as a random sample from the population of all possible measures of distance and capacity. The X_i and the Y_i are put into one of three classes to give the cell frequencies shown in fig. 2.34.

Size class	*Distance-to-market class*			
	A	*B*	*C*	*Total*
A	25	8	32	65
B	14	11	8	33
C	24	16	26	66
Total	63	35	66	164

2.34 Contingency table showing capacity class and distance-to-market class of the mills in fig. 2.33.

The null hypothesis is that the two classifications are independent. There is certainly no logical necessity for association. The null hypothesis can be put into terms that relate clearly to the calculation of the test statistic.

H_0 : the probability, P (size class i, Market class j) =
the probability, P (size class i) P (Market class j) for all i, j.

The alternative to H_0 if we decide to reject this null hypothesis is that H_0 is not true for *some* i, j. We shall use the same rejection region as in the previous example with $(r-1)(c-1) = (3-1)(3-1) = 4df$. In other words once we have made 4 cell entries in this 3×3 table with these marginal totals, the remaining cell entries are determined (fig. 2.35).

With $df = 4$, the 0·95 quantile of the χ^2 *cdf* is 9·488 and so we accept H_0 if $X^2 \leqslant 9{\cdot}488$.

The statement of the H_0 reflects the way the test statistic is calculated, in that we argue that the frequency of allocations to each cell, a_{ij}, is in proportion to the marginal totals of the ith row and the jth column. The expected values and the contributions to the test statistic of each cell are given in figs. 2.36 and 2.37.

				Total
		8	32	65
	14			33
			26	66
Total	63	35	66	

2.35 Calculating the degrees of freedom in an $r \times c$ contingency table.

Size class	*Distance-to-market class*			
	A	*B*	*C*	*Total*
A	24·970	13·872	26·159	65
B	12·677	7·043	13·280	33
C	25·354	14·085	26·561	66
Total	63	35	66	164

2.36 Expected values under H_0 for fig. 2.34.

Size class	*Distance-to-market class*			
	A	*B*	*C*	
A	0·000	2·486	1·304	
B	0·138	2·223	2·099	
C	0·072	0·260	0·012	
				8·594

2.37 X^2 contributions of each cell for fig. 2.34.

$X^2 = 8{\cdot}594$, we accept H_0.

Our general enquiry may well have concerned whether a firm's production capacity is related to its location with respect to market, and our supposition of association may well have been strengthened by the map. This establishes our proposition. Refining this to a hypothesis involves measuring the attributes and then selecting a test statistic. The (X_i, Y_i) were measured on a continuous scale, but the values were put into categories and so treated apparently as measurements on a nominal scale. Market was defined as the nearest market able to absorb the output of a mill of a given size. In fact this was a median-sized mill; such mills produced over three-quarters of the total UK output. Size was defined in terms of machine width. Definitions such as these must be made, but we must always be aware that our terms such as market and capacity are then identified by the particular definitions, and we must beware of showing a relationship for defined terms and then generalizing the relationship to other definitions of the same terms without substantiation. In the above case the definition of market can be criticized for ignoring variations in market capacity or the exclusiveness of one market. Similarly the output of a mill is not a simple function of machine size. Thus when the null hypothesis is accepted it is accepted for those definitions of terms, for those particular classifications and for the test statistic used. Some geographers may be disappointed by these conditions and feel that these restrictions weaken the proposition intolerably. On the contrary, recognizing these restrictions strengthens the statements by giving precise meaning to them. Vagueness is inimical to progress in understanding.

2.5.2 Partitioning the degrees of freedom

Chi-square used as a general test for independence in an $r \times c$ contingency table is not directed to particular sorts of departure from independence. A total value that lies outside a selected rejection region leads to acceptance of the null hypothesis even though there may be reason to suppose that the whole table is not homogeneous. The rejection region depends on the test statistic value and the degrees of freedom. A similar value of X^2 with reduced *df* could lie *within* the rejection region defined at the same quantile of the *pdf*. This can be illustrated by the last example in which the total value of 8·594 lies outside the 0·95 quantile with 4 *df*, but any value $\geqslant$ 7·815 lies within the 0·95 quantile with 3 *df*, and any value $\geqslant$ 3·84 lies within the 0·95 quantile with 1 *df*. An inspection of the table of contributions to the test statistic shows that four cells make substantial contributions, while the remainder have small values. How can the $r \times c$ table be divided in an acceptable way?

When expected frequencies are estimated from the marginal totals, the total chi-square value may be partitioned exactly into single degrees of freedom so that we can associate a particular portion of the total value with each of the $(r-1)$ $(c-1)$ subtables of the full table. In order to express the general formula for calculating the chi-square values associated with each of the subtables it is convenient to put the general $r \times c$ table in the following form.

$$
\begin{array}{cccc|ccc||c}
a_{11} & a_{12} & \cdots & a_{1j-1} & a_{1j} & \cdots & a_{1c} & a_{1.} \\
a_{21} & a_{22} & \cdots & a_{2j-1} & a_{2j} & \cdots & a_{2c} & a_{2.} \\
\cdot & & & & & & & \\
\cdot & & & & & & & \\
\cdot & & & & & & & \\
a_{i-11} & a_{i-12} & \cdots & a_{i-1j-1} & a_{i-1j} & \cdots & a_{i-1c} & a_{i-1.} \\
\hline
a_{i1} & a_{i2} & \cdots & a_{ij-1} & a_{ij} & \cdots & a_{ic} & a_{i.} \\
\cdot & & & & & & & \\
\cdot & & & & & & & \\
\cdot & & & & & & & \\
a_{r1} & a_{r2} & \cdots & a_{rj-1} & a_{rj} & \cdots & a_{rc} & a_{r.} \\
\hline\hline
a_{.1} & a_{.2} & \cdots & a_{.j-1} & a_{.j} & \cdots & a_{.c} & a_{..}
\end{array}
$$

This table uses the dot notation with

$$a_{i.} = \sum_{j=1}^{c} a_{ij} \; ; a_{.j} = \sum_{i=1}^{r} a_{ij} \; ; a_{..} = \sum_{i=1}^{r} \sum_{j=1}^{c} a_{ij}$$

To show the fourfold table under consideration is

$\sum_{i=1}^{i-1} \sum_{j=1}^{j-1} a_{ij}$	$\sum_{i=1}^{i-1} a_{ij}$
$\sum_{j=1}^{j-1} a_{ij}$	a_{ij}

2.38 The general $r \times c$ contingency table in a notation for partitioning the degrees of freedom.

The $(r-1)(c-1)$ fourfold subtables can be formed in a number of ways. A straightforward procedure is to start from the a_{22} cell and systematically form tables for the a_{ij} cells with i going from $i = 2, 3, \ldots, r$ and j from $j = 2, 3, \ldots, c$. Thus we can write fig. 2.34 as below, fig. 2.39.

	$j=1$	$j=2$	$j=c=3$
$i=1$	a_{11} 25	a_{12} 8	a_{1c} 32
$i=2$	a_{21} 14	a_{22} 11	a_{2c} 8
$i=r=3$	a_{r1} 24	a_{r2} 16	a_{rc} 26

2.39 Fig. 2.34 in the format of fig. 2.38.

From this we can derive four subtables, with a_{22} as the bottom right-hand-side cell of the first 2×2 table. The second table has a_{23} as its bottom right-hand-side cell and $a_{11} + a_{12}$, a_{13}, $a_{21} + a_{22}$ as the other entries in this fourfold table. The third 2×2 table has entries $a_{11} + a_{21}$, a_{31}, $a_{12} + a_{22}$, with a_{32} as its bottom right-hand-side cell. The final table has entries $a_{11} + a_{12} + a_{21} + a_{22}$, $a_{31} + a_{32}$, $a_{13} + a_{23}$, with a_{33} as the bottom right-hand-side cell. Each is a 2×2 table and has 1 degree of freedom. The fourfold table frequencies are given in fig. 2.40.

(i)

25	8
14	11

(ii)

33	32
25	8

(iii)

39	19
24	16

(iv)

58	40
40	26

2.40 Four 2×2 subtables of fig. 2.39 corresponding to the four degrees of freedom.

A test statistic calculated for each of these tables would be based on their individual marginal totals and would give expected values for the cell entries different from the expected values calculated from the marginal totals of the full $r \times c$ table. The totals will also be different. The intention is to split the full table into these subtables *and* allocate to each subtable its contribution to the total value. The general formula must relate each subtable cell entries to the correct marginal totals of the complete contingency table. This formula for the subtable with the a_{ij}th entry as its bottom right-hand-side cell is given as equation 2.5.2.

$$X^2 = \frac{a_{..}\left\{a_{.j}\left[a_{i.}\left(\sum_{i=1}^{i-1}\sum_{j=1}^{j-1} a_{ij}\right) - \left(\sum_{i=1}^{i-1} a_{i.}\right)\left(\sum_{j=1}^{j-1} a_{ij}\right)\right] - \sum_{j=1}^{j-1} a_{.j}\left[a_{.j}\left(\sum_{i=1}^{i-1} a_{ij}\right) - \left(\sum_{i=1}^{i-1} a_{i.}\right)(a_{ij})\right]\right\}^2}{a_{.j}\left(\sum_{j=1}^{j-1} a_{.j}\right)\left(\sum_{j=1}^{j} a_{.j}\right)(a_{i.})\left(\sum_{i=1}^{i-1} a_{i.}\right)\left(\sum_{i=1}^{i} a_{i.}\right)} \tag{2.5.2}$$

This appears far more formidable than it is, and substitution in this equation for the four 2×2 subtables (fig. 2.40) reveals its straightforward structure and use. Shortcut formulae are available for 3×3 tables (Kimball 1954). The availability of electronic desk calculators in most geography departments reduces the need for such shortcut formulae and such formulae are restricted to arrays of particular dimension. In many undergraduate dissertations 4×5 and 5×5 contingency tables arise often enough to justify the availability in an introductory text of the general formula for partitioning.

Example
It is worthwhile the reader's making sure he can follow these substitutions and can associate each value with the cell in the main table (fig. 2.39).

Subtable (i)

a_{i-1j-1} 25	a_{i-1j} 8	a_{i-1c}	$a_{i-1.}$ 65
a_{ij-1} 14	a_{ij} 11	a_{ic}	$a_{i.}$ 33
a_{rj-1}	a_{rj}	a_{rc}	$a_{r.}$
$a_{.j-1}$ 63	$a_{.j}$ 35	$a_{.c}$	$a_{..}$ 164

$$X_1^2 = \frac{164\{35[33(25) - 65(14)] - 63[33(8) - 65(11)]\}^2}{35\,.\,63\,.\,98\,.\,33\,.\,65\,.\,98}$$

$$= \underline{2{\cdot}336}$$

2.41 Calculating the X^2 contribution of subtable (i) in fig. 2.40.

Subtable (ii)

a_{i-11} 25	a_{i-1j-1} 8	a_{i-1j} 32	$a_{i-1.}$ 65
a_{i1} 14	a_{ij-1} 11	a_{ij} 8	$a_{i.}$ 33
a_{r1}	a_{rj-1}	a_{rj}	$a_{r.}$
$a_{.1}$ 63	$a_{.j-1}$ 35	$a_{.j}$ 66	$a_{..}$ 164

$$X_2^2 = \frac{164\{66[33(33)-65\,(25)]-98[33(32)-65(8)]\}^2}{66\,.\,98\,.\,164\,.\,33\,.\,65\,.\,98}$$

$$= \underline{5{\cdot}683}$$

2.42 Calculating the X^2 contribution of subtable (ii) in fig. 2.40.

Subtable (iii)

a_{1j-1} 25	a_{1j} 8	a_{1c}	$a_{1.}$ 65
a_{i-1j-1} 14	a_{i-1j} 11	a_{i-1c}	$a_{i-1.}$ 33
a_{ij-1} 24	a_{ij} 16	a_{ic}	$a_{i.}$ 66
$a_{.j-1}$ 63	$a_{.j}$ 35	$a_{.c}$	$a_{..}$ 164

$$X_3^2 = \frac{164\{35[66(39)-98(24)]-63[66(19)-98(16)]\}^2}{35\,.\,63\,.\,98\,.\,66\,.\,98\,.\,164}$$

$$= \underline{0{\cdot}543}$$

2.43 Calculating the X^2 contribution of subtable (iii) in fig. 2.40.

Subtable (iv)

a_{11} 25	a_{1j-1} 8	a_{1j} 32	$a_{1.}$ 65
a_{i-11} 14	a_{i-1j-1} 11	a_{i-1j} 8	$a_{i-1.}$ 33
a_{i1} 24	a_{ij-1} 16	a_{ij} 26	$a_{i.}$ 66
$a_{.1}$ 63	$a_{.j-1}$ 35	$a_{.j}$ 66	$a_{..}$ 164

$$X_4^2 = \frac{164\{66[66(58)-98(40)]-98[66(40)-98(26)]\}^2}{98\,.\,66\,.\,164\,.\,66\,.\,98\,.\,164}$$

$$= \underline{0{\cdot}033}$$

2.44 Calculating the X^2 contribution of subtable (iv) in fig. 2.40.

$X^2 = X^2_{(i)} + X^2_{(ii)} + X^2_{(iii)} + X^2_{(iv)} = 8{\cdot}594$ with $1 + 1 + 1 + 1 = 4$ *df*. Subtable (ii) shows the greatest evidence of independence and there is reason to suppose that there is some association between a classification of size in category $j = 2$ and of location in category $i = 3$. Inspection of the values shows that this is an inverse dependence as there are fewer entries than expected of median size mills at greater than median distances from market.

The same set of mills could provide a sample of a different two-dimensional random variable $(Z_i Y_i)$ in which Z_i is the distance from a different set of origins: say sources of raw material defined precisely on some criteria. These $(Z_i Y_i)$ could be treated in a similar way. Comparisons could also be made between $(Z_i Y_i)$ for different years $T_1, T_2, \ldots, T_k$, under the null hypothesis that the probabilities of the i,jth classification will not vary over the k samples (Lewis 1969).

Measurement scales can be changed too. Location relations measured on a cost scale may yield values that coincide more closely with the effective constraints on papermills. Cost or distance scales provide ordered relationships and this increase in information can be exploited by different test statistics that are derived on the assumption that this information is available. When it is meaningful to arrange the categories of a nominal scale in an order, then the chi-square distribution can be exploited in various ways by directing the test statistic to a specified pattern of expected values (article 2.5.4). Our consideration of these adaptations is deferred until we have discussed and illustrated procedures for testing for independence in incomplete contingency tables.

2.5.3 Independence in incomplete contingency tables

In the previous example we examined the contribution of each subtable to the total test statistic. Now we shall examine the possibility of *reducing* the total $r \times c$ by ignoring particular cells or subtables. An example motivates the need for such procedures and illustrates their use. The reader who wishes to understand the theoretical basis of these procedures can refer to Cochran (1954), Goodman (1968) and Lancaster (1969).

The previous example can be extended by ascertaining whether factories of one size class at one point in time tend to remain in the same size class at a subsequent point in time. It may be that there is complete size-class mobility such that the size class of a factory at the later time, T_2, is independent of its size class at T_1. There is certainly no necessary restriction on such size-class mobility. Indeed some ideas of regional changes in production capacity imply that such changes principally result from changes in the size class of existing factories rather than from regional changes in the number of factories.

The contingency table (fig. 2.45) summarizes the results of this dual classification of papermills which survived from 1915 to 1965, and for which size classes are defined equivalently in the two years by putting each mill's capacity in each year as a multiple of the median machine size for that year. We put

H_0 : the size class of a mill at t_1 is independent of its size class at t_0.

The rejection region is defined at the 0·95 quantile of a χ^2 random variable with 4 degrees of freedom. $X^2 = 57{\cdot}8424$, the 0·999 quantile is 18·47 and so we reject

	Size class			
	t_1			
t_0	*C*	*B*	*A*	
C	41	3	4	48
B	10	8	4	22
A	7	14	31	52
	58	25	39	122

t_0 = initial period is 1910–15

t_1 = final period is 1960–5

Size classes range from the smallest, *A*, to the largest, *C*, and are symmetric about the median.

2.45 Contingency table of papermaking mills according to capacity class in 1915 and 1965.

H_0 in favour of the alternative hypothesis that the classifications are dependent. Our inference is that a mill's size class is dependent on its size class at the preceding time.

We shall now examine two separate procedures for analysing reduced contingency tables. These reductions arise because:

1. one or more cells of the original table is deleted. The need may occur because one cell dominates the frequencies and, as X^2 is proportional to size, a particular row, column dependence may have an overwhelming effect on the whole table.
2. the entries in the principal diagonal are deleted; that is for a_{ij}, $i = j$, giving $a_{11}, a_{22}, \ldots, a_{nn}$.

Case 1

In fig. 2.45 cell a_{11} is dominant and we may feel that this affinity between classification in the first row and the first column obscures the independence of the remaining classes. This cell can be deleted from the table and the reduced table analysed as if the data were missing (fig. 2.46). The approach developed by Goodman (1968) to meet such incomplete tables is to put them in the general form of subtables of different dimensions as in fig. 2.47.

-	3	4	7
10	8	4	22
7	14	31	52
17	25	39	81

2.46 Contingency table of fig. 2.45 reduced by one cell entry.

–	X
Y	Z

2.47(a) Format for analysis of reduced contingency tables.

–	x_1 3	x_2 4	$x_.$ 7
y_1 10	z_{11} 8	z_{12} 4	$z_{1.}$ 12
y_2 7	z_{21} 14	z_{22} 31	$z_{2.}$ 45
$y_.$ 17	$z_{.1}$ 22	$z_{.2}$ 35	$z_{..}$ 57

2.47(b) Fig. 2.46 in the format of fig. 2.47(a).

In this notation the x_j are the elements of the row which contains the empty cell, and the y_i are the elements of the column that contains the empty cell. The z_{ij} are the elements of the remaining subtable (fig. 2.47). The marginal symbols retain the dot notation, so that $x_.$ is the sum of the row of x_j over the jth column, $y_.$ is the sum of the column of y_i over the ith row and $z_{i.}$ and $z_{.j}$ are the row and column sums of the z_{ij} entries.

As the intention is to compare observed with expected values for the whole incomplete table the expected values must sum to the observed marginal totals. Straightforward modifications of the usual calculation ensure this, as is shown by the following equations for the expected values.

1. Expected value of X for the jth column is

$$\hat{x}_j = x_.(x_j + z_{.j})/(x_. + z_{..}) \tag{2.5.3}$$

2. Expected value of Y for the ith row is

$$\hat{y}_i = y_.(y_i + z_{i.})/(y_. + z_{..}) \tag{2.5.4}$$

3. Expected value of Z for the ijth cell is

$$\hat{z}_{ij} = (y_i + z_{i.})(x_j + z_{.j})\, z_{..}/(y_. + z_{..})(x_. + z_{..}) \tag{2.5.5}$$

Substituting from table (ii) in fig. 2.47 into these equations gives the expected values with respect to the appropriate marginal totals and N. Thus in equation 2.5.3 the Y are ignored, in equation 2.5.4 the X are ignored and in equation 2.5.5 the formula puts the effect of the additional marginals more succinctly than words. The expected frequencies for each cell are calculated in the following way.

$$\hat{x}_1 = 7(3+22)/(7+57) = 2{\cdot}734$$
$$\hat{x}_2 = 7(4+35)/(7+57) = 4{\cdot}266$$
$$\hat{y}_1 = 17(10+12)/(17+57) = 5{\cdot}054$$
$$\hat{y}_2 = 17(7+45)/(17+57) = 11{\cdot}946$$
$$\hat{z}_{11} = (10+12)(3+22)57/(7+57)(17+57) = 6{\cdot}620$$
$$\hat{z}_{12} = (10+12)(4+35)57/(7+57)(17+57) = 10{\cdot}326$$
$$\hat{z}_{21} = (7+45)(3+22)57/(7+57)(17+57) = 15{\cdot}646$$
$$\hat{z}_{22} = (7+45)(4+35)57/(7+57)(17+57) = 24{\cdot}408$$

These expected values and their X^2 contributions are tabulated in fig. 2.48.

—	2·734	4·266	7
5·054	6·620	10·326	16·946
11·946	15·646	24·408	40·054
17	22·266	34·734	57

2.48(a) Expected values under H_0 in the reduced table of fig. 2.46.

—	0·0259	0·0166	
4·8403	0·2877	3·8755	
2·0478	0·1732	1·7803	
			13·0473

2.48(b) X^2 contributions of each cell for fig. 2.46.

The total $X^2 = 13{\cdot}0473$ is the sum of these individual values with $(r-1)(c-1)-k = (3-1)(3-1)-1 = 3$ degrees of freedom, when k is the number of deleted cells. Using the 0·95 quantile to define the rejection region, we reject the null hypothesis of independence in this incomplete table. Goodman refers to *quasi-independence* in such incomplete tables and extends the treatment to more cells, to row and cell permutations and to alternative estimation procedures. The probability models basic to these procedures are dealt with succinctly by Bishop and Fienberg (1969) and the reader interested in both the extensions and their justification should begin by consulting those articles. We shall proceed to illustrate the case in which a complete diagonal of cells is deleted as this seems to be the commonest requirement after the one-cell case. It is also the case to which the one-cell deletion procedure cannot be extended directly. The procedure is stated and used in an example, but the reader will appreciate the approach more once he has read the use of contingency tables for three or more variables (article 2.6).

Case 2

In studying the size mobility of factories we may expect most factories to stay in the same size category. But for those factories that do change from one category to another we may be interested to see whether there is any association of classes. Thus we arrange the two classifications in the same sequence and delete cells, a_{ij} for $i = j$, giving a table in which entries along a principal diagonal are empty. Such a table has marginal totals based on the off-diagonal cells. For the same data used in case 1 we have 42 factories which changed size class in the following manner (fig. 2.49).

Size class				
		t_1		
t_0	*C*	*B*	*A*	
C	–	3	4	7
B	10	–	4	14
A	7	14	–	21
	17	17	8	42

2.49 Fig. 2.45 modified to show papermaking mills changing capacity class between 1915 and 1965.

Our null hypothesis is one of independence in these 6 cells. Independence can be envisaged as involving no interaction between categories such that knowledge of the class of one attribute gives no information about the class of the other attribute. In such a case, the cross-product ratio in the fourfold table is unity. This idea of interaction being related to independence can be extended to $r \times c$ tables as well as to multiple contingency tables. It is the basis for estimating whether there is sufficient dependence in the table to reject the null hypothesis of independence in arrangements where the principal diagonal entries are omitted.

In any contingency table the frequencies can be expressed as proportions of the total, that is

$$p_{ij} = f_{ij}/N \tag{2.5.6}$$

In a standard 2×2 table the cross-product ratio is 1, if $p_{11} \cdot p_{22} = p_{12} \cdot p_{21}$. This is interpreted as there being no interaction in the table. The cross-product ratio can be extended to the general $r \times c$ table as well as to multiple contingency tables as the basis for measuring the amount of interaction. The interaction provides the basis for judging the acceptability of the hypothesis of independence.

In a contingency table with $r = c = 3$, such that there are entries in off-diagonal cells, the analogy to the cross-product ratio is $p_{12}p_{23}p_{31}/p_{13}p_{32}p_{21}$ which equals 1 if there is no interaction. For the data in fig. 2.49, we have $p_{12} = f_{12}/N = 3/42 = 0{\cdot}0714$ and so on, giving a ratio of $0{\cdot}15$ which indicates some interaction. The question is whether this is sufficiently different from 1 for us to reject the null hypothesis of independence.

A standard model for such tables expresses the proportions as natural logarithms, and as ln 1 = 0 the model shows the condition of no interaction as:

$$L = \ln p_{12} + \ln p_{23} + \ln p_{31} - \ln p_{13} - \ln p_{32} - \ln p_{21} = 0 \qquad (2.5.7)$$

If $L \neq 0$ then there is some interaction modifying the proportions systematically by column and by row. This interaction can be expressed in terms of coefficients, α, such that

$$L_\alpha = \sum_{i,j} a_{ij} \ln p_{ij} \qquad \text{with } \sum_i \alpha_{ij} = 0 \text{ and } \sum_j \alpha_{ij} = 0 \qquad (2.5.8)$$

because the α modify values *within* the table, but they do not change the marginal totals. The problem is to estimate this interaction. Estimated values are shown conventionally by a $\wedge$ over the symbol representing the value that is being estimated and we have, for equation 2.5.8, the estimates, $\hat{L}_\alpha$, given by

$$\hat{L}_\alpha = \sum_{i,j} \alpha_{ij} \ln f_{ij} \qquad (2.5.9)$$

The variance, S^2, of these estimates is given by

$$S_\alpha^2 = \sum_{i,j} \alpha_{ij}^2 g_{ij} \quad \text{with } g_{ij} = 1/f_{ij} \qquad (2.5.10)$$

Substituting the data into these equations gives

$$\begin{aligned} \hat{L} &= \ln f_{12} + \ln f_{23} + \ln f_{31} - \ln f_{13} - \ln f_{32} - \ln f_{21} \\ &= \ln 3 + \ln 4 + \ln 7 - \ln 4 - \ln 14 - \ln 10 \\ &= 1{\cdot}8971 \end{aligned}$$

and

$$\begin{aligned} S^2 &= g_{12} + g_{23} + g_{31} - g_{13} - g_{32} - g_{21} \\ &= 1{\cdot}1476 \end{aligned}$$

Now the statistic $X^2 = \hat{L}^2/S^2$ is distributed approximately as the χ^2 random variable, thus

$$\begin{aligned} X^2 &= \hat{L}^2/S^2 \\ &= 3{\cdot}1361 \end{aligned} \qquad (2.5.11)$$

Referring this to χ^2 with $(r-1)(c-1)-k = (3-1)(3-1)-3 = 1 df$ we accept the null hypothesis of independence in the incomplete table for the rejection region defined on 0·95 quantile of χ^2. Thus we state that there is evidence in the table (fig. 2.49) to support the assertion that when factories change size class, the final size class is independent of the initial size class.

2.5.4 Chi-square and ordered categories

In this section so far we have considered hypotheses and test statistics appropriate to maps which give information about one or two attributes of a set of objects. These attributes are location class and one other attribute for both of which two or more categories are recognized. A great many maps are contained in these conditions that require nominal measurement only. However in a substantial subset of these maps it is reasonable to put the categories of one or both of the attributes in an order which corresponds to a logical sequence. For example, the location classes could be distance classes from some origin or from a number of datum points. Equally the non-locational attribute could consist of measurements on a continuous scale for which an ordered sequence of increasing or decreasing values is rational. An increase in the usable information increases the subtlety of the hypothesis that can be tested. We shall consider the case of maps that show two properties, one in an ordered sequence, the other as a dichotomy, and illustrate the use of chi-square by providing

1. a procedure for testing for homogeneity of the proportion of the dichotomy in subsets of the sequence;
2. a measure of association between change in the proportions of the dichotomy and position in the sequence.

For both these procedures we shall begin by arranging the results as frequencies in a contingency table and then calculate a test statistic, $X^2 = \Sigma\ (O_{ij} - E_{ij})^2/E_{ij}$ and refer this to values of χ^2 in a rejection region defined on a quantile of the χ^2 distribution. We noted earlier that X^2 calculated for the whole table is not sensitive to particular sorts of deviation that may occur if the null hypothesis of independence or of homogeneity is false. The two situations stated above can be met by adaptations to the usual X^2 statistic which makes it sensitive to the particular patterns of deviation entailed. These two situations are common in initial stages of geographic enquiry using point symbol maps.

2.5.4.1 Homogeneity in an r × 2 contingency table

The example illustrating the procedure puts location classes in an ordered sequence and uses survival and failure as the non-locational dichotomy. The method is equally applicable to two mutually exclusive location classes which contain all the objects for which another attribute such as capacity, temperature, flow and so on is put into ordered classes of intensity.

For the papermills already considered, define X as the distance from a source of fibrous raw material, Y as its survival or failure over a given time period. Record $(X_i Y_i)$ for each mill. Then $[(X_1 Y_1), \ldots, (X_N Y_N)]$ can be seen as a random sample from the population of all possible distance, survival measures. Put these N observations into an $r \times 2$ contingency table corresponding to a suitable classification of X and arrange these classes in a sequence, as in fig. 2.50.

Set the null hypothesis as

H_0 : the proportion of mills surviving is independent of distance class.

Now, if the distance classes had been established initially such that each distance class was treated as a homogeneous location class for n_i observations in the ith location class on survival and failure, then the appropriate null hypothesis would

Ordered location classes	Attribute class								
	Do survive			Do not survive			Total	X^2	
	x_i	$\hat{x}_i = n_i\hat{p}$	$(x_i - \hat{x}_i)^2/\hat{x}_i$	y_i	$\hat{y}_i = n_i(1-\hat{p})$	$(y_i - \hat{y}_i)^2/\hat{y}_i$	n_i	$(x_i - \hat{x}_i)^2/\hat{x}_i + (y_i - \hat{y}_i)^2/\hat{y}_i$	$p_i = x_i/n_i$
	29	24·802	0·7106	8	12·198	1·4448	37	2·1554	0·78378
	25	22·121	0·3747	8	10·879	0·7619	33	1·1366	0·75758
	30	26·813	0·3788	10	13·187	0·7702	40	1·1490	0·75000
	19	22·791	0·6306	15	11·209	1·2822	34	1·9128	0·55882
	19	25·473	1·6449	19	12·527	3·3448	38	4·9897	0·50000
	122	122·000	3·7396	60	60·000	7·6039	182	11·3435	$\hat{p}=0{\cdot}67033$
									$(1-\hat{p})=\hat{q}=0{\cdot}32967$

2.50 $r \times 2$ contingency table with r ordered classes.

be that the proportion of survivors (or of failures to survive) is homogeneous. In either case we calculate $X^2 = \Sigma_i (O_i - E_i)^2/E_i$ for each of the $r \times c = 5 \times 2 = 10$ cells. The correspondence between this and the X^2 calculated already is seen clearly from fig. 2.50, and consolidated by considering some of the steps in the following manner.

There are $\Sigma\, x_i$ survivors altogether out of $\Sigma\, n_i$ factories. The mean proportion surviving regardless of distance class is, then,

$$\hat{p} = \Sigma\, x_i/\Sigma\, n_i \tag{2.5.12}$$

The expected number of survivors in the ith distance class is $\hat{x}_i = n_i\hat{p}$. Thus for fig. 2.50 we have

$$\hat{p} = 122/182 = 0{\cdot}67033$$
$$x_1 = 37(0{\cdot}67033) = 24{\cdot}80220$$

Alternatively, the marginal totals can be used directly by

$$\hat{x}_1 = \frac{n_i\, \Sigma x_i}{\Sigma\, n_i} = \frac{37.122}{182} = 24{\cdot}80220 \tag{2.5.13}$$

These expected values provide the basis for the usual calculation of X^2. Thus we have the contributions to the total X^2 value for x_1 and y_1.

$$\text{For } x_1 : (29-24{\cdot}80220)^2/24{\cdot}80220 = 0{\cdot}7106$$
$$\text{For } y_1 : (8-12{\cdot}198)^2/12{\cdot}198 = 1{\cdot}4448$$

The separate contributions to X^2 for all x_i, y_i are given in fig. 2.50. The total value is $X^2 = 11{\cdot}3435$, with $(r-1)(c-1) = (5-1)(2-1) = 4$ *df.* Define the rejection region as the $0{\cdot}95$ quantile of X^2 which is $11{\cdot}143$ with 4 *df* : consequently, as $X^2 > 11{\cdot}143$, we reject the null hypothesis and infer that in this table there is evidence that the number of surviving factories is dependent on the distance class from a source of raw material.

This statement does not direct attention to any of the distance classes and the inference may be able to be refined by modifying the test to see whether the proportion of survivors in the first n_k rows is different from the proportion in the remaining $n_N - n_k$ rows. This refinement in interpretation is accomplished by allocating the total X^2 value to each row in relation to its deviation from the overall proportion.

The mechanism of the allocation is more easily grasped by establishing some identities and substituting these in the X^2 formula already used. These identities also emphasize that the binomial distribution is the underlying exact probability model for which X^2 is being used as an approximation (see article 2.3.2).

Define	$p_i = x_i/n_i$	then	$x_i = n_i p_i$
	$q_i = (1-p_i)$	then	$y_i = n_i(1-p_i)$
	$\hat{p} = \Sigma x_i/\Sigma n_i$	then	$\hat{x}_i = n_i\hat{p}$
		and	$\hat{y}_i = n_i(1-\hat{p})$

Giving $$(x_i - \hat{x}_i) = (n_i p_i - n_i \hat{p}) = n_i (p_i - \hat{p}) \tag{2.5.14}$$

and $$(y_i - \hat{y}_i) = n_i (1 - p_i) - n_i (1 - \hat{p}) = n_i (p_i - \hat{p}) \tag{2.5.15}$$

For one term of X^2 contributions we have

$$\frac{(x_i - \hat{x}_i)^2}{\hat{x}_i} + \frac{(y_i - \hat{y}_i)^2}{\hat{y}_i} = \frac{n_i (p_i - \hat{p})^2}{n_i \hat{p}} + \frac{n_i (p_i - \hat{p})^2}{n_i \hat{q}}$$

$$= \frac{n_i (p_i - \hat{p})^2 \hat{q} + n_i (p_i - \hat{p})^2 \hat{p}}{n_i \hat{p}\, \hat{q}} = \frac{(\hat{p} + \hat{q}) n_i (p_i - \hat{p})^2}{n_i \hat{p}\, \hat{q}}$$

$$= \frac{1 \,.\, n_i (p_i - \hat{p})^2}{n_i \hat{p}\, \hat{q}}$$

summing over i to give the total X^2 we have

$$X^2 = \frac{\Sigma n_i (p_i - \hat{p})^2}{\hat{p}\, \hat{q}} \tag{2.5.16}$$

which shows that X^2 is a weighted sum of squares of the individual proportions of success, p_i, from their mean, $\hat{p}$, in which the weights are the $n_i / \hat{p}\, \hat{q}$.

This total sum of squares can be subdivided and allocated to subsets of the distance classes. The procedure is best understood by rearranging this formula, equation 2.5.16. First substitute $p_i = x_i / n_i$ and $\hat{p} = \Sigma x_i / \Sigma n_i$ to give, after cancellation,

$$X^2 = \frac{\sum \frac{x_i^2}{n_i} - \frac{(\Sigma x_i)^2}{\Sigma n}}{\hat{p}\, \hat{q}} \tag{2.5.17}$$

Then replace $$\sum \frac{x_i^2}{n_i} \text{ by } \sum x_i \,.\, \frac{x_i}{n_i} = \sum x_i p_i$$

and replace $$\frac{(\Sigma x_i)^2}{\Sigma n_i} \text{ by } \Sigma x_i \frac{\Sigma x_i}{\Sigma n_i} = \Sigma x_i \hat{p}$$

to give

$$X^2 = \frac{\Sigma x_i p_i - \hat{p}\, \Sigma x_i}{\hat{p}\, \hat{q}} \tag{2.5.18}$$

This formula enables the total X^2 to be partitioned into independent components which sum to the correct total for the initial table. These components can be arranged conveniently as below (fig. 2.51).

In the case of the survival of papermills, factories in the median distance class or less may have a survival propensity which is different from that of mills in the remaining distance classes. Some such division recognizes that there could be a response to costs of assembling the raw material, and as these increase in the sequence of distance classes these costs may approach a threshold which affects the likelihood of survival. We form the following division of the k distance classes:

1. Differences between classes (1, 2, 3) and (4, 5)
2. Differences within classes (1, 2, 3)
3. Differences within classes (4, 5)

X^2 *component due to*		*Degrees of freedom*
(i)	Difference between p's in first N_1 and last $N-N_1$ rows	1
(ii)	Variation among p's within the first N_1 rows	N_1-1
(iii)	Variation among p's within the last $N-N_1$ rows	$N-N_1-1$
		$N-1$

2.51 Summary format for analysis of $r \times 2$ contingency tables with ordered categories.

For each of these partitions the numerator of the formula (equation 2.5.18) is used to calculate a sum of squared deviations, and this sum is converted to an X^2 value on division by the denominator, $\hat{p}\hat{q}$. The preliminary arithmetic is established in fig. 2.52.

	x_i	y_i	n_i	$p_i = x_i/n_i$	$x_i p_i$	
−2	29	8	37	0·78378	22·72962	
−1	25	8	33	0·75758	18·93950	N_1
0	30	10	40	0·75000	22·50000	
1	19	15	34	0·55882	10·61758	$N-N_1$
2	19	19	38	0·50000	9·50000	
Total	122	60	182	$\hat{p}=0{\cdot}67033$	84·28670	$= \sum_i x_i p_i$
				$\hat{q}=0{\cdot}32967$		
				$\hat{p}\hat{q}=0{\cdot}22099$	$\hat{p}\sum_i x_i$	= 81·78026

2.52 Initial details required for analysing tables like fig. 2.50.

The total sum of squares is $\Sigma x_i p_i - \hat{p}\,\Sigma x_i = 84{\cdot}28670 - 81{\cdot}78026 = 2{\cdot}50644$, which on division by $\hat{p}\hat{q} = 0{\cdot}22099$ gives the total $X^2 = 11{\cdot}34187$.

In order to ascertain what part of that sum of squares results from differences in proportions *between* the two subsets, we perform a similar operation. Again the arithmetic can be seen clearly from a table that links the procedure to the initial table (fig. 2.53).

The x_i and p_i now refer to the first n_i rows and the remaining $n-n_i$ rows of the original table. Substituting these combined values in the same formula, equation 2.5.18, we get

$$\text{Sum of squares } [(1, 2, 3); (4, 5)] = 64{\cdot}145424 + 20{\cdot}055556 - 81{\cdot}78026 = 2{\cdot}42072$$

	x_i	y_i	n_i	p_i	$x_i p_i$	
(−2, −1, 0)	84	26	110	0·763636	64·145424	N_1
(1, 2)	38	34	72	0·527778	20·055556	$N-N_1$

2.53 Partitioning fig. 2.50 into two subsets giving X^2 contribution for appropriate degrees of freedom.

$$\begin{aligned} X^2\ [(1, 2, 3); (4, 5)] &= 2{\cdot}42072/\hat{p}\,\hat{q} \\ &= 10{\cdot}95398 \end{aligned}$$

with 1 *df* because we are comparing two things. It is plain that most of the variation in the proportions is contained in this comparison. The difference between the overall X^2 and this component is 0·387890 and is the sum of the contributions to X^2 of variations *within* each subset. The variation within the subset containing the first three distance classes (1, 2, 3) is calculated from a comparison of their separate proportions with the mean proportion for all three together. Again using equation 2.5.18 we have

$$\begin{aligned} \text{Sum of squares } (1, 2, 3) &= 29(0{\cdot}78378) + 25(0{\cdot}75758) + 30(0{\cdot}7500) - 84(0{\cdot}763636) \\ &= 64{\cdot}169120 - 64{\cdot}145424 \\ &= 0{\cdot}023696 \\ X^2\ (1, 2, 3) &= 0{\cdot}023696/\hat{p}\,\hat{q} \\ &= 0{\cdot}107227 \end{aligned}$$

with 2 *df* because three things are being compared. The remaining component is that representing the variation within the subset (4, 5) distance classes.

$$\begin{aligned} \text{Sum of squares } (4, 5) &= 19(0{\cdot}55882) + 19(0{\cdot}5) - 38(0{\cdot}527778) \\ &= 20{\cdot}117580 - 20{\cdot}055556 \\ &= 0{\cdot}062024 \\ X^2\ (4, 5) &= 0{\cdot}062024/\hat{p}\,\hat{q} \\ &= 0{\cdot}280664 \end{aligned}$$

with 1 *df* because two things are being compared. These results are summarized in fig. 2.54 and indicate how the important difference in survival rates can be isolated, whereas the differences *within* each subset are negligible in comparison. By this relatively straightforward procedure attention is directed to a particularly important difference in survival rate within a table that showed overall heterogeneity in proportions surviving. The reduction of the degrees of freedom enables inferences to be made with increased confidence.

2.5.4.2 Regression in an r × 2 contingency table

In many cases where location classes correspond to increases in distance or cost from some origin, we may expect that the proportions, in the *i*th location class, p_i, on the dichotomized property will vary systematically with location class. In such cases the appropriate question is whether these proportions increase or decrease linearly as distance class increases. In order to provide a measure of

X^2 component due to	X^2 contribution	Degrees of freedom	χ^2 α: 0·05	χ^2 α: 0·001
(i) Difference between (−2, −1, 0)(1, 2)	10·95398	1	3·841	10·828
(ii) Variation within (−2, −1, 0)	0·107227	2	5·991	
(iii) Variation within (1, 2)	0·280664	1	3·841	
Total	11·341871	4	9·488	18·467

2.54 Fig. 2.50 in the format of fig. 2.51.

association between the proportions, p_i, and distance class, these distance classes must be assigned a value, as it is to this value that the proportions may show a linear relation. As these are distances, we are quite properly able to assume a more refined underlying scale than the nominal one used previously. Indeed we are surjectively mapping measurements on a continuous scale to measurements on an ordinal scale such that all assignments in a given interval on the continuous scale are given the same assignment on the ordinal scale. The supposition is that the new scale, less restricted than the continuous scale and more restricted than the nominal scale, bears a closer relationship to changes in the proportions surviving than either of the other scales. The only necessary restriction on the numbers given to the location classes is that the numerals indicate the correct position of the class in the ordered sequence of classes. If the classes are defined on equal increments of a continuous scale then it seems reasonable in the first instance to imitate this regularity. If the increments change systematically in size of interval on the continuous scale then this could be matched by weighting the ordinal measurements to correspond to these intervals.

Let us call these location-class measurements, z_i, then we wish to ascertain what change in z_i corresponds to a unit increase in the proportions, p_i. If the relationship were exact we should have a unit increase in p_i such that $p_i = a + bz_i$, where a and b are respectively the intercept value of the line and the slope coefficient of the line. However the relationship is unlikely to be exact and the standard procedure is to estimate a straight line, known as the regression line, that is nearest to the values of (p_i, z_i). In this case the b coefficient is referred to as the regression coefficient of P or Z. Since the p_i are assigned a weight $n_i/p\,q$, as in equation 2.5.16, we use the formula for calculating a weighted regression coefficient.

$$b = \frac{\Sigma n_i\,(p_i - \hat{p})(z_i - \bar{z})}{\Sigma n_i\,(z_i - \bar{z})^2} \qquad (2.5.19)$$

Although this can be used directly to calculate X^2, it is usually more convenient to replace the n_iP_i by X_i and separate the terms to give

$$b = \frac{\Sigma x_i z_i - \dfrac{\Sigma x_i(\Sigma n_i z_i)}{\Sigma n_i}}{\Sigma n_i z_i^2 - \dfrac{(\Sigma n_i z_i)^2}{\Sigma n_i}} \qquad (2.5.20)$$

The total sum of squares is again partitioned, as in article 2.5.4.1, but this time the component due to the linear regression of P and Z is isolated, converted to an X^2 value with 1 *df*, leaving the difference between the total X^2 and this component X^2 to be attributed to non-linear regression with the remaining degrees of freedom.

Example

This procedure can be illustrated by applying it to the data of the previous article. A linear scale of Z values centred on zero was assigned to the distance classes as shown in fig. 2.55. In this particular case, the papermills are put into five classes centred on the median distance; this is the average class and is given a zero score. The extremes in both directions are given the same scores with a change in sign. The mills in the distance class nearest to a source of raw material are given a score of -2, those in the distance class furthest from a source of raw material are given a score of $+2$. Any set of numeral assignments that preserves the order of the classes provides a valid test, but the closer the given scores are to the actual values of the variable on the true, but unknown, underlying scale the more sensitive the test is.

z_i	x_i	n_i	$x_i z_i$	$n_i z_i$	$n_i z_i^2$	y_i	$y_i z_i$
−2	29	37	−58	−74	148	8	−16
−1	25	33	−25	−33	33	8	−8
0	30	40	0	0	0	10	0
1	19	34	19	34	34	15	15
2	19	38	38	76	152	19	38
	122	182	−26	3	367	60	29

2.55 Regression analysis of ordered categories in $r \times 2$ contingency tables; details shown for fig. 2.50.

We wish to determine the truth of the assertion that the proportion of mill survivals decreases linearly with an increase in distance class from a source of raw materials. This becomes H_1; H_0 is then that there is no decrease in survival with an increase in distance class. Again the rejection region is defined by the 0·95 quantile of χ^2 with the appropriate *df*. The test statistic to be referred to χ^2 is given as

$$X^2 = \frac{1}{\hat{p}\,\hat{q}} \cdot \frac{(\text{Numerator of equation 2.5.20})^2}{(\text{Denominator of equation 2.5.20})} \qquad (2.5.21)$$

with 1 degree of freedom.

Substituting the details from fig. 2.55 in equations 2.5.20 and in 2.5.21 we have

$$\text{Numerator} = -26 - \frac{122(3)}{182}$$

$$= -28{\cdot}01099$$

$$\text{Denominator} = 367 - \frac{(3)^2}{182}$$

$$\mp 366{\cdot}950549$$

Consequently the X^2 for regression with 1 *df* is

$$X^2 = \frac{(-28{\cdot}01099)^2}{(366{\cdot}950549)(0{\cdot}22099)}$$

$$= 9{\cdot}675574$$

The results can be put in a convenient table form as in fig. 2.56.

X^2 *component due to*	X^2	*Degrees of freedom*
Regression of p_i on z_i	9·675574	1
Departure from linear regression	1·666297	3
	11·341871	4

2.56 Results of regression analysis of fig. 2.50.

The X^2 for regression lies well within our rejection region and we accept H_1 with some confidence. Again we have directed our test statistic to a particular pattern of deviations and as a result we have both refined the nature of our assertion and increased our confidence in it.

A diagram (fig. 2.57) may show more clearly the way in which the test statistic has been directed. In essence we have drawn a graph of the proportions surviving,

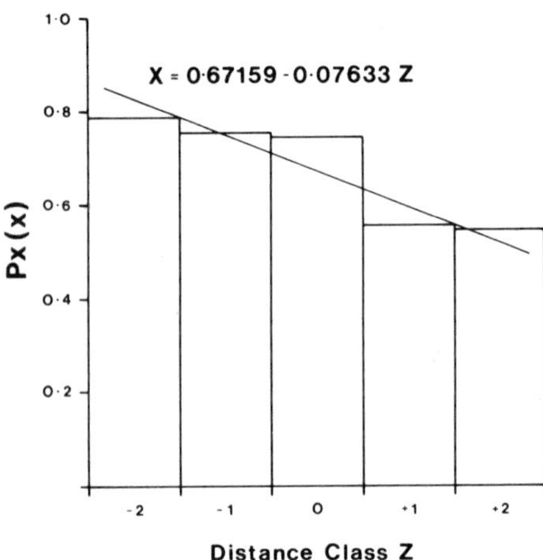

2.57 Graphical comparison of the observed and expected values under the regression hypothesis for fig. 2.50.

p_i, on the vertical axis against distance class, z_i, on the horizontal axis. In the *undirected* case we used X^2 against the H_0 of uniformity; i.e. about a line parallel to the horizontal axis through the overall proportion of survivors. In the *directed* case of *partitions* we replaced this one line by two lines parallel to the horizontal axis; i.e. one line for each subset passing through the overall proportion of survivors in each subset. In the *directed* case of *regression* line we used X^2 against the H_0 of decreasing proportions; i.e. about a line *not parallel* to the horizontal axis. In each case we could calculate the expected numbers of survivors and failures by weighting the *expected proportions* by the number of objects in each class. In the undirected case the expected proportion, p, was the same for all classes, whereas in the regression case the expected proportions correspond to the regression line. The slope of that line is $b = -0{\cdot}0763$. The minus sign implies that the line slopes downwards from left to right such that decreasing proportions correspond to increasing distance classes. The value shows what change in proportion arises from a change of one distance class; of course the full range of proportions is the interval (0, 1).

In order to superimpose the regression line on the graph we need to know where the line intersects the vertical axis when that axis passes through $z = 0$. This number is known as the intercept value and is often denoted by a in equations. Thus if X is the proportion or number of mills surviving and Z is the distance class, then $X = a + bZ$ is the equation of the line : in the present case b is $-b$ and so the equation is $X = a + (-b)Z$ which is $X = a - bZ$. By rearranging the terms we have $a = X - bZ$ in general, and as the regression line always passes through $(\bar{x}, \bar{z})$ we can calculate a simply by substituting the average values of X and Z in the equation. Both means can be calculated directly from fig. 2.55. The mean value of Z, $\bar{z}$, might seem to be 0, however the distance classes have to be weighted by the frequency of occurrences in them, thus

$$\bar{z} = z_0 + \left(\frac{\Sigma n_i z_i}{\Sigma n_i}\right) k$$

Where z_0 is an arbitrary origin (i.e. a guessed mean) which in this case is 0, and k is the size of the step in the classes, in this case k is 1. Substituting from 2.5.19 in 2.5.22 we get

$$\begin{aligned} \bar{z} &= 0 + \frac{3}{182} \cdot 1 \\ &= 0{\cdot}01648 \end{aligned}$$

For $\bar{x}$ we use the same approach. We call the survivors 1 and the failures 0, $k = 1$ to give

$$\begin{aligned} \bar{x} &= 0 + \left(\frac{122}{182}\right) \cdot 1 \\ &= 0{\cdot}67033 \end{aligned}$$

Thus we find a as

$$\begin{aligned} a &= 0{\cdot}67033 - (-0{\cdot}07633)0{\cdot}01648 \\ &= 0{\cdot}67159 \end{aligned}$$

And the equation of the regression line is

$$X = 0{\cdot}67159 - 0{\cdot}07633\ Z$$

Substituting in this equation for values of $Z = -2, -1, 0, 1, 2$ we get $X =$ 0·8243, 0·7479, 0·6716, 0·5953, 0·5189. These proportions can be plotted on the graph and the regression line drawn. To calculate X^2 from these proportions involves weighting them by the number in each class, the n_i, and using these expected numbers for survivors and failures in the original formula for X^2. The formulae we have developed avoid these intermediate steps and graphical constructions.

2.6 Propositions of independence (3)

Properties: three or more
Measurement scale: nominal or ordinal
Statistical procedure: chi-square with log-likelihood estimates

We shall now discuss some of the procedures available for the analysis of maps in which more than two variables are depicted. There is no change in the point symbolism or the constraints on the maps : the change is in the increase in information contained in the map which is accompanied by a rapid increase in the number of hypotheses and the complexity of the statistical tests. The information may not all be put on to one map, instead there may be a number of maps of the same objects with separate maps representing one or more attributes. Such maps are amongst the most common maps drawn by geographers and are often not restricted to three variables, but commonly show four or five variables each at two or more levels of classification. Such maps entail many comparisons. Although these comparisons are obviously worthwhile we must acknowledge the complexities of analysis they pose. The information in such maps can be seen as a straightforward extension of the two-variable maps and the information can be presented in an $r \times c \times l$ contingency table which is a simple extension of the $r \times c$ format to include layers. The methods of analysis cannot be so simply extended. It is worth emphasizing that analytical procedures for many-variable contingency tables are rarely discussed in statistics textbooks and yet the conceptual complexities of the hypotheses make such a treatment imperative. The importance of such tables is acknowledged by the number of articles devoted to them in recent publications (Bishop 1969; Goodman 1970, 1971; Ku and Kullback 1968; Plackett 1962, 1969, 1971) which extend the useful work contained in Birch 1963, Dyke and Patterson 1952 and Lewis 1962.

The purpose of including such material in this introductory text is:

1. to exploit the structural similarities of the ideas presented already with those of the principal model used in these more complex cases, and so to enable the reader to follow the more advanced reference material from a secure background;
2. to reveal the analytical implications of these standard geographic maps;
3. to illustrate how many of the map problems with which geographers are customarily concerned require for their statistical treatment quite intricate

procedures of hypothesis formulation, model selection and significance testing.

Many undergraduate geography dissertations contain substantive problems that require three or more – variable maps and yet it is rare that the statistical models appropriate to their analysis are taught. Let us encourage such ambitious maps at the same time as we enlarge the technical awareness of the student so that more positive and more precisely defined answers can be provided to these interesting problems. It is always a useful exercise to equate map statements to their statistical-model implications and this should be a necessary preliminary to proceeding to more involved map statements.

2.6.1 The log-linear model, 2 × 2 case

In maps showing two variables at each of two levels we have used chi-square to test the null hypothesis of independence. Fig. 2.58 is a straightforward illustration of this sort of map in which we distinguish two location classes, L_1, L_2, and two levels of response, such as survival and failure, say R_1, R_2. Inspection of fig. 2.58 reveals that each location class contains half the objects, say factories, that survive and half the factories that do not survive. We should infer immediately that survival is independent of location. The corresponding contingency table is shown as diagram (a) in fig. 2.61.

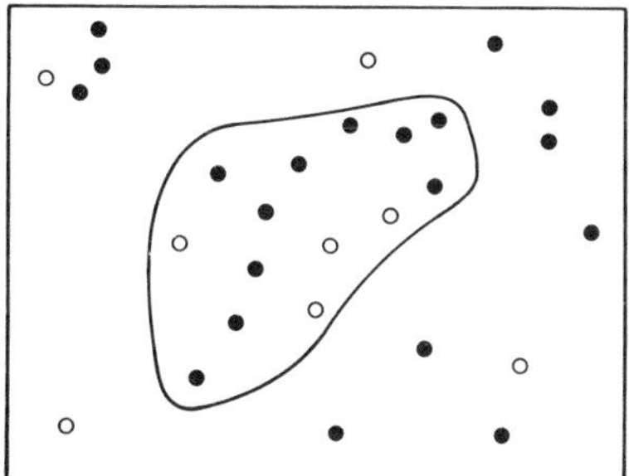

2.58 Objects with two attributes each at two levels, satisfying H_0 of independence; no two-factor interaction in which the two levels of location (attribute A) are equally probable given the level of response (attribute B).

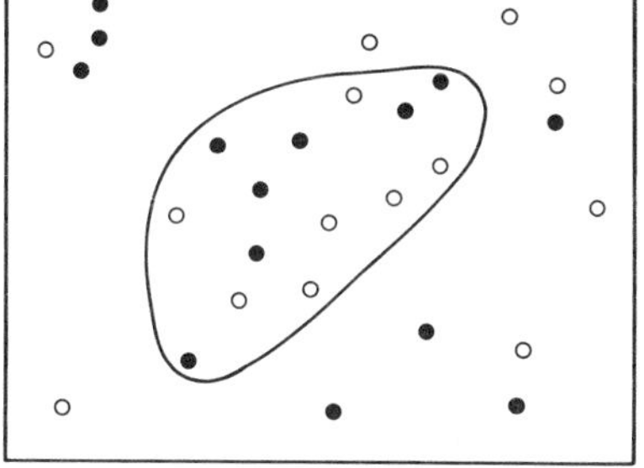

2.59 Objects with two attributes each at two levels, satisfying H_0 of independence; a uniform distribution in which the joint variable AB, defined as in fig. 2.58, is equally probable.

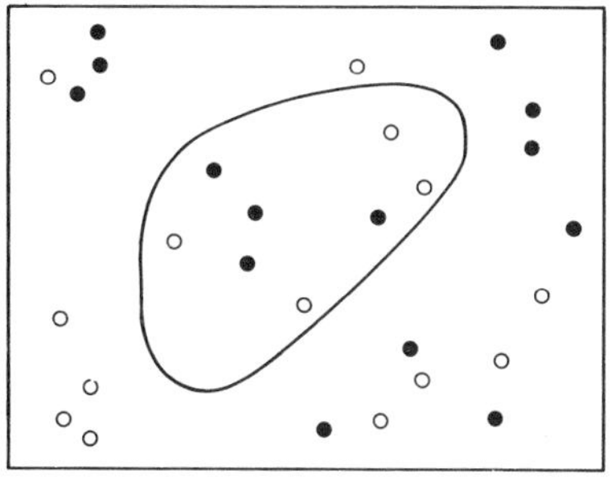

2.60 Objects with two attributes each at two levels, satisfying H_0 of independence; no two-factor interaction in which the two levels of response (B) are equally probable given the level of location (A).

(a)

	R_1	R_2	
L_1	10	4	14
L_2	10	4	14
	20	8	28

$H_0 : A \otimes B$

$H_0 : A = \Phi | B$

(b)

	R_1	R_2	
L_1	7	7	14
L_2	7	7	14
	14	14	28

$H_0 : A \otimes B$

$H_0 : AB = \Phi$

(c)

	R_1	R_2	
L_1	4	4	8
L_2	10	10	20
	14	14	28

$H_0 : A \otimes B$

$H_0 : B = \Phi | A$

(d)

		B_j		
		B_1	B_2	
A_i	A_1	f_{ij}		f_{1j}
	A_2			f_{2j}
		f_{ij}	f_{i2}	$f_{..}$

R = response, non-locational attribute

L = location class

A, B = general designations for attributes

2.61 (a) Contingency table corresponding to fig. 2.58.
(b) Contingency table corresponding to fig. 2.59.
(c) Contingency table corresponding to fig. 2.60.
(d) General contingency table for such maps.

In such a contingency table when the two properties are independent we say that there is no two-factor interaction, or that the first-order interaction is zero. When the frequencies are put as proportions the condition of independence is defined as $P_{11}P_{22} = P_{12}P_{21}$ which relates directly to the earlier discussion of the use of the cross-product ratio as a measure of association (article 2.4.2). The null hypothesis of independence is accepted when all the cell frequencies are the same. Independence is also said to exist when such a uniform table is multiplied by row or column constants. Thus diagram (b) in fig. 2.61 is equivalent to diagram (c) in fig. 2.61 in the sense that for each the null hypothesis of independence is accepted : rows 1 and 2 of diagram (b) in fig. 2.61 are multiplied by 4/7 and 10/7 respectively and the cross-product ratio is 1. Figs. 2.59 and 2.60 correspond to these tables and represent no interaction between survival and location. The calculation in terms of diagram (d) in fig. 2.61 is of course $f_{1j}\, f_{i1}/f_{..} = 14.20/28$ $= 10$ and so on. The maps are not identical and other hypotheses can be stated. In fig. 2.58 it is plain that survival is 2.5 times as likely as failure regardless of location class, whereas either location class is as likely when we know the response class. In these cases our calculations are $f_{i1}/I = 20/2 = 10$ and so on where I

represents the number of rows, i.e. location classes. The hypothesis that the classes of response, the B_j, are equiprobable given the location class, the A_i, is evidently not the case. Whether the difference is important is a matter of a significance level. If they were equiprobable the expected value would be $f_{ij}/J = 14/2 = 7$ and so on. In fig. 2.60 and table (c) in fig. 2.61, it is equally plain that knowing the level of location of an object does not help in deciding whether it survives or fails. Finally in fig. 2.59 and diagram (b) in fig. 2.61 we have a uniform distribution for which we can say, first, that the properties of location and response are independent, secondly, that knowing response does not alter the level of location, and thirdly, tht knowing location does not alter the level of response. Indeed it is enough if we know the total, $f_{..}$, and the number of row and column classes, I, J, thus the expected frequencies in the cells are $f_{..}/IJ = 28/4 = 7$.

There are three distinct types of hypothesis.

1. In the first we are asserting that the two attributes, location class and response, which we now designate as A and B, are independent.
2. In the second that either a) the classes of A are equiprobable given the level of B, that is if we know that the response is survival then we know that the object is as likely to be in one location class as in the other, or b) the classes of B are equiprobable given the level of A.
3. In the third hypothesis we are saying that the joint variable, AB, is equiprobable.

Figs. 2.58, 2.59 and 2.60 illustrate the three types of hypothesis, and it is worth emphasizing that even in this straightforward 2×2 case there are four distinct hypotheses that can be tested. These hypotheses are a nested sequence in which the amount of information used to estimate the expected cell frequencies diminishes. In the first type of hypothesis both pairs of marginal totals are used. In the second type of hypothesis, only one pair of marginal totals is used, either f_{1j}/J or f_{i1}/I. In the third, only the grand total, $f_{..}$ or N, is used with IJ.

If we have as our problem the task of measuring the association between classifications of objects, then we see that there is a structure to the idea of association that is being measured. The cross-product ratio, or some function of it, is a suitable basis for measuring association. When the association in a set of frequencies is characterized in this way we can use these measures of association to estimate the cell values. A linear model of the effects of successive types of association is convenient. A useful function of the cross-product ratio which leads to a linear model of hierarchic association is $\lambda = \ln CPR$.

The frequencies, f_{ij}, or the proportions, $P_{ij} = f_{ij}/N$ can provide a basis for the model. Using the P_{ij} the function of the cross-product ratio is

$$\begin{aligned}\lambda^{AB} &= \ln (P_{11}P_{22}) - \ln (P_{12}P_{21}) \\ &= \ln P_{11} + \ln P_{22} - \ln P_{12} - \ln P_{21}\end{aligned}$$

This equation measures the association between the two variables, A and B, which are denoted by superscripts of λ. This association is termed *either* two-factor interaction *or* first-order interaction. When the variables are independent the cross-product ratio is 1, and as $\ln 1 = 0$, when $\lambda^{AB} = 0$, we say there is no first-order interaction. In all the maps (figs. 2.58, 2.59, 2.60), $\lambda^{AB} = 0$. In the first map, the classes of A are equiprobable given the class of B, but the classes of B are not equiprobable given the class of A. In the third map the converse is true. Let us designate these restricted interactions by

$$\lambda^A = \frac{1}{4}\left(\ln\frac{10}{28} - \ln\frac{4}{28} - \ln\frac{10}{28} + \ln\frac{4}{28}\right) = 0$$

$$\lambda^B = \frac{1}{4}\left(\ln\frac{10}{28} + \ln\frac{10}{28} - \ln\frac{4}{28} - \ln\frac{4}{28}\right) \neq 0$$

for map 1, and $\lambda^A \neq 0$, $\lambda^B = 0$ for map 3. Finally we can recognize that to say that the joint variable, AB, is equiprobable is to say that each of the IJ cells has N/IJ objects and we designate this as μ, the mean number. This number is modified by the classification that is not equiprobable.

Now we can postulate that the two classifications are associated and that we wish to provide a model that will reveal these associations if the rows and columns are subject to multiplication by constants: namely we wish to retain the properties of association we deemed important in the cross-product ratio. We have defined the interactions that can affect the classifications and have put as our model that

$$f_{ij} = M\, L^A\, L^B\, L^{AB} \tag{2.6.1}$$

in which M is an average expected value, and the L terms are log-likelihood terms. The model is multiplicative and contrasts with the usual analysis of variance model which is additive in the same terms.

$$f_{ij} = M + V^A + V^B + V^{AB} \tag{2.6.2}$$

Thus the multiplicative relationship of the cross-product ratio is retained deliberately. Now equation 2.6.1 is equivalent to

$$f_{ij} = e^{\mu}\, e^{\lambda_i^A}\, e^{\lambda_j^B}\, e^{\lambda_{ij}^{AB}} \tag{2.6.3}$$

and this is readily made into an additive model

$$f_{ij} = \mu + \lambda_i^A + \lambda_j^B + \lambda_{ij}^{AB} \tag{2.6.4}$$

with $\mu = \frac{1}{N}\,\Sigma \ln f_{ij}$ and the λ as defined previously.

Thus our model says that the number in any cell is given as the mean number modified by the set of variable-interactions when these interactions are effected multiplicatively. Some of the variable-interactions can be zero. In the two-variable case:

1. $\lambda^{AB} = 0$ if and only if AB are independent;
2. $\lambda^B = 0$ if and only if the two classes of B are equiprobable;
3. $\lambda^A = 0$ if and only if the two classes of A are equiprobable.

When $\lambda^{AB} = \lambda^A = \lambda^B = 0$, then $f_{ij} = \mu$. The hypotheses of interaction are nested in a hierarchy in the sense that a hypothesis of a term entails the inclusion of all lower-order terms. Thus λ^{AB} entails λ^A and λ^B. This notion becomes more important as the number of possible interactions increases and this occurs as the number of variables increases. The data used in fig. 2.62 lead us to doubt independence. This doubt is vindicated by a test statistic and our model must include λ^{AB} and all lower-order terms, λ^{A}, λ^{B}, and μ. The hypotheses, the equations and the calculations are given below the map in fig. 2.62.

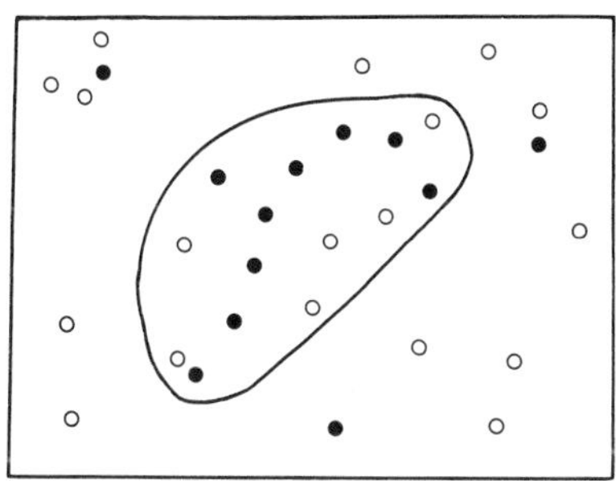

1. $H_0: A \otimes B \longrightarrow \hat{F}_{ij} = f_i^A f_j^B \,|\, n = \begin{matrix} 14 \cdot 1 \,|\, 29; & 14 \cdot 17 \,|\, 29 \\ 15 \cdot 1 \,|\, 29; & 15 \cdot 17 \,|\, 29 \end{matrix}$

2(a) $H_0: B = \Phi | A \longrightarrow \hat{F}_{ij} = f_i^A \,|\, J = \begin{matrix} 12|2\ ; & 17|2 \\ 12|2\ ; & 17|2 \end{matrix} \qquad \begin{array}{c|c} 6 & 8{\cdot}5 \\ \hline 6 & 8{\cdot}5 \end{array}$

(b) $H_0: A = \Phi | B \longrightarrow \hat{F}_{ij} = f_j^B \,|\, I = \begin{matrix} 14|2\ ; & 14|2 \\ 15|2\ ; & 15|2 \end{matrix} \qquad \begin{array}{c|c} 7 & 7 \\ \hline 7{\cdot}5 & 7{\cdot}5 \end{array}$

3. $H_0: AB = \Phi \longrightarrow \hat{F}_{ij} = n \,|\, IJ = \begin{matrix} 29|4\ ; & 29|4 \\ 29|4\ ; & 29|4 \end{matrix} \qquad \begin{array}{c|c} 7{\cdot}25 & 7{\cdot}25 \\ \hline 7{\cdot}25 & 7{\cdot}25 \end{array}$

2.62 Reappraisal of the data in figs. 2.18, 2.19.

We shall adopt the notation conventions introduced by Goodman (Goodman 1970) and use ⊗ to represent independence, | to represent conditional probability, the letters $A, B, C, \ldots$ to represent variable 1, variable 2, variable 3, . . . and subscript i for variable 1, j for variable 2, k for variable 3, . . . with $I, J, K, \ldots$ meaning the number of classes in each variable respectively. Fig. 2.63 illustrates the possible hypotheses, using these conventions for two-variable maps and tables, together with the marginal totals used in the calculation of the expected value and the λ assumed zero under each hypothesis.

Hypothesis rank	*Number of* H_0 *at such rank*	$\lambda = zero$	*Fitted marginals*	*Notation*
1	1	(AB)	$\{A\}\{B\}$	$A \otimes B$
2	2	$(AB)(A)$	$\{B\}$	$A = \Phi \| B$
3	1	$(AB)(A)(B)$	n	$AB = \Phi$

2.63 General notation for $r \times c$ contingency table analysis in terms of hypotheses and marginal totals implied (after Goodman).

The values below fig. 2.62 show that neither $H_0 : B = \Phi|A$, nor $H_0 : A = \Phi|B$ is likely and the equiprobability of the joint variable, AB, is even less tenable, i.e. $H_0 : AB = \Phi$. When we test the hypothesis of independence we are testing whether $\lambda^{AB} = 0$ in the sense defined earlier. If $\lambda^{AB} = 0$, we can proceed, in general, to test whether the classes of variable A are equiprobable given the level of variable B, that is $H_0 : A = \Phi|B$, or analogously the $H_0 : B = \Phi|A$. If $\lambda^A = \lambda^B = 0$ then we can test the third type of hypothesis that $AB = \Phi$. In this simplest case, there are three different types of hypothesis and four testable hypotheses.

The choice of model is determined by the hypothesis tested and a model is considered to be hierarchical if the λ terms are omitted only in descending order of dimensionality (Bishop 1969). Thus if our model includes λ^{AB} then the more elementary terms, λ^A, λ^B, must be included. However, a model in some cases could be

$$f_{ij} = \mu + \lambda^B$$

This model would be satisfactory for fig. 2.58 but not for fig. 2.62. In each hypothesis some λ are *assumed* zero and the expected values under that assumption are compared with the actual values and the test is based on a function of those differences. The function is the log-likelihood statistic referred to χ^2 tables. The λ assumed zero, or the fitted marginals used, will be given in each case.

2.6.2 The log-linear model, $r \times c \times l$ case

The simplest extension of two dichotomous variables is to three dichotomous variables yielding a $2 \times 2 \times 2$ contingency table. The general form of such a table is illustrated in fig. 2.64.

	Variable B_j			
Variable A_i	B_1 *Variable C_k*		B_2 *Variable C_k*	
	C_1	C_2	C_1	C_2
A_1	$A_1B_1C_1$	$\boxed{A_1B_1C_2}$	$\boxed{A_1B_2C_1}$	$A_1B_2C_2$
A_2	$\boxed{A_2B_1C_1}$	$A_2B_1C_2$	$A_2B_2C_1$	$\boxed{A_2B_2C_2}$

2.64 General format for a $2 \times 2 \times 2$ contingency table.

The condition for no three-factor interaction suggested by Bartlett (Bartlett 1935) for such a table is

$$\frac{A_1B_1C_1 \,.\, A_2B_1C_2}{A_1B_1C_2 \,.\, A_2B_1C_1} = \frac{A_1B_2C_1 \,.\, A_2B_2C_2}{A_1B_2C_2 \,.\, A_2B_2C_1} \tag{2.6.5}$$

giving by cross-multiplication

$$A_1B_1C_1 \,.\, A_2B_1C_2 \,.\, A_1B_2C_2 \,.\, A_2B_2C_1 \;=\; A_1B_2C_1 \,.\, A_2B_2C_2 \,.\, A_1B_1C_2 \,.\, A_2B_1C_1$$

and putting these frequencies as proportions we get

$$P_{111}P_{212}P_{122}P_{221} \;=\; P_{121}P_{222}P_{112}P_{211} \qquad (2.6.6)$$

which is the equivalent of the cross-product ratio form of the two-way table. In the general case of the $r \times c \times l$ contingency table with r rows, c columns and l layers the condition of no three-factor interaction is defined as

$$\frac{P_{rcl} \cdot P_{ijl}}{P_{icl} \cdot P_{rjl}} = \frac{P_{rck} \cdot P_{ijk}}{P_{ick} \cdot P_{rjk}} \text{ with } \begin{cases} 1 \leqslant i \leqslant r-1 \\ 1 \leqslant j \leqslant c-1 \\ 1 \leqslant k \leqslant l-1 \end{cases} \qquad (2.6.7)$$

of course in the dichotomous case $i = j = k = 1$ and $r = c = l = 2$ (Roy and Kastenbaum 1956).

However, the independence, no-interaction hypothesis of the two-way table can be generalized in other ways. For example, variables A and B could be independent in each layer of variable C, or the three variables could be mutually independent; for further discussion of these possibilities see Lewis (1962). In fact there are 8 different types of hypothesis that can be tested for the three-variable case, and 18 distinct hypotheses (Goodman 1970). These 8 types of hypothesis can be arranged in a nested sequence that reveals the implications of each hypothesis in terms of the interactions that are assumed zero in each case. When the hypothesis is of no three-factor interaction the λ^{ABC} is assumed to be zero. The information equivalent to that of the two-variable case is presented in fig. 2.65.

Hypothesis rank	*Number of such rank*	λ = *zero*	*Fitted marginals*	*Notation*
1	1	(ABC)	$\{CA\}\{CB\}\{AB\}$	
2	3	$(ABC)(AB)$	$\{CA\}\{CB\}$	$A \otimes B \mid C$
3	3	$(ABC)(AB)(CB)$	$\{CA\}\{B\}$	$AC \otimes B$
4	3	$(ABC)(AB)(CB)(B)$	$\{CA\}$	$B = \Phi \mid AC$
5	1	$(ABC)(AB)(CB)(CA)$	$\{A\}\{B\}\{C\}$	$A \otimes B \otimes C$
6	3	$(ABC)(AB)(CB)(CA)(A)$	$\{B\}\{C\}$	$B \otimes C \cap A = \Phi \mid BC$
7	3	$(ABC)(AB)(CB)(CA)(A)(C)$	$\{B\}$	$CA = \Phi \mid B$
8	1	$(ABC)(AB)(CB)(CA)(A)(C)(B)$	$\{n\}$	$ABC = \Phi$

The symbol $\cap$ denotes the intersection of two events.

2.65 General notation for $r \times c \times l$ contingency table analysis in terms of hypothesis and marginal totals implied (after Goodman 1970).

The way in which the hypotheses are set up, their implications and a verbal statement of them are probably most conveniently illustrated by an example.

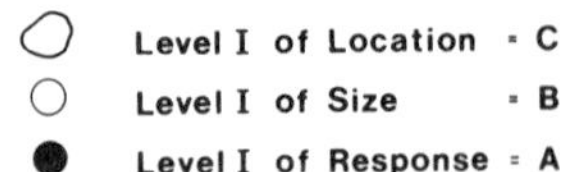

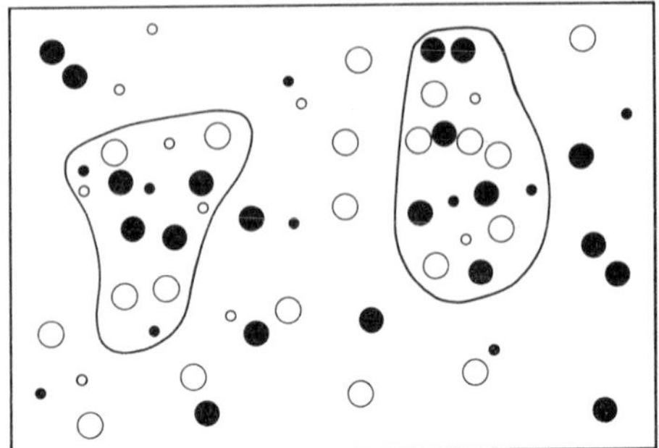

2.66 Map showing objects in terms of three properties each at two levels corresponding to simplest $r \times c \times l$ contingency table.

Example

In the map (fig. 2.66) a third variable, size, has been added to the previous variables, location class and response. Each variable is given at two levels, although location class 1 is no longer contiguous. It is not easy even in this simple case merely to look at the map and state some hypothesis. The usual procedure associated with such an inspection is to count the occurrences in the class combinations. These frequencies could be arranged as in fig. 2.67.

	Variable C or 3 Location class			
Variable A or 1 Response class	C_1		C_2	
	Variable B or 2 Size class			
	B_1	B_2	B_1	B_2
A_1	ABC_{111} 12	ABC_{121} 9	ABC_{112} 9	ABC_{122} 3
A_2	ABC_{211} 4	ABC_{221} 3	ABC_{212} 15	ABC_{222} 5

Variable 1 = A = Response class, two categories
Variable 2 = B = Size class, two categories
Variable 3 = C = Location class, two categories

ABC_{ijk} are shown for convenience and for reference to the general Bartlett condition of no three-factor (or three-variable, or three-attribute) interaction.

2.67 Contingency table corresponding to fig. 2.66.

Perhaps the initial confusion of the map is reduced in this tabulated form, as inspection of the frequencies shows that there is no three-factor interaction as defined in Bartlett's sense because

$$12 \,.\, 15 \,.\, 3 \,.\, 3 = 4 \,.\, 9 \,.\, 9 \,.\, 5$$

In general there will be no such contrived simplicity of arrangement and estimates of the expected values for this hypothesis depend on an iterative procedure that is considered subsequently.

As this hypothesis of no three-factor interaction is accepted, we can proceed hierarchically until we meet a hypothesis we reject. We can ask whether any pair of variables is independent given the level of the third variable. For example, is response, A, independent of size, B, so long as we know whether the objects are in location class 1 or location class 2. In the notation of fig. 2.65 we are setting $H_0 : A \otimes B \mid C$. From that figure we see that this is equivalent to asking whether we can estimate the cell frequencies acceptably well from the marginal totals $\{CA\}\,\{CB\}$ and this is equivalent to asking whether $\lambda^{AB} = 0$ in addition to $\lambda^{ABC} = 0$ which is entailed by hypothesis 1. As we are proceeding through the hierarchy of hypotheses to illustrate their implications in simple cases before introducing some actual data, a standard tabulation is advisable. Each table refers to a distinct hypothesis and contains similar details in a common arrangement as shown in fig. 2.68(a). In fig. 2.68(b)–(j) this format is used to show the details of each of the 8 kinds of hypothesis for the three-variable case. The details for the estimating equation and the degrees of freedom are given completely so that they can be adapted easily to other circumstances.

If we had chosen to ask whether size and location were independent given the level of response on the basis of the map (fig. 2.66), then we should have set up $H_0 : B \otimes C \mid A$ as in fig. 2.68(c). If our interest had been such that we expected survival to be independent of location in each size class then we should have set up $H_0 : A \otimes C \mid B$ (fig. 2.68(d)). These tables indicate exactly how the hypotheses imply particular marginal totals as the information necessary and sufficient (Birch 1963) to calculate the cell frequencies for the complete table if and only if the appropriate $\lambda = 0$ (Birch 1963; Goodman 1970; Plackett 1971).

The information in fig. 2.66 supports the proposition that size is independent of survival in each location class and that size is independent of location for each response, i.e. survivors and failures. We should reject the assertion that there was no association between response and location if we knew the size class. Suppose we had put as our proposition the hypothesis $H_0 : A \otimes B \mid C$, then we should proceed to the hypothesis that the joint-variable, survival-size, was independent of location class. The details are set out for $H_0 : BA \otimes C$ in fig. 2.68(e). As with all subsequent hypotheses we should reject this at the 0·95 quantile of χ^2. However the remainder are well worth working through to grasp the mechanical implications of the hypotheses, their appropriate marginal totals and the $\lambda = 0$. The difference between $\lambda^{ABC} = 0$ and $\lambda^{ABC} = \lambda^{AB} = \lambda^{AC} = \lambda^{BC} = 0$ is instructive; both are, loosely, hypotheses of independence, With $\lambda^{ABC} = 0$ we require the marginal totals $\{AB\}\,\{AC\}\,\{BC\}$, with $H_0 : A \otimes B \otimes C$ we require the marginal totals $\{A\}\,\{B\}\,\{C\}$.

(a)

Hypothesis description
Fitted marginals
Values of fitted marginals

Estimating function

Degrees of freedom

Cell	Actual values	Fitted values	Likelihood ratio
ABC_{ijk}	f_{ijk}	$\hat{F}_{ijk}$	X^2

$\chi^2_{.95}$ at particular degrees of freedom

(b)

$H_0 : A \otimes B|C$
$\{AC\}\ \{BC\}$

	R_1	R_2	
L_1	21	7	28
L_2	12	20	32
	33	27	

	S_1	S_2	
L_1	16	12	28
L_2	24	8	32
	40	20	

$\hat{F}_{ijk} = f_{ik}^{AC}\, f_{jk}^{BC} / f_k^{C}$

$df = K(I-1)(J-1) = 2$

ABC_{ijk}	f_{ijk}	$\hat{F}_{ijk}$		X^2
$L_1S_1R_1$	12	12	= 21.16/28	
$L_1S_1R_2$	4	4	= 7.16/28	
$L_1S_2R_1$	9	9	= 21.12/28	
$L_1S_2R_2$	3	3	= 7.12/28	
$L_2S_1R_1$	9	9	= 12.24/32	
$L_2S_1R_2$	15	15	= 20.24/32	
$L_2S_2R_1$	3	3	= 12.8/32	
$L_2S_2R_2$	5	5	= 20.8/32	

$\chi^2_{(2)\,.95}$ = 6·0   0

(c)

$H_0 : B \otimes C|A$
$\{BA\}\ \{CA\}$

	R_1	R_2	
S_1	21	19	40
S_2	12	8	20
	33	27	

	R_1	R_2	
L_1	21	7	28
L_2	12	20	32
	33	27	

$\hat{F}_{ijk} = f_{ij}^{BA}\, f_{ik}^{CA} / f_i^{A}$

$df = I(J-1)(K-1) = 2$

ABC_{ijk}	f_{ijk}	$\hat{F}_{ijk}$		X^2
$L_1S_1R_1$	12	13·36	= 21.21/33	0·138
$L_1S_1R_2$	4	4·93	= 7.19/27	0·175
$L_1S_2R_1$	9	7·64	= 21.12/33	0·242
$L_1S_2R_2$	3	2·07	= 7.8/27	0·418
$L_2S_1R_1$	9	7·64	= 12.21/33	0·242
$L_2S_1R_2$	15	14·07	= 20.19/27	0·061
$L_2S_2R_1$	3	4·36	= 12.12/33	0·424
$L_2S_2R_2$	5	5·93	= 20.8/27	0·146

$\chi^2_{(2)\,.95}$ = 6·0   1·846

(d)

$H_0 : A \otimes C|B$
$\{AB\}\ \{CB\}$

	R_1	R_2	
S_1	21	19	40
S_2	12	8	20
	33	27	

	S_1	S_2	
L_1	16	12	28
L_2	24	8	32
	40	20	

$\hat{F}_{ijk} = f_{ij}^{AB}\, f_{jk}^{CB} / f_j^{B}$

$df = J(I-1)(K-1) = 2$

ABC_{ijk}	f_{ijk}	$\hat{F}_{ijk}$		X^2
$L_1S_1R_1$	12	8·4	= 21.16/40	
$L_1S_1R_2$	4	7·6	= 19.16/40	
$L_1S_2R_1$	9	7·2	= 12.12/20	
$L_1S_2R_2$	3	4·8	= 12.8/20	
$L_2S_1R_1$	9	12·6	= 21.24/40	
$L_2S_1R_2$	15	11·4	= 24.19/40	
$L_2S_2R_1$	3	4·8	= 8.12/20	
$L_2S_2R_2$	5	3·2	= 8.8/20	

$\chi^2_{(2)\,.95}$ = 6·0   8·441

(e)

$H_0 : BA \otimes C$
$\{BA\}\{C\}$

	R_1	R_2		L_1	L_2
S_1	21	19	40	28	32
S_2	12	8	20		
	33	27			

$\hat{F}_{ijk} = f_{ij}^{AB} f_k^C / n$

$df = (IJ - 1)(K - 1) = 3$

ABC_{ijk}	f_{ijk}	$\hat{F}_{ijk}$	X^2
$L_1S_1R_1$	12	9·8 = 21.28/60	
$L_1S_1R_2$	4	8·87 = 19.28/60	
$L_1S_2R_1$	9	5·6 = 12.28/60	
$L_1S_2R_2$	3	3·73 = 8.28/60	
$L_2S_1R_1$	9	11·2 = 21.32/60	
$L_2S_1R_2$	15	10·13 = 19.32/60	
$L_2S_2R_1$	3	6·4 = 12.32/60	
$L_2S_2R_2$	5	4·27 = 8.32/60	

$\chi^2_{(3)\,.95} = 7{\cdot}8$ 9·977

(f)

$H : C = \phi | AB$
$\{AB\}\{K\}$

	R_1	R_2		K
S_1	21	19	40	2
S_2	12	8	20	
	33	27		

$\hat{F}_{ijk} = f_{ij}^{AB} / K$

$df = IJ(K - 1) = 4$

ABC_{ijk}	f_{ijk}	$\hat{F}_{ijk}$	X^2
$L_1S_1R_1$	12	10·5 = 21/2	
$L_1S_1R_2$	4	9·5 = 19/2	
$L_1S_2R_1$	9	6 = 12/2	
$L_1S_2R_2$	3	4 = 8/2	
$L_2S_1R_1$	9	10·5 = 21/2	
$L_2S_1R_2$	15	9·5 = 19/2	
$L_2S_2R_1$	3	6 = 12/2	
$L_2S_2R_2$	5	4 = 8/2	

$\chi^2_{(4)\,.95} = 9{\cdot}5$ 10·296

(g)

$H_0 : A \otimes B \otimes C$
$\{A\}\{B\}\{C\}$

R_1	33	S_1	40	L_1	28
R_2	27	S_2	20	L_2	32

$\hat{F}_{ijk} = f_i^A f_j^B f_k^C / n^2$

$df = IJK - I - J - K + 2 = 4$

ABC_{ijk}	f_{ijk}	$\hat{F}_{ijk}$	X^2
$L_1S_1R_1$	12	10·27 = 28.40.33/60²	
$L_1S_1R_2$	4	8·4 = 28.40.27/60²	
$L_1S_2R_1$	9	5·13 = 28.20.33/60²	
$L_1S_2R_2$	3	4·2 = 28.20.27/60²	
$L_2S_1R_1$	9	11·73 = 32.40.33/60²	
$L_2S_1R_2$	15	9·6 = 32.40.27/60²	
$L_2S_2R_1$	3	5·87 = 32.20.33/60²	
$L_2S_2R_2$	5	4·8 = 32.20.27/60²	

$\chi^2_{(4)\,.95} = 9{\cdot}5$ 10·895

(h)

$H_0 : B \otimes C \cap A = \Phi | BC$
$\{B\}\{C\}\{I\}$

S_1	40	L_1	28	I	2
S_2	20	L_2	32		

$\hat{F}_{ijk} = f_j^B f_k^C / I n$

$df = IJK - J - K + 1 = 5$

ABC_{ijk}	f_{ijk}	$\hat{F}_{ijk}$	X^2
$L_1S_1R_1$	12	9·33 = 28.40/2.60	
$L_1S_1R_2$	4	9·33 = 28.40/2.60	
$L_1S_2R_1$	9	4·67 = 28.20/2.60	
$L_1S_2R_2$	3	4·67 = 28.20/2.60	
$L_2S_1R_1$	9	10·67 = 32.40/2.60	
$L_2S_1R_2$	15	10·67 = 32.40/2.60	
$L_2S_2R_1$	3	5·33 = 32.20/2.60	
$L_2S_2R_2$	5	5·33 = 32.20/2.60	

$\chi^2_{(5)\,.95} = 11{\cdot}1$ 11·496

(i)				(j)			
$H_0 : CA = \Phi \| B$				$H_0 : ABC = \Phi$			
$\{B\}\{IK\}$				$n \quad IJK$			
S_1 40, S_2 20	IK 4						
$\hat{F}_{ijk} = f_j^B / IK$				$\hat{F}_{ijk} = n / IJK$			
$df = J(IK - 1) = 6$				$df = IJK - 1 = 7$			
ABC_{ijk}	f_{ijk}	$\hat{F}_{ijk}$	X^2	ABC_{ijk}	f_{ijk}	$\hat{F}_{ijk}$	X^2
$L_1S_1R_1$	12	10 = 40/4		$L_1S_1R_1$	12	7·5 = 60/8	
$L_1S_1R_2$	4	10 = 40/4		$L_1S_1R_2$	4	7·5 = 60/8	
$L_1S_2R_1$	9	5 = 20/4		$L_1S_2R_1$	9	7·5 = 60/8	
$L_1S_2R_2$	3	5 = 20/4		$L_1S_2R_2$	3	7·5 = 60/8	
$L_2S_1R_1$	9	10 = 40/4		$L_2S_1R_1$	9	7·5 = 60/8	
$L_2S_1R_2$	15	10 = 40/4		$L_2S_1R_2$	15	7·5 = 60/8	
$L_2S_2R_1$	3	5 = 20/4		$L_2S_2R_1$	3	7·5 = 60/8	
$L_2S_2R_2$	5	5 = 20/4		$L_2S_2R_2$	5	7·5 = 60/8	
$\chi^2_{(6)\,.95} = 12{\cdot}6$			11·763	$\chi^2_{(7)\,.95} = 14{\cdot}1$			18·559

2.68(a) General format for hypothesis description, fitted marginals and their expected values for three-variable map contingency table analysis.
(b) Results of $H_0 : A \otimes B | C$ for fig. 2.67.
(c) Results of $H_0 : B \otimes C | A$ for fig. 2.67.
(d) Results of $H_0 : A \otimes C | B$ for fig. 2.67.
(e) Results of $H_0 : BA \otimes C$ for fig. 2.67.
(f) Results of $H_0 : C = \Phi | AB$ for fig. 2.67.
(g) Results of $H_0 : A \otimes B \otimes C$ for fig. 2.67.
(h) Results of $H_0 : B \otimes C \cap A = \Phi | BC$ for fig. 2.67.
(i) Results of $H_0 : CA = \Phi | B$ for fig. 2.67.
(j) Results of $H_0 : ABC = \Phi$ for fig. 2.67.

The map (fig. 2.66) implies that $\lambda^{ABC} = \lambda^{AB} = \lambda^{BC} = 0$, and therefore our model can omit these terms, leaving $\mu + \lambda^A + \lambda^C + \lambda^{AC}$ as an acceptably efficient estimator of the frequencies.

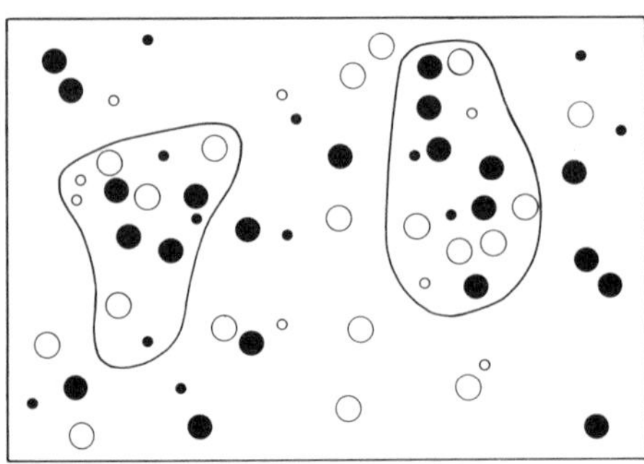

2.69 Modification of fig. 2.66 to satisfy $H_0 : BA \otimes C$ exactly.

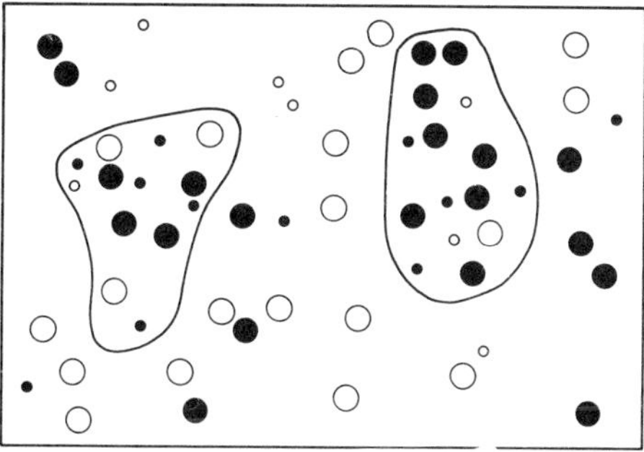

2.70 Modification of fig. 2.66 to satisfy $H_0 : A \otimes B \otimes C$ exactly.

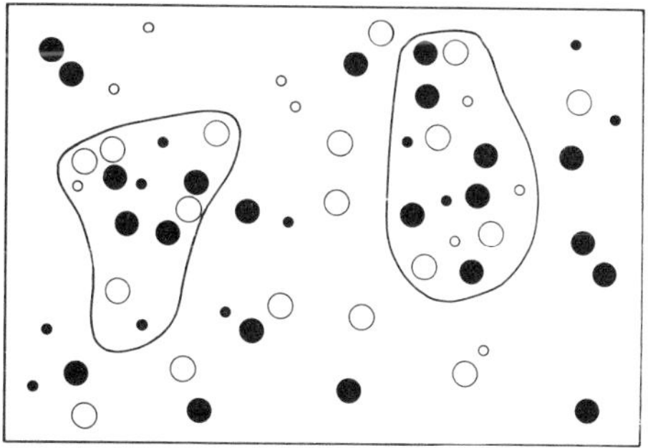

2.71 Modification of fig. 2.66 to satisfy $H_0 : CA = \Phi|B$ exactly, but marginal totals are no longer all the same as in fig. 2.66.

The other maps, figs. 2.69, 2.70, 2.71, are included for $H_0 : BA \otimes C$ (fig. 2.69), $H_0 : A \otimes B \otimes C$ (fig. 2.70), and $H_0 : CA = \Phi|B$ (fig. 2.71) for as nearly the exact map situation as the marginal totals allow, to illustrate the difficulty of visually comparing what we as geographers would consider traditionally as straightforward maps containing only three variables, each at only two levels. It is rare to find proportional and divided symbol maps with as few as three dichotomous variables.

Now that we have established a structural connection between the two- and three-variable cases, we can consider a more involved situation.

Example: age, decay and land-use of buildings

One notion of urban rent is that it is some monotonic decreasing function of distance from the city centre, and that this reduction in land value is accompanied by systematic changes in land use from the central city business area to the more extensive manufacturing and warehousing establishments, through a broad belt of residential suburbs in which building density diminishes out to the commuter zone. As cities change, the rent-surface changes and, with expansion, we can postulate a successive erosion of zones of distinct land-uses by their inner neighbours. Some mechanism like this is a convenient rationalization of the sort of impression that many people gain from travelling into a city.

Most people are aware of the qualities of buildings in towns and references to land-use zones are common. Land-use descriptions are usually made in terms of the activity, such as retailing, manufacturing or residential, rather than in terms of the age of the buildings, their level of deterioration or their form. Some association is assumed to exist and is often given tacit recognition by phrases like high-rise city-centre office developments, Victorian terrace housing, the industrial twilight zone. However, examples are easy to provide to counter the belief that industry is always an area of dereliction, or that the high street or suburban areas are uniform in terms of age or form. In the belief that it is worth examining existing urban building before making assertions about relationships or venturing to undertake major alterations to a city, a number of geographers and planners have considered samples of city building (Medhurst and Parry-Lewis 1969; White 1973). These studies have classified buildings according to three variables: age, decay and use.

One straightforward notion is that building age, building decay and building use are associated in some way. This is a weaker idea than that building use has some particular relationship with a cost surface and thus, if cost is a monotonic function of distance, with distance from a city centre. However we now appreciate that there is more than one sort of relationship that could exist between the three variables. Our impressions of particular towns might encourage us to suppose that all three variables are jointly associated, or we may imagine that only age and decay are related and land use is largely superfluous in characterizing buildings.

Such ideas are tested by reference to part of northeast London from the Bank of England to St Leonard's Hospital, Shoreditch, a transect 1·5 miles long by 0·4 miles wide (White 1973). The buildings in this sample transect are classified according to one of five age-classes: 1 is post-1939; 2 is 1914–39; 3 is 1870–1914; 4 is 1830–70; 5 is pre-1830. Their state of decay is also coded on a scale devised by Medhurst and Parry-Lewis (1969). The coding is done on the basis of scores on a number of criteria such as façade, woodwork, roof sag and so forth. The brief class descriptions for this data from White's study are as follows: *Level 1*, very little or no decay; *Level 2*, some slight decay; *Level 3*, much decay but largely superficial; *Level 4*, substantial decay; *Level 5*, severe decay with major structural weakness. The function of each building refers to the use to which it is currently put. The original detailed classes are compressed to three major uses: *Level 1*, residential; *Level 2*, manufacturing and warehousing; *Level 3*, offices. In this case function of a building is used rather than land-use zones. But as simple land-use zones imply a tendency to homogeneous building use, maps by building can be subjected to repeated analysis according to changes in the bases for designating location class. We can use maps as data sources without insisting that location class is considered each time.

The original maps are complicated and are not able to be reduced sufficiently clearly to be included in this text. There seem to be three belts of comparatively homogeneous building use, but the superimposition of age and decay details on these 1407 buildings is hardly to be resolved by inspection. To try to confine more closely the nature of the association we shall tabulate the data by classifications. There are 5.5.3 = 75 cells in the contingency table which is arranged by age – decay arrays in each of three building-use classes.

Age level	Land-use level I, Decay level 1	2	3	4	5	Land-use level II, Decay level 1	2	3	4	5	Land-use level III, Decay level 1	2	3	4	5	Total
1	160	9	0	0	0	101	11	0	0	0	168	4	0	0	0	453
	158·8	10·2	0	0	0	99·5	12·3	0	0	0	170·7	1·5	0	0	0	
2	3	3	1	0	0	1	35	1	1	0	34	15	0	0	0	94
	1·4	5·4	0	0	0	4·2	31·1	1·7	1	0	32·3	16·5	0	0	0	
3	3	43	20	2	0	6	49	65	40	4	112	152	24	3	2	525
	4·0	38·9	21·6	2·9	1	2·9	54·7	63·2	38·9	4·4	114·1	150·3	24·2	3·2	1	
4	0	4	9	6	1	0	14	51	102	24	20	6	8	4	2	251
	1	3·8	9·4	4·2	1·8	1·3	10·1	51·5	104·7	23·4	17·8	10	7·1	3·1	1·9	
5	0	2	2	0	3	0	0	0	4	2	48	18	1	1	3	84
	1	2·6	1	1	1·6	0	1	1	2·4	2·2	46·9	16·7	1·7	1·7	4·2	
Total	166	61	32	8	4	108	109	117	147	30	382	195	33	8	7	1407

Age level	*Decay level* 1	2	3	4	5	*Total*
1	429	24	0	0	0	453
2	38	53	2	1	0	94
3	121	244	109	45	6	525
4	20	24	68	112	27	251
5	48	20	3	5	8	84
Total	656	365	182	163	41	1407

Age level	*Land-use level* 1	2	3	*Total*
1	169	112	172	453
2	7	38	49	94
3	68	164	293	525
4	20	191	40	251
5	7	6	71	84
Total	271	511	625	1407

Decay level	*Land-use level* 1	2	3	*Total*
1	166	108	382	656
2	61	109	195	365
3	32	117	33	182
4	8	147	8	163
5	4	30	7	41
Total	271	511	625	1407

2.72 Contingency tables (three-way and implied two-way) summarizing a map of land-use, age, decay of buildings in East London, 1973.

There seems to be a basic association between age and decay but there is sufficient irregularity to make us reluctant to simplify. Most buildings are at decay level 1, but age level 3. The decline in residential decay frequencies is similar to the overall pattern whereas manufacturing land-use increases with decay to level 4 and offices have a distinct pattern which is readily observed. Speculation is good and we might expect there to be a three-factor interaction and, of the two-factor interactions, we may suspect that age-decay will be the important one. Others may suppose however that the frequencies by cell are very close to the values expected under the hypothesis of mutual independence: namely that to estimate the number of buildings expected in cell ABC_{111} we use the marginal totals of age, decay and location each at level 1.

Hypothesis	*Fitted marginals*	*Likelihood ratio*	*df*	$\chi^2_{.95}$
H_0: no three-factor interaction	$\{AB\}\{AC\}\{BC\}$	39·636	32	
$H_0: B \otimes C \mid A$	$\{AB\}\{AC\}$	435·548	40	
$A \otimes C \mid B$	$\{AB\}\{BC\}$	344·426	40	55·8
$A \otimes B \mid C$	$\{AC\}\{BC\}$	1140·144	48	55·8
$H_0: A \otimes B \otimes C$	$\{A\}\{B\}\{C\}$	2079·680	66	
$H_0: ABC = \Phi$	n	3504·977	74	

A_i = Variable 1 = Age = 5 levels
B_j = Variable 2 = Decay = 5 levels
C_k = Variable 3 = Land use = 3 levels

We can infer that $\lambda^{ABC} = 0$, but that all λ_{ij} are non zero. Thus our model is

$$\hat{f}_{ijk} = \mu + \lambda_i^A + \lambda_j^B + \lambda_k^C + \lambda_{ij}^{AB} + \lambda_{ik}^{AC} + \lambda_{jk}^{BC}.$$

The estimates for this model are given beneath the actual values in the original table.

Hypotheses of lower dimension than the two 2-way marginals would not normally be calculated. Two are included here to reinforce understanding.

2.73 Acceptable and unacceptable hypotheses from fig. 2.72.

From the summary table (fig. 2.73) we see that there is no significant three-factor interaction at $\alpha = 0{\cdot}05$ and so we could set $\lambda^{ABC} = 0$; however at a more stringent level our decision may be different. The closeness of the estimates from the three two-way marginals $\{AB\}\{AC\}\{BC\}$ can be judged from the italicized values given in fig. 2.72.

Given that we accept $\lambda^{ABC} = 0$ we might now suppose that age and decay will show strong association and be most unlikely to yield $\lambda^{AB} = 0$. We are left with age and function, or decay and function; of these I imagine most people would suppose there to be less prior reason to anticipate an association between age and function, but as we have supposed age and decay to be associated there is unlikely to be much confidence in selecting $H_0: A \otimes C \mid B$ rather than $H_0: B \otimes C \mid A$.

We choose $A \otimes C|B$ and recall that this implies $\lambda^{AC} = 0$ using the marginal totals {age, decay} and {age, function} to estimate the full table of frequencies. We reject this hypothesis and we reject $H_0 : B \otimes C|A$. The values for this latter hypothesis are given in fig. 2.74, and from these we note how the particularly discrepant values are for age level 3. In the table relating to $A \otimes C|B$ the discrepant values were in decay levels 1 and 2. Treating age and decay as independent given functions led to the likelihood-ratio statistic $X^2 = 1140{\cdot}144$ with $K(I-1)(J-1) = 3.4.4. = 48$ degrees of freedom; again this was firmly rejected.

We conclude that only $\lambda^{ABC} = 0$ can be assumed. From this evidence it seems as if we can discard the idea that there is any important increase in information gained by assuming $\lambda^{ABC} \neq 0$. All the important associations seem to be implicit in the two-factor interactions. We cannot reduce this further and use only one-factor information, nor can we discard any one of the three two-factor interactions. Any reduction beyond $\lambda^{ABC} = 0$ gives too great a loss of information and so none of the conditional hypotheses is supportable. In detail then we cannot say that age and decay are sufficiently associated to enable us to discard interactions entailed by land use; building use modifies age and decay individually.

The next stage in this sort of study is to specify land-use classes and characterize them as some function of distance from the city centre, or if cost in the form of urban rent is not a monotonic transformation of distance, we should use a different classification based on costs. An extension of the study to other sectors of London and to other cities would show whether there was consistency in the association of variables. There may well be close correspondence between cities revealed by using standardized tables which would result in a more parsimonious description of urban land-use. Cities in Africa or Asia may show different patterns, but even though the age and decay class criteria will be different they can be ordered similarly and comparisons made. As a general principle, the number of distinctions of level can be increased substantially with great benefit as indicated by Mosteller (Mosteller 1968; Bishop 1969). The ability of such models to cope with a greater fineness of classification will appeal to many geographers who firmly believe in approaching the individual circumstance more closely than usually seems possible in many statistical studies. Cross classifications that result in a very few cases or even no cases in some cells in a particular study can be set against the context of the greater detail of the rest of the table. Zero cell entries have also been contained within the model used in this example as will have been noted from the last set of data.

Such classificatory approaches to geographic problems are a regular feature of the early stages of undergraduate geography learning and research. This model, and its likelihood equations for estimating the expected values when the information used is reduced as interactions are progressively assumed zero, will help to guide interpretation of involved map patterns and clarify the nature of the statements that can be made about those maps. The basic algorithm providing a log-linear fit to contingency tables is published (Haberman 1972) and is readily adapted to particular computer installations. Other models exist and reference to the cited works in this section will enable you to deal with the majority of problems of many-variable maps, in terms of hypotheses of independence and tests of association. It is in these situations of greater than two variables that

Age level	Land-use level I Decay level					Land-use level II Decay level					Land-use level III Decay level				
	1	2	3	4	5	1	2	3	4	5	1	2	3	4	5
1	160·1	9	0	0	0	106·1	5·9	0	0	0	162·9	9·1	0	0	0
2	2·8	4	0·2	0·1	0	15·4	*21·4*	0·8	0·4	0	*19·8*	*27·6*	1	0·5	0
3	*15·7*	*31·6*	*14·1*	*5·8*	0·8	*37·8*	*76·2*	*34·1*	*14·1*	1·9	*67·5*	*136·2*	*60·8*	*25·1*	3·3
4	1·6	1·9	5·4	8·9	2·2	15·2	18·3	51·7	*85·2*	20·5	*3·2*	3·8	10·8	*17·8*	4·3
5	4	1·7	0·3	0·4	0·7	3·4	1·4	0·2	0·4	0·6	40·6	16·9	2·5	4·2	6·8

This table shows the values estimated under the assumption that $\lambda^{ABC} = \lambda^{BC} = 0$; namely $H_0 : B \otimes C \mid A$.
The hypothesis would be rejected.
The italicized values are the most severely discrepant. It is interesting to note the consistently poor estimation of age level 3. For $H_0 : A \otimes C \mid B$ it is decay level 1 and 2 that are consistently poorly estimated.

2.74 Expected values under $H_0 : B \otimes C \mid A$ for fig. 2.72.

assertions concerning the non-locational attributes arise inevitably from maps. As the number of attributes considered is increased so the number of assertions increases and so does the number of assertions involving conditional independence between non-locational properties. Restricting attention to locational attributes excludes important relationships as well as the opportunity to relate the importance of location as an attribute to that of other properties of the objects. Two actual data sets are given in the exercises to illustrate these points.

2.7 Propositions of equality of location and identity of distribution

Properties: two or more
Measurement scale: ordinal or continuous
Statistical procedure: Wilcoxon rank-sum; Mann-Whitney U; Kruskal-Wallis; Friedman

Most of the statistical tests introduced so far are suitable for measurements on nominal scales. The elaboration of the basic point symbol was a response, first, to an increase from one attribute to two attributes, secondly to an increase from dichotomous variables to variables with many classes. All of these tests can be used for measurements on more restricted scales and in two cases we made use of the increase in information of such scales by modifying a chi-square statistic to ordered categories. The proportional and divided symbol maps may be based on ordinal or continuous measurement scales but for a number of reasons this extra information may be ignored deliberately. This sacrifice of information often reduces the sensitivity of a test. A number of tests that make use of the extra information of more restricted measurement scales are discussed next.

The corresponding maps are proportional and divided point-symbol maps. The maps may be based on direct rankings from assessment of order, say, from most preferred to least preferred, or on measurements from a continuous scale which have been converted to ranks. Of course if these measurements represent a random variable with a normal distribution, the extra information should be exploited by using the appropriate t or F tests. Even if this information is not used, the equivalent rank tests are very nearly as efficient in indicating whether a result lies within a particular rejection region. The best-known of the distribution-free tests are known as rank-randomization tests because the original measurements are replaced by ranks and the test statistic is the probability distribution of the ranks for particular events. The tests presented for the propositions of equality and identity for point symbol maps are two-sample or multi-sample tests for ordinal measurements. Other similar tests applicable to the same proposition are introduced in the section on line symbols. This division of tests is one of convenience of presentation not necessity of application.

2.7.1 Wilcoxon rank-sum test

In a typical proportional symbol map the symbols are proportional to the non-locational attribute, but the objects occur in two different location classes or under two sets of conditions. Objects from each location class are treated as separate samples from two, possibly different, populations of measurements of the non-locational attribute. The geographer wishes to know whether he can assert that the two populations of measurements are different on the basis of the evidence shown on the map. In particular he may be interested in whether one population engenders larger values than the other. Thus, if the distributions are not identical, does one distribution have a different location parameter, say the median. The null hypothesis is that the two populations of measurements are identical.

The basis of the test is straightforward. Let us denote one sample of n objects as X, and the other sample of m objects as Y. There are $n + m$ objects altogether. If the null hypothesis is true, then the two samples, X, Y, are simply one of many pairs of subsets of size, n, m, from the total of $n + m$ objects. There are $\binom{n+m}{n}$ ways of forming two samples of sizes n, m from the total of $n + m$ objects, and, under the assumptions of the null hypothesis, each of these ways is equally likely. Any one pair was as likely as any other to have arisen as the actual pair.

If each measure in the combined sample is given its rank, with rank 1 given to the largest numerical value and rank $n + m$ to the smallest, then a set of ranks can be associated with each of the $\binom{n+m}{n}$ combinations. Let us take as the basis of the test statistic the sum of the ranks in each n, m, subset. Then to each of the $\binom{n+m}{n}$ samples there is a corresponding sum of ranks. The frequency of these values is the probability distribution of the test statistic and is the distribution the sum of ranks will have if the null hypothesis is true. This is known as the randomization distribution. The H_0 is accepted or rejected on the basis of where the actual sum lies in this randomization distribution. If the two populations differ in location only, then we expect larger values for one population than for the other and, as a consequence, we anticipate a smaller sum of ranks, because large values are given low ranks. The test based on this sum of ranks is known as the Wilcoxon-rank-sum test.

Example 1

In Kent, nineteen farms which, five years ago, had similar acreage, capital equipment and income, sold their produce entirely to tradespeople in the same market town. Twelve of these farms continue to rely on the local market, the remaining seven farms rely solely on metropolitan sales with higher prices plus higher transfer costs. The difference between initial and present income was calculated and presented as a map (fig. 2.75). We wish to test whether there is any difference in farm income increment between the two subsets of farms. It could be that neither subset of farms was drawn as a random sample from a population of farms, but the measures are the random variable and it is reasonable to suppose that these values are unbiased.

The Wilcoxon test is chosen to test

H_0 : the distribution of farm income increments is the same for farms that sell locally as for farms that sell to the London Market.

H_1 : farms selling to London have larger increments.

H_0 will be rejected in favour of H_1 for values of the test statistic that are less than the value of the test statistic, W_n, that defines the 0·95 quantile of the randomization distribution.

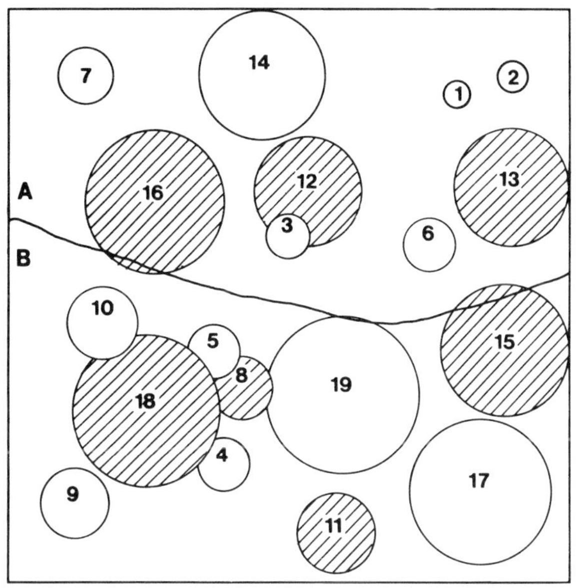

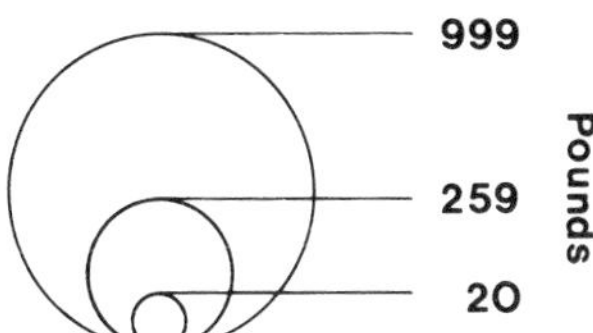

Farms that changed to supplying the metropolitan market

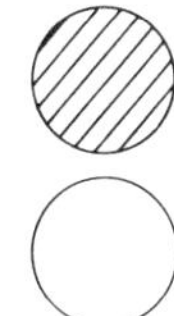

Farms that continued to supply local tradespeople

1–19 Farm index numbers ranked on increment

A Chalk
B Lower Greensand

2.75 Increment in farm income.

Arrange the increments as follows (fig. 2.76). The test statistic, W_n, is the sum of ranks of X,

$$W_n = \Sigma R_x = 47 \quad \text{with } n = 7, m = 12 \tag{2.7.1}$$

This result is looked up in tables. Entering the appropriate table (table 6 in appendix 2), with $n = 7$, $m = 12$, a value of $W_n = 49$ defines the critical region and

X	946, 791, 691, 536, 501, 259, 141		$n = 7$
Y	999, 801, 646, 179, 164, 118, 104, 101, 96, 72, 36, 20		$m = 12$
Rank X, X_{r_i}	2, 4, 5, 7, 8, 9, 12	$\sum_{i=1}^{n+m} X_{r_i} = 47 = W_n$	
Rank Y, Y_{r_i}	1, 3, 6, 10, 11, 13, 14, 15, 16, 17, 18, 19	$\sum_{i=1}^{n+m} Y_{r_i} = 143$	

2.76 Wilcoxon rank-sum calculations for fig. 2.75.

a value of $W_n = 46$ defines the 0·975 quantile. We reject H_0 in favour of the alternative that the increments are not distributed equally and that the median income is higher in farms selling solely to the London market. There are $\binom{19}{7} = 50\,388$ ways in which the rankings could arise out of the 19! ways all the rankings could occur. It is the distribution of the sums of these ranks that yields the probability distribution for the decision. This is too unwieldy a task to illustrate with these figures, but the exact distribution and decision rule are enumerated for a subset of these farms.

Example 2

For this purpose we note that the farms can be separated into one of two location classes. Nine farms are on the Chalk, ten on the Lower Greensand. Let us construct the same test, the same H_0, H_1, and the same rejection region for the farms on the Chalk.

Arrange the increments as follows (fig. 2.77).

X	791, 536, 501		$n = 3$
Y	648, 118, 104, 72, 36, 20		$m = 6$
Rank X, X_{r_i}	1, 3, 4	$\sum_{i=1}^{n+m} X_{r_i} = 8 = W_n$	
Rank Y, Y_{r_i}	2, 5, 6, 7, 8, 9	$\sum_{i=1}^{n+m} Y_{r_i} = 37$	

2.77 Wilcoxon rank-sum calculations for a subset of fig. 2.75.

Let us examine the sense in which $W_n = 8$ is a basis for accepting or rejecting H_0.

1. There are $\binom{9}{3} = 84$ combinations.
2. Each has three ranks
   (1,2,3), (1,2,4), (1,2,5), . . ., (1,3,9), . . ., (2,5,6), . . ., (7,8,9).

3. Each has a sum of ranks
   (6), (7), (8), . . ., (13), . . ., (13), . . ., (24).

Thus the minimum value of W_n is 6, the maximum is 24. Some occur once, most occur more often.

4. The distribution of these sums for $n = 3$, $m = 6$ is given in fig. 2.78.

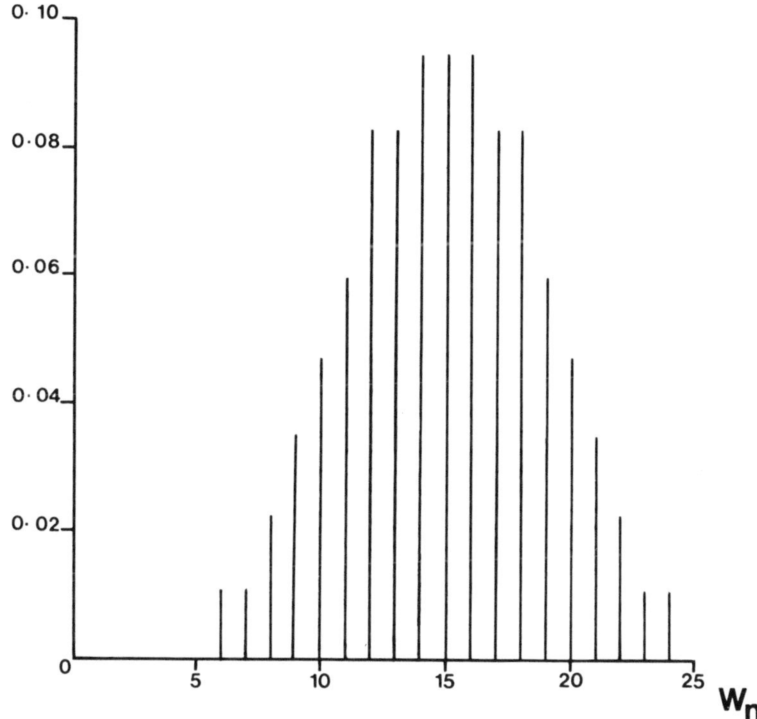

2.78 Probability distribution of Wilcoxon rank-sum for n=3, m=6.

5. Each point in the sample space is equally likely and has probability 1/84. The point probability of each event $W_n = 6, 7, \ldots, 24$ is given as the product of the frequency of that sum and 1/84.
6. The *cdf* is given in fig. 2.79.
7. Thus our decision to reject H_0 is made because the cumulative probability of a sum as small as or smaller than the actual sum is 0·0476192 and falls in the rejection region defined at the nominal α : 0·05, for which there is no rank-sum.

The distribution of W_n is symmetric about its mean, which is 15 in this example. For a two-tailed test, that is for an H_1 that the distributions are not identical rather than an H_1 that their lack of identity is in a particular direction, the one-tailed value is doubled because of this symmetry. However, in many cases

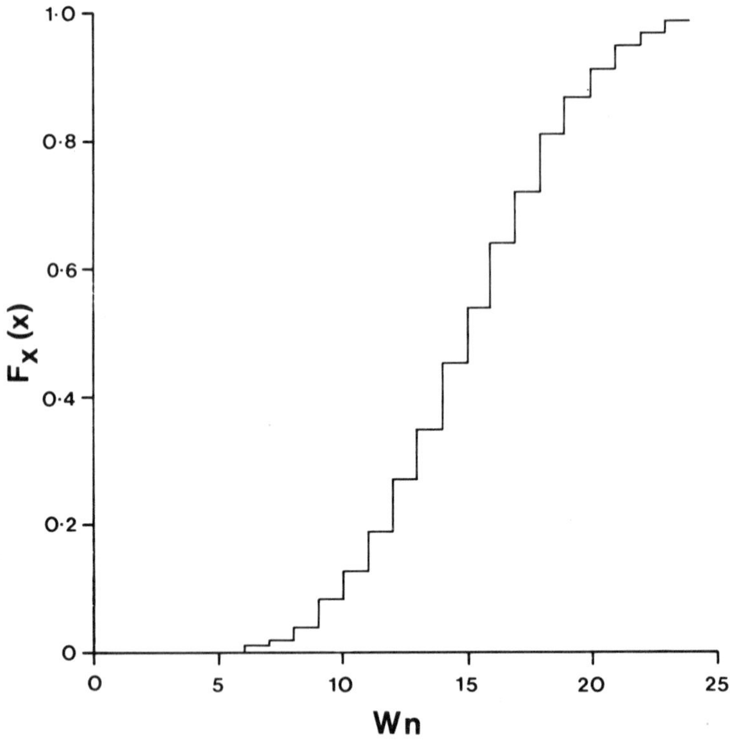

2.79 *cdf* for Wilcoxon rank-sum for n=3, m=6.

the test is chosen because a particular directed alternative is envisaged as an outcome of the conditions.

The symmetry of the distribution means that an upper-tail test can be made from the tabulated lower-tail values simply by including an extra column in the tables giving the mean. Suppose we give a rank of 1 to the smallest values in the last example and a rank of $n + m = 9$ to the largest value, then the sum of ranks for $X = 22$ and for $Y = 23$. From fig. 2.79 we can see that $W_n' = 22$ is placed symmetrically to $W_n = 8$ with respect to the mean. Thus $W_n = W_n' - 2\,(W_n' - \bar{W}_n) = 2\bar{W}_n - W_n'$. In this case $W_n = 22 - 2\,(22 - 15) = 30 - 22 = 8$.

The outcome of the test indicates that the two distributions are unequal generally, i.e. have unequal averages, probably unequal variances and so on. The test indicates ways in which the study can be elaborated.

2.7.2 Mann-Whitney *U*

An alternative, equivalent form of the Wilcoxon test is known as the Mann-Whitney *U* test. The order of presentation was chosen because the randomization distribution of the Wilcoxon form seems to be grasped more readily and to have a more spontaneous appeal.

In the Mann-Whitney test the two subsets are combined in increasing size and the number of times an X precedes a Y is counted. Consider the Chalk location class sample as a line of numbers (fig. 2.80). In the diagram the X and Y values are put above and below the line for ease of understanding the counting procedure.

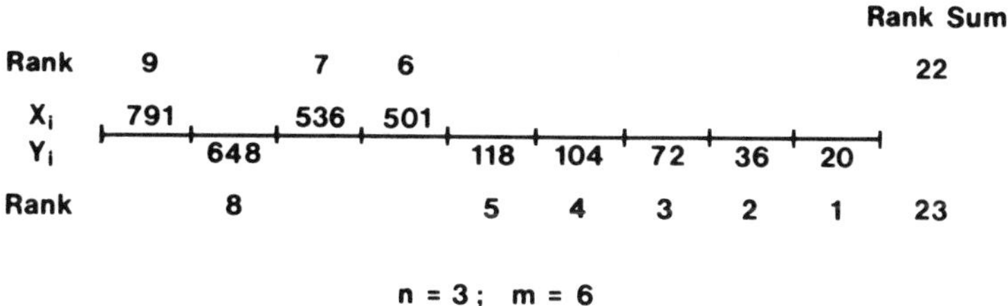

2.80 Number-line format for Mann-Whitney U statistic applied to the subset of fig. 2.77.

The sum of ranks is of course the same as the second, upper-tail count for Wilcoxon not the first, because the ranks increase as the magnitudes increase. In the above example the X value, 501, precedes one Y value, and the X value, 536, precedes one Y value, so $U = 2$. This is referred to published tables, for the values at nominal quantiles (see table 7 in appendix 2).

In other cases the enumeration is more intricate and error prone, and the value of U can be obtained from the formula given as equation 2.7.2.

$$U = \frac{n\,(n + 2\,m + 1)}{2} - \Sigma R_x \qquad (2.7.2)$$

Substituting from fig. 2.80 we have

$$U = \frac{3\,(3 + 2.6 + 1)}{2} - 22$$

$$= \frac{48}{2} - 22$$

$$= 2$$

From this formula we note that ΣR_x, the basis of the Wilcoxon test, is used and the remaining terms are simply sums of the integers involved in ranking the $n + m$ values. U is a linear function of W_n and this is why the two statistics yield identical decisions.

2.7.3 Kruskal-Wallis test

The direct generalization of the two-sample Wilcoxon test to the many-sample situation is known as the Kruskal-Wallis test. It is a distribution-free analogue of the parametric analysis of variance and F test. The test is suited to geographical and map situations which are $k-$ condition or $k-$ location-class extensions of the situations stated previously for Wilcoxon. It is essential to emphasize that the separate samples are assumed independent in the sense that a measurement in the kth sample does not affect a measurement in a different sample. This is an additional requirement to independence of measurements within each sample. The map form is no different. The test uses the same amount of information as the Wilcoxon test and is therefore applicable to ordinal measurements. As in many rank tests, tied values can disturb the sensitivity of the test and an implicit assumption is that the measurements are made on a continuous scale which precludes the possibility of tied values if there is sufficient refinement of the measuring procedure. In many practical cases some tied values do occur; so long as there are comparatively few such pairs the test may be used, but for a large number of tied values a test such as the Median test may be preferable. Thus although it is ostensibly a test that requires an ordinal scale, its derivation excludes tied values and implies a continuous scale.

The general null hypothesis is that all the k populations of measurements on the same attribute are identical. The general alternative is that they are not all identical and at least one of the populations of measurements leads to larger values than at least one of the other populations. However the test does not isolate these pairs, nor does accepting the null hypothesis preclude the possibility that for one pair of samples the null hypothesis could be rejected by the Wilcoxon test at the same α level.

As with the Wilcoxon test the exact distribution of the test statistic is derived under the assumption that the measurements in each sample all come from identical populations of measurements. There are N values altogether with n_1 values in the first sample, n_2 values in the second sample, and so on to n_k values in the kth sample. There are then 1 to N ranks arranged in a particular way over the k samples. If the assumption is correct then each of the $N!/n_1!n_2! \ldots n_k!$ different arrangements of the rankings among the k samples is equally likely. For each of these arrangements there will be a sum of ranks for each sample and a mean sum of ranks for all samples. This is the mean of the rank randomization distribution of all the sample means if the assumption of the null hypothesis is true. The distribution of this mean is the basis of the test statistic, equation 2.7.3.

$$KW = \frac{12}{N(N+1)} \Sigma n_i (\bar{R}_i - \frac{N+1}{2})^2 \qquad (2.7.3)$$

where $\bar{R}_i$ is the mean of the ranks in the ith sample with n_i elements.

$\frac{(N+1)}{2}$ is the mean of the first N integers. The sum of the first N integers is given by $N(N+1)/2$, consequently their mean is $\frac{1}{N}$ of this sum.

The variance of the first N integers is $N(N+1)/12$ and this is used to divide the weighted, n_i, squared differences of the actual mean rank, $\bar{R}_i$, and the expected mean rank, $N(N+1)/2$.

An equivalent form, easier to calculate is

$$KW = -3(N+1) + \frac{12}{N(N+1)} \Sigma \frac{R_i^2}{n_i} \qquad (2.7.4)$$

Example

The organic content of soil at six inches was determined for eight sites in each of three location classes. The first location class was high forest of mixed woodland undisturbed for over 100 years, this was used as control. The second location class was oak forest planted as replacement in a section of the high forest 100 years ago. The third location class was pine forest also planted in cutover high forest 100 years ago. The aim was to test the hypothesis that the measurement populations were identical. The data is tabulated showing actual values and the ranks over the *combined* twenty-four site results. In this case $n_1 = n_2 = n_3$ but this is *not* a requirement of the test (fig. 2.81).

Location class					
1		2		3	
High forest	R_1	*Oak forest*	R_2	*Pine forest*	R_3
7·1	5	9·5	11	14·0	22
6·3	3	10·3	14	6·0	2
7·8	7	19·1	24	7·5	6
13·4	20	12·0	17·5	6·4	4
9·6	12	10·2	13	10·7	15
14·2	23	13·6	21	8·1	8
12·0	17·5	8·6	9	5·2	1
11·7	16	12·2	19	8·8	10
$n_1 = 8$ ΣR_1	103·5	$n_2 = 8$ $\Sigma R_2 = 128·5$		$n_3 = 8$ $\Sigma R_3 = 68$	
$\bar{R}_1 = 12·9375$		$\bar{R}_2 = 16·0625$		$\bar{R}_3 = 8·5$	

$N = n_1 + n_2 + n_3 = 24$

$(N+1)/2 = 12·5$

2.81 Organic content of soil at 6″ under three woodland types in format for the Kruskal-Wallis test statistic.

Substitution in equation 2.7.3 gives

$$KW = \frac{12}{24(25)}\left[8\left(\frac{103·5}{8} - 12·5\right)^2 + 8\left(\frac{128·5}{8} - 12·5\right)^2 + 8\left(\frac{68}{8} - 12·5\right)^2\right]$$

$$= 4·62125$$

and, equivalently, substitution in equation 2.7.4 gives

$$KW = -3\ (25) + \frac{12}{24(25)}\left[\frac{(103·5)^2}{8} + \frac{(128·5)^2}{8} + \frac{(68)^2}{8}\right]$$

$$= -75 + \frac{1}{50}\ [1339·03 + 2064·03 + 578]$$

$$= 4·62125$$

Tables for exact probabilities are published for $n_1 \leqslant n_2 \leqslant n_3 = 5$ (see table 8 in appendix 2). For larger n_i the test statistic can be referred to χ^2 with $k-1$ degrees of freedom. In this case $k = 3$, $df = 2$ and the 0·95 quantile of χ^2 is 5·991. Thus our decision is to accept H_0. This decision means that for all the possible combinations of n_i measures from N measures, discrepancies as larger as those observed in their rankings would arise sufficiently often for us to accept that the measurements came from a common population of measurements or from $k = 3$ populations of identical measurements.

2.7.4 Friedman's related samples test

Finally in this section we consider a situation similar to the previous one except that the objects are related. The objects are related in the sense that they are qualitatively similar: this stipulation avoids comparing things which are different in kind. However, the objects may be different in degree. Such quantitative differences may be in terms of size, age or other object-characteristics which are considered irrelevant by the geographer in the particular circumstances. These R_n objects are subject to the same set of conditions which induce C_k responses in each object. The test is designed to see whether the objects reveal the same response structure.

In geography we often have a number of qualitatively similar objects which are related in the sense that they are subject to certain common conditions, but whose response to those conditions varies. The traditional cartographic representation is to depict the objects by point symbols and to subdivide these symbols to show the structure of each object's response. The map of divided symbols is then used as the basis of some assertion about object response to conditions.

A general tabulated format for this test by object response is given in fig. 2.82.

Object	*Response attribute*			
	C_1	C_2	...	C_k
R_1	X_{11}	X_{12}	...	X_{1k}
R_2	X_{21}	X_{22}	...	X_{2k}
·				
·				
·				
R_n	X_{n1}	X_{n2}	...	X_{nk}

The objects can be seen as separate samples, rows, and the columns as outcomes. The X_{ij} is a measure of the response (outcome) at each level of the response.

2.82 General format for Friedman's related samples test statistic.

There are k responses shown as columns and n objects shown as rows. For each object a measurement is made for each response, and the random variable is the set of response measurements for each object. Thus the set $(X_{11}, X_{12}, \ldots, X_{1k})$

constitutes the random variable for the first object. There are n such sets, such that the results of measurements on the k responses of one object do not influence the measurements on those k responses of any other object. In other words the n, k— variate random variables are independent. The null hypothesis is that the n random variables are taken from the same population of measurements or from identical populations of measurements.

Each measure is replaced by its rank. Rank 1 is given to the smallest value and rank k to the largest. Thus these ranks go from 1 to k for each of n objects. If the null hypothesis is true, the original values will be distributed randomly among the k responses for each object. Consequently the ranks of these values will be random permutations of the first k integers for each object under the essentially similar conditions. For each row, i.e. each object, the ranks will be arranged in one of the $k!$ permutations. Each of these $k!$ permutations is equally likely, and can occur for each of the n objects. Thus there are $(k!)^n$ distinguishable arrangements of the ranks for the $R_n C_k$ table. Each of these is equally likely under the null hypothesis. How can this set of arrangements be characterized in such a way that a fairly straightforward characteristic can be used for a test statistic?

The sum of the k ranks in each row is $\frac{k(k+1)}{2}$, thus the mean rank of each row is $\frac{1}{k} \cdot \frac{k(k+1)}{2} = (k+1)/2$. There are n such rows, thus the expected sum of ranks of each column for the complete table is $n(k+1)/2$. The actual sum of ranks is $T_j = \sum_i^n r_{ij}$. For each of the $(k!)^n$ tables that are equally likely under H_0 there will be a value corresponding to the statistic

$$F = \Sigma \left(T_j - \frac{n(k+1)}{2} \right)^2 \tag{2.7.5}$$

The frequency distribution of these sums, F, is the randomization distribution which defines the probability of any particular value of F for n, k.

Exact tables for F are given for small values of N and K. For values of n and k not covered by exact tables the statistic, F', can be referred to χ^2 with k—1 degrees of freedom.

$$\begin{aligned} F' &= \frac{12}{nk(k+1)} \Sigma \left(T_j - \frac{n(k+1)}{2} \right)^2 \\ &= \frac{12F}{nk(k+1)} \end{aligned} \tag{2.7.6}$$

Thus the general null and alternative hypotheses are:

H_0 : each of the $k!$ rankings in each row is equally likely. In other words the responses are equally likely.

H_1 : at least one of the responses tends to give larger values.

Example

Twelve papermills in Kent in 1865 made varying tonnages of each of three grades of paper:

1. Specialist papers
2. Printing papers
3. Packaging papers.

The aim is to decide whether their responses were the same. The outputs of paper are ranked in fig. 2.84 and mapped in fig. 2.83. In some cases the same value arises for two or more responses. Such tied values are given an average rank as in fig. 2.84. A large number of tied values affects the sensitivity of the test statistic.

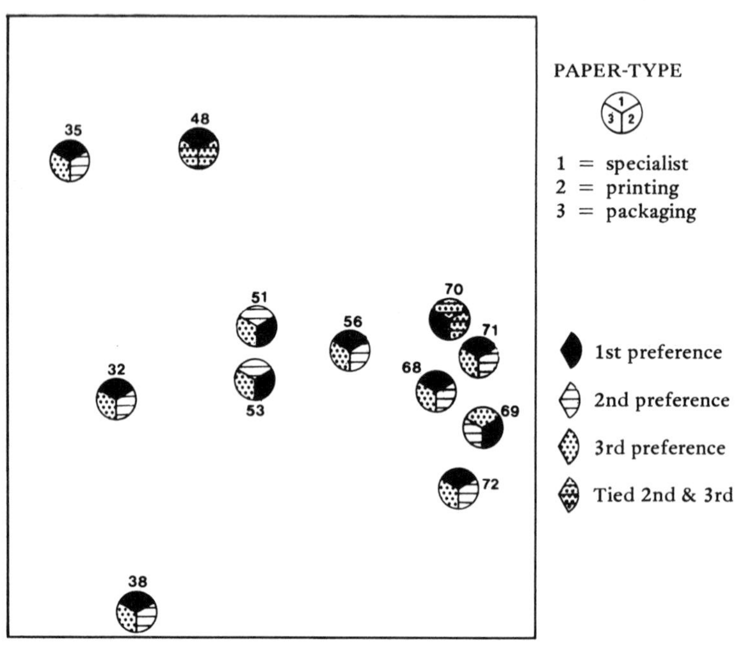

2.83 Divided symbol representation of paper-types made in a sample of mills in Kent, 1860–5.

H_0 : the responses are identical

H_1 : at least one of the responses tends to give higher values.

Rejection region defined at values of F that lie within α : 0·05, and of F' that lie within 0·95 quantile of χ^2 with 2 *df*.

$$F = \Sigma\ [(30{\cdot}5 - 24)^2 + (22{\cdot}0 - 24)^2 + (15{\cdot}5 - 24)^2]$$

$$= 118{\cdot}5$$

$F = 114$ for $n = 12$, $k = 3$ has probability 0·0080 (see table 9 in appendix 2). And

$$F' = \frac{12.118{\cdot}5}{12.3.4}$$

$$= 19{\cdot}75 \qquad \chi^2_{0{\cdot}999} \text{ quantile} = 13{\cdot}82$$

Mills		Preferred output ranked		
		Specialist	*Printing*	*Packaging*
32	Sundridge	3	2	1
35	Cray	3	2	1
38	Chafford	3	2	1
48	Darenth	3	1·5	1·5
51	Basted	2	3	1
53	Roughway	2	3	1
56	East Malling	3	2	1
68	Lower Tovil	3	2	1
69	Upper Tovil	1	3	2
70	Medway	1·5	1·5	3
71	Springfield	3	2	1
72	Hayle	3	2	1
	T_i	30·5	26	15·5
	$T_i - \bar{T}$	6·5	2	−8·5
	$(T_i - \bar{T})^2$	42·25	4·0	72·25

2.84 Details of fig. 2.83 in tabular form.

We accept H_1 and infer that these mills responded to the local and national conditions by producing specialist papers. This inference is consistent with the sort of assertion most geographers would make from the map. The inference is not always so plain. It is worth emphasizing that Friedman's test does not assume homogeneity of individuals, but it does require qualitative similarity. The mills varied in size, investment, power source and machinery, but the test yields useful information because the mills all made paper under qualitatively similar production conditions. A test that can ignore differences between individuals and be sensitive to differences in response to common conditions is often useful: also this often matches the circumstances depicted by divided point symbols.

2.7.4.1 *Friedman's test with partial or complete order prediction*

Suppose we had constructed a cost surface for the industry and assembled various structural details too, such that on this evidence we had the confidence to predict that there would be a preferential emphasis in response, namely that the column mean ranks were ordered. Put $\bar{T}_1$ for specialist papers, $\bar{T}_2$ for printing papers and $\bar{T}_3$ for packaging papers and predict the order as $\bar{T}_1 > \bar{T}_2 > \bar{T}_3$. Of course this must be done before seeing the data.

Regardless of the order in which our column values occur, the column totals would remain the same, and the value of F as the sum of squared differences of these totals from the expected total rank would remain the same. The column totals, that is the whole column of responses, can be rearranged in $k! = 3! = 6$

ways. Thus the probability of predicting any particular order is $1/k!$ which in this case is $1/6 = 0{\cdot}1667$. The total table remains unchanged and has a probability of, say, α' which may be different from the α that defines the rejection region. As there are $k!$ ways of getting F with this particular α', there is a probability of $1/k!\alpha'$ of predicting the particular order of the columns in a table with probability α'. In the previous example to have predicted the order $\bar{T}_1 > \bar{T}_2 > \bar{T}_3$ for the table with $\alpha' = 0{\cdot}008$ is given as $1/6 \,.\, 0{\cdot}008 = 0{\cdot}0013$. Thus α' is the probability of getting a table with those means and $1/k!$ is the probability of getting the particular table.

By including a prediction of order we have enlarged our rejection region and when we can combine a partial or full ordering of deviations from the average we have a useful and effective test. If the outcome is an order *almost* the same as the predicted order then we cannot reject, no matter how small the probability is, because our rejection region refers only to the F value for the predicted order. The use of complete ordering should be restricted to those cases where the order seems almost logically entailed by other considerations. The expedient of using a partial ordering rather than a complete order prediction is often more appropriate to map situations. Once $k > 3$, $k! \geqslant 24$ and to predict the full order implies a probability of at least $1/24 = 0{\cdot}04167$ and H_0 can be rejected without calculating F. Partial ordering in such circumstances seems advisable. For example we could predict that only λ of the means will have the lowest ranks or ν of them will have the highest ranks. Thus of the k means, λ can be placed in $\lambda!$ ways and for each of these ways, ν of them can be placed in the highest ranking positions in $\nu!$ ways. The remaining $(k - \lambda - \nu)$ can be put in the $(k - \lambda - \nu)$ positions in $(k - \lambda - \nu)!$ ways. Thus there are $(k - \lambda - \nu)!\lambda!\nu!$ ways of picking the specified orders out of the $k!$ possible arrangements of column means. The overall probability of H_0 of given F and given order under this circumstance is $[(k - \lambda - \nu)\,!\,\lambda\,!\,\nu\,!\,\alpha']/k!$ (exercise 2.37).

The tests introduced in this section seem particularly appropriate to the propositions and measurement scales that arise in point symbol maps as they are used in the initial stages of geographic enquiry. We proceed to consider tests that seem particularly appropriate to line symbol maps even though the same propositions and the same measurement scales are involved. Considerable interchange is possible. The division is one of convenience. The presentation becomes progressively more concise to match what I hope is your increasing understanding of hypothesis testing.

Exercises

2.1 Are the following mappings bijective, injective, surjective, none of these?

   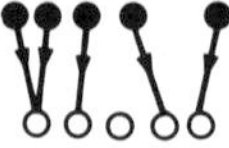 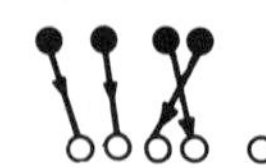

2.2 What sort of mapping is a mapping from the sphere to the plane?

2.3 If cost is a monotonic function of distance what sort of mapping is involved?

Is this mapping retained in the plane? In fig. (ii) accompanying exercise 1.29 join the positions of the towns. What sort of mapping is it?

2.4 Why can there be no standard map for a distribution which is hypothesized to be a random scatter of objects from some probability distribution? Does this apply to a particular random process?

2.5 Is the distribution of objects unusual under the assumption that their frequency is a binomial random variable?

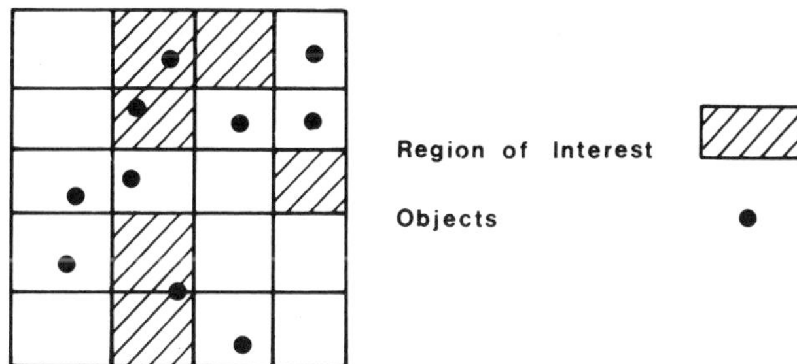

2.6 Is the distribution of objects unusual under the assumption that their frequency is a multinomial random variable?

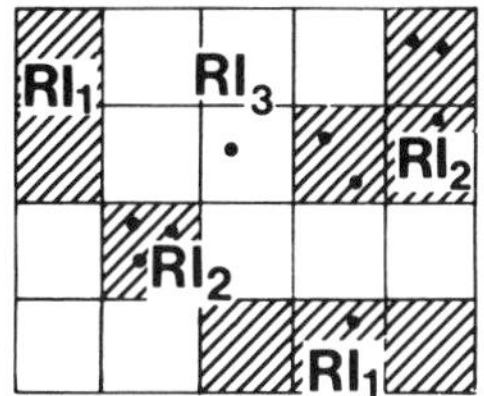

a) What is the point probability of (1, 8, 1)?
b) What is the cumulative probability for a result at least as extreme as this?
c) Can you use the chi-square approximation?

2.7 The 19 waterfall sites were all occupied by papermills at the beginning of the nineteenth century. Economic pressures encouraged specialization in

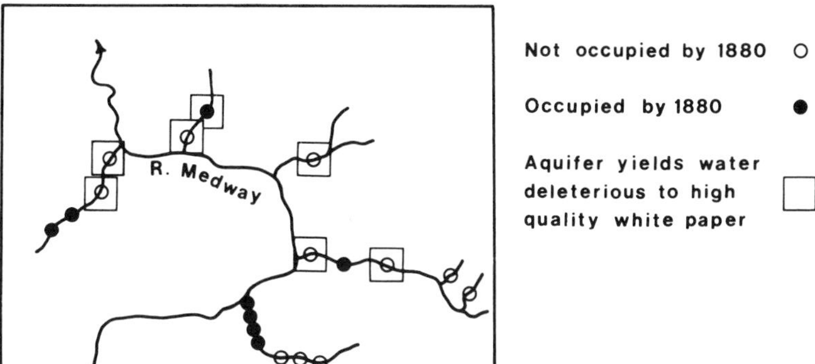

high-grade white papers and was accompanied by a progressive reduction in occupied sites. The region of interest consists of those waterfall sites that had aquifers yielding water that was unable to be used then to make high-quality white papers. Is there evidence that the distribution of successful occupation is concentrated to an unexpected extent on sites with favourable aquifers?

a) Set up H_0, H_1.
b) What is the random variable?
c) What is the probability distribution?

2.8 Suppose $\lambda = 2$. Evaluate $\frac{\lambda^n}{n!} . e^{-\lambda}$.

2.9 Suppose $\lambda = 1{\cdot}68$. Evaluate $\frac{\lambda^n}{n!} . e^{-\lambda}$.

2.10 In a random sample of firms in a large industrial city, each of 73 firms listed its most important trading partners. We have n, the number of trading partners within the same unit area, and Nn, the number of firms with n partners. $N = 73$.

n	Nn
0	8
1	16
2	18
3	15
4	9
$\geqslant 5$	7

Do firms' trading contacts (linkages) with other factories tend to be clustered in the same part of the city? Does the evidence provide a satisfactory basis for testing the hypothesis indicated?

2.11 Suppose we are concerned with the occurrence or non-occurrence of some event, A, under particular conditions. Keep the assumptions of the binomial random variable that $P(A) = p$, $P(\tilde{A}) = (1-p) = q$ remain constant for all repetitions of the conditions. Now suppose that the number of repetitions is a random variable; this is unlike the binomial case where N was fixed. Define the random variable, X, to be the number of repetitions needed to get A, thus $X = 1, 2, \ldots$. Now $X =$ some number k, if and only if the first $(k-1)$ repetitions do not result in A (i.e. they result in $\tilde{A}$), and if the kth repetition gives A. We have

$$P(X=k) = q^{k-1}p \text{ for } k = 1, 2, \ldots .$$

Such a random variable is said to have a *geometric distribution.* Say the event is that a river floods during the Spring (1 January – 30 April) with probability 0·1. What is the probability that the first flood occurs on 1 February? Let $X =$ number of days, starting with 1 January, until the first flood, so that we require $P(X = k = ?) = q^{k-1}p$. Calculate this value.

This distribution is interesting in that if A has not occurred during the first t repetitions of the conditions, then the probability that it will not occur during the next t repetitions is the same as for the first t repetitions. It is said to have no memory.

2.12 If we define our random variable, X, as the number of repetitions needed for A to occur r times, again with $P(A) = p$, $P(\tilde{A}) = 1-p$ we have a random variable that is said to have a *negative binomial* distribution. Suppose $X = k$ repetitions to get r occurrences of the event, A. Then

$$\begin{aligned} P(X = k) &= P\{(r-1)\ A\text{'s in } (k-1) \text{ repetitions} \cap \text{one further} \\ &\qquad A \text{ in the } k\text{th repetition}\} \\ &= \binom{k-1}{r-1}(1-p)^{k-r}p^{r-1}p \\ &= \binom{k-1}{r-1}(1-p)^{k-r}p^{r} \end{aligned}$$

Suppose we have an RI with $p = 0{\cdot}2$ and we wish to determine the probability that $X=k=15$ before $r=3$ events occurred in RI.

2.13 Those of you whose mathematics and interest is strong may care to look up in a more advanced text the *exponential* and *gamma* distributions which are the continuous analogues of the geometric and negative binomial discrete distributions. Consider their applicability to point pattern problems.

2.14 In a study of urban recreational facilities, parks were classed as being in areas of high or low residential density. They were also classified as being used heavily or not heavily. The following frequencies were observed in a study of the 24 parks.

Residential density	*Intensity of use*		
	Heavy	*Not heavy*	*Total*
High	11	4	15
Low	3	6	9
Total	14	10	24

Given the hypothesis that parks in areas of high residential density show a greater intensity of usage,

a) state H_0, H_1,
b) set an α, for one- or two-tailed rejection region,
c) comment on choice of tests available.

2.15 In equation 2.3.5 substitute for equal column totals (i.e. $2R=N$) and equal row totals (i.e. $2n=N$) and show why in these cases the two-tailed test is twice the one-tailed test.

2.16 For exercise 2.14 calculate the value that is as extreme for the lower tail as for the upper tail (see equations 2.3.4, 2.3.5).

2.17 'Nothing propinks like propinquity' (*Diamonds are Forever*). Do the following figures justify Ian Fleming's view? The figures are a random sample of married couples living in Hull.

Geographic origin of the male	*Geographic origin of the female*		
	Hull and East Riding	Elsewhere	Total
Hull and East Riding	196	11	207
Elsewhere	36	7	43
Total	232	18	250

2.18 In a study of immigration into the Glamorgan coalfield in the late nineteenth century (Jones 1969) the following frequencies were obtained.

Residence	*Nationality*		
	Welsh	*Not Welsh*	*Total*
Coalfield Glamorgan	98,569	71,687	170,256
Not-coalfield Glamorgan	24,396	57,597	81,993
Total	122,965	129,284	252,249

H_1 : Welsh immigrants show a greater preference for the coalfield. What is H_0? Is H_1 justified?

2.19 From the same study the question is asked whether the Welsh show an unusually strong preference for the Rhondda Valley. Use the following frequencies to set H_0, H_1 and determine a judgement with respect to an α.

Residence	*Nationality*		
	Welsh	*Not Welsh*	*Total*
Rhondda	12,684	7,079	19,763
Not Rhondda	25,325	31,871	57,196
Total	38,009	38,950	76,959

2.20 Calculate measures of association for exercises 2.17, 2.18, 2.19.

2.21 Chi-square with fixed marginal totals may be used to test for independence in two variables. This could be exploited in a distribution of symbols on a graph or map where (X, Y) coordinates were distances along each axis from the origin. Equally it can be applied to the observed value along X and along Y for other non-distance scales. In the graph of exercise 1.32 divide the 63 X values into 3 groups of equal size and similarly divide them up with respect to the Y axis, giving 9 cells with an expected value of 63/9 in each cell. Test $H_0 : X$ and Y are independent. What happens if you increase the number of cells?

2.22 In a study of the use made of local parks and recreational areas the following data were observed for Ashdown Forest.

Summer bank-holiday Sunday	*Distance class*			
	1	2	3	4
Regular user	14	9	21	3
Adventitious user	4	4	6	9

Non-bank-holiday Summer Sunday	1	2	3	4
Regular user	26	14	3	0
Adventitious user	9	4	11	3

Distance classes increase from category 1, within 15 miles, to category 4, over 44 miles.

a) Which chi-square model is implicit?
b) Set up H_0 (article 2.5) and test.
c) Combine the tables and test a similar H_0.

2.23 Set up appropriate null hypotheses for the frequencies given in the following tables and, if the total X^2 value could give significant results with reduced degrees of freedom, partition the tables as in article 2.5.2.

a)

Size class of mill	*Distance class to market (1865)*			
	−1	0	+1	Total
−1	25	18	40	83
0	17	9	22	48
1	50	21	21	92
Total	92	48	83	223

b)

Size class of mill	*Distance class to source of raw material (1865)*			
	−1	0	+1	Total
−1	19	19	45	83
0	17	13	18	48
1	55	16	21	92
Total	91	48	84	223

2.24 Calculate X^2 for each of the four 2 X 2 subtables given in article 2.5.2, fig. 2.40. These test statistics are based on the marginal totals of the individual tables and give expected values different from the expected values calculated from the marginal totals of the full table.

2.25 Calculate X^2 for each of the four 2 X 2 subtables you considered appropriate for exercise 2.23.

2.26 The 108 papermills which survived from 1865 to 1915 and which made paper by machine in both periods are shown in the following table in terms of their size class in each period (see article 2.5.3).

a) Does the data support the assertion that size class in 1915 is dependent on size class in 1865?

b) Does the data support the assertion that changes in size class are independent of size class in the initial period?

t_1 *Size class in 1915*	t_0 *Size class in 1865*			
	−1	0	1	Total
−1	17	16	7	40
0	9	6	9	24
1	3	4	37	44
Total	29	26	53	108

2.27 Does the following table contain evidence to support the proposition that papermills tend to be oriented to sources of raw material (Lewis 1969)?

Distance class	*Did survive*	*Did not survive*	*Total*
−2	34	36	70
−1	29	35	64
0	26	44	70
1	33	32	65
2	21	50	71
Total	143	197	340

a) Consider the assertion for the whole table.
b) Consider the assertion for row subsets [−2, −1] [0, 1, 2] and [+2 + 1] [0, −1, −2] (article 2.5.4.1).
c) Consider the assertion for regression (article 2.5.4.2).
d) Draw the histogram for the table.
e) Calculate the regression line.

2.28 From the map set up your own hypotheses of independence and conditional probabilities and then test by the methods of article 2.6.2. The frequencies have been contrived to lead to easy arithmetic in case you do not have access to an electronic calculator.

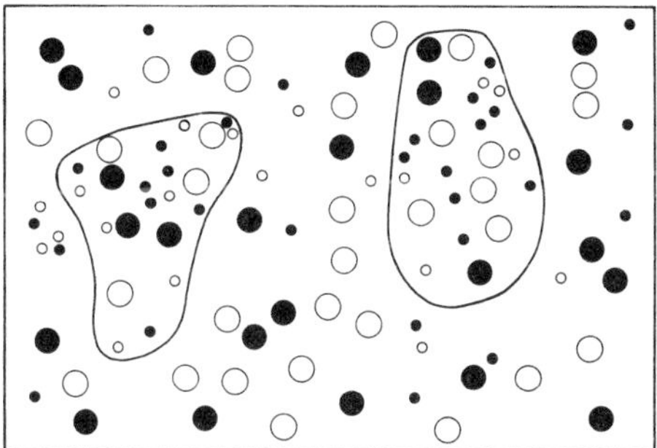

2.29 The industries of Hull are shown in the map below and the details summarized as a three-dimensional contingency table. The variables and their levels are given below the table together with eight hypotheses and the details appropriate to these hypotheses.

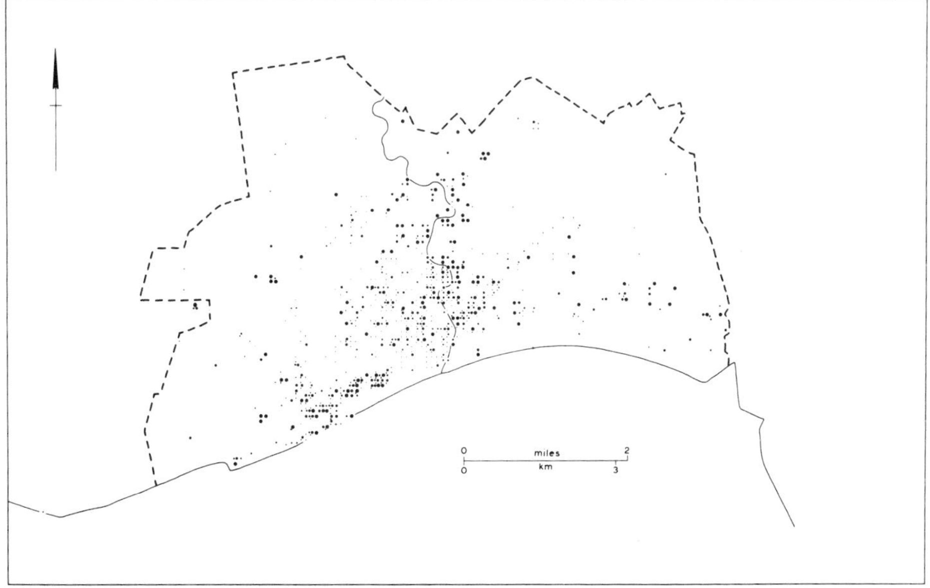

	S_1			S_2			S_3			
	C_1	C_2	C_3	C_1	C_2	C_3	C_1	C_2	C_3	*Total*
L_1	24 17·4	15 14·4	25 26·1	19 17·11	11 14·4	20 26·1	14 17·11	16 14·4	30 16·1	174
L_2	28 27·2	19 22·6	40 40·9	28 27·7	27 22·6	41 40·9	21 27·2	23 22·6	45 40·9	272
Total	52	34	65	47	38	61	35	39	75	446

Variable 1, A_i, size, 3 levels
Variable 2, B_j, category, 3 levels
Variable 3, C_k, location, 2 levels, L_1 access to waterfront

	Fitted marginals	*Likelihood ratio*	*d.f.*	$\chi^2.95$
H_0: no three-factor interaction	$\{BC\}\{BA\}\{AC\}$	1·039	4	9·5
$H_0: A \otimes C \mid B$	$\{BC\}\{BA\}$	3·318	6	12·6
$H_0: CB \otimes A$	$\{BC\}\{A\}$	8·674	10	18·3
$H_0: A = \Phi \mid CB$	$\{BC\}$	8·759	12	21·0
$H_0: A \otimes B \otimes C$	$\{A\}\{B\}\{C\}$	9·677	12	21·0
$H_0: B \otimes C \cap A = \Phi \mid BC$	$\{B\}\{C\}$	9·762	14	23·7
$H_0: AC = \Phi \mid B$	$\{B\}$	31·472	15	25·0
$H_0: ABC = \Phi$	n	60·015	17	27·6

a) Which hypothesis is the least demanding in terms of the necessary fitted marginals at a nominal α : 0·05?
b) What does this H_0 imply?
c) What are the marginal totals used?
d) Can you use these to provide the expected values given in the table?

2.30 New and existing factories were compared with respect to location and category of production with the following outcome. This shows how much two-variable interaction there can be.

Waterfront location	*Factory*		
	New	*Existing*	*Total*
Category 1	8	151	159
Category 2	6	88	94

Non-waterfront location	*Factory*		
	New	*Existing*	*Total*
Category 1	5	25	30
Category 2	43	195	238
Total	62	459	521

Category 1 represents industries closely dependent on raw materials.
Category 2 represents industries related more to markets.

Test a) H_0: no three-factor interaction.

b) H_0: Factory age is independent of factory category is independent of factory location, i.e. $A \otimes B \otimes C$.

c) $H_0: B \otimes C \mid A$.

Using for

a) all the marginal totals

	C_1	C_2
L_1	159	94
L_2	30	238

	C_1	C_2
F_1	13	176
F_2	49	283

	F_1	F_2
L_1	14	239
L_2	48	220

giving as the fitted table with degrees of freedom = 1

		New	Existing
Waterfront	Category 1	8·253	150·732
	Category 2	5·747	88·268
Non-waterfront	Category 1	4·726	25·291
	Category 2	43·274	194·709

b) the marginal totals

F_1	62	C_1	189	L_1	253
F_2	459	C_2	332	L_2	268

and calculate the fitted table with degrees of freedom = 4

c) the marginal totals

	F_1	F_2
C_1	13	176
C_2	49	283

	F_1	F_2
L_1	14	239
L_2	48	220

and calculate the fitted table with degrees of freedom = 2.

2.31 The data shown in the following diagrams represent a sample of 14 farms, Zellwood, Lake Apopka, Florida. In 1967, 5 farmers decided that cooperative organization of their farm production would permit more efficient

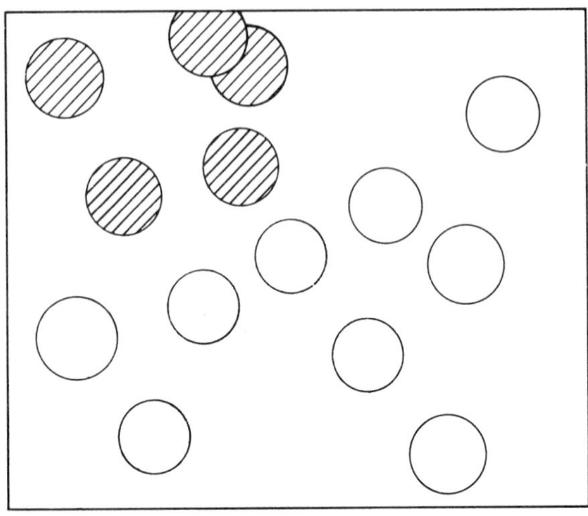

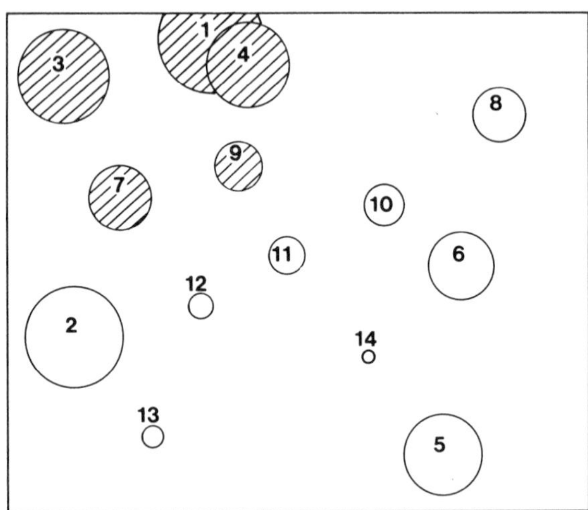

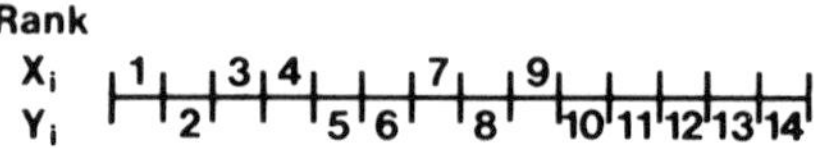

use of resources, particularly in enabling them to install railway-truck freezing-plant for access to New York market.

	Farm income (\$ in 1970)
* 1	17,323
* 2	15,239
* 3	15,692
4	13,252
5	12,388
* 6	12,753
* 7	13,817
8	14,196
9	11,681
10	11,663
11	16,859
12	11,350
13	11,444
14	14,807

* represents a cooperative member's farm.

In 1970 all 14 farmers examined their farm income to decide whether to join in a larger cooperative venture.

a) Does the evidence as given here support the viewpoint of the 5 cooperative farmers that their income is greater than that of their 9 non-cooperative farm neighbours?
b) What important details would you include in any test that attempted to determine a significant discrimination?
c) How many combinations of 5 farms are there from 14 farms?
d) How many such combinations would give a sum less than or equal to the actually obtained value?
e) Can you justify the use of symbols that are proportional to rank rather than to magnitude?

2.32 In a study of historic towns, rateable value per foot frontage was used as an index of the value of building site. Buildings scheduled as of historic importance are subjected to severe planning restrictions over changes in use or appearance. In order to ascertain whether these restrictions are reflected in their site value a random selection of buildings from such an historic town was made and is given below in £/foot frontage.

Building scheduled X_S : 132, 157, 201, 214, 305, 374, 410, 700

Building not scheduled X_U : 84, 150, 263, 280, 293, 534, 620, 680, 810, 830, 921, 1020, 1090, 1260

Test $H_0 : X_S, X_U$ come from identical populations.

2.33 Use Mann-Whitney's U test statistic with exercises 2.31 and 2.32.

2.34 In a rain-making experiment there were 15 storms, of which 6 were selected by a random procedure for seeding. The rainfall collected over the regional rain gauges was recorded as:

Control: 0·07, 0·15, 0·17, 0·23, 0·39, 0·45, 0·73, 0·91, 1·01.
Treated: 0·01, 0·09, 0·24, 0·31, 0·35, 0·81.

Does seeding increase rainfall?

2.35 For the data in fig. 2.81 compare, using Wilcoxon,
a) high forest with oak;
b) high forest with pine;
c) oak with pine.
Use high forest as control in (a) and (b).
Devise one-tailed tests carefully with respect to expected direction of differences.

2.36 For the data in fig. 2.81, round *down* to integers and find only 5, 11, 19 remain untied. Use Kruskal-Wallis, giving a value of 4·77375. Why have the tied values made so little difference?

2.37 By 1960/5 the greatest diversity of opportunity for papermaking was concentrated in southeastern UK whereas in northwestern UK the constraints were more severe. The conditions within both regions were remarkably homogeneous. A random sample of size $n = 10$ was taken of mills in the southeast and of size $n = 11$ in the northwest and for each mill the production was ordered by preferred output as shown below.

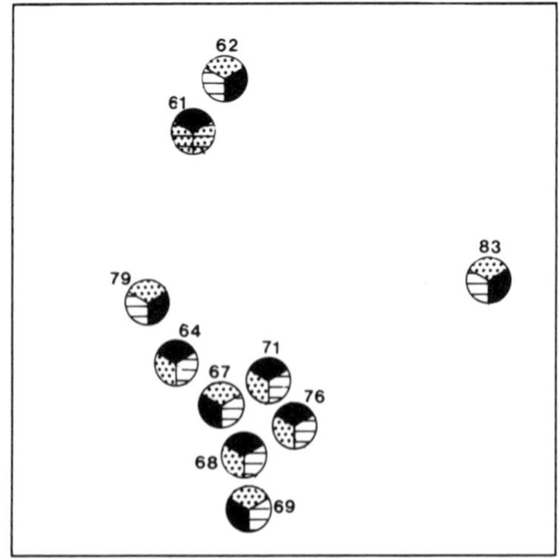

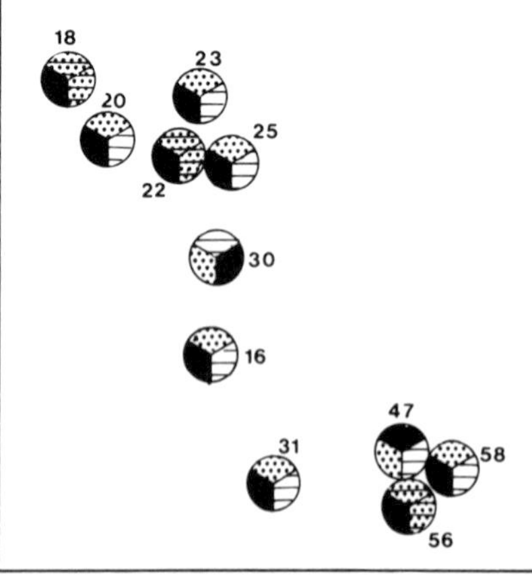

a)

Mill	*Specialist high grade*	*Printings*	*Wrappings*
83	1	3	2
61	3	1·5	1·5
62	1	3	2
64	3	2	1
67	1	2	3
68	3	2	1
69	1	2	3
71	3	2	1
76	3	2	1
79	1	3	2

b)

Mill	*Specialist high grade*	*Printings*	*Wrappings*
16	1	2	3
18	1·5	1·5	3
20	1	2	3
22	1·5	1·5	3
23	1	2	3
25	1	2	3
30	2	3	1
31	1	2	3
47	3	2	1
56	1·5	1·5	3
58	1	2	3

H_0 : mills show no preferred order.
For Lancashire we suggest *in advance of seeing* the data that the mills will tend to concentrate on wrappings and expect no discrimination between the other two categories. Thus our alternative hypothesis is

H_1 : (northwest) the column sums will be ordered as specialist $\leqslant$ printings $<$ wrappings

H_1 : (southeast) the mills will show a preferred order but it is unspecified.

Test these hypotheses.

Section 3/Analysing line symbol maps

Analysing line symbol maps

3.1 Propositions of randomness (1)

Property: non-locational
Measurement scale: nominal, dichotomy
Statistical procedure: runs tests; using number of runs, length of run, number of pairs.

In the article considering randomness in point patterns we devised test statistics that used the probability of getting r objects in k homogeneous location classes. In this section we shall use some of the same probability distributions for deciding how likely a particular pattern of attribute-states is in terms of a line sequence. The objects are characterized by one of two mutually exclusive categories of one attribute. We are often interested in such a sequence of objects by attribute state. These sequences are one-dimensional in time or distance, and the randomness which we test is randomness in order of appearance of the attribute-states. Typical examples in geography are:

1. the increase or decrease of adjacent values of, say, pollution level in a river;
2. the incidence of wet or dry days;
3. a dichotomy on the median or other quantile as, say, with sediment characteristics at different sites;
4. survival, failure along a transect.

Tests for sequential randomness are not limited to dichotomies. The attribute may be measured by many categories such as:

1. categories of land use on each side of a road or across a particular geomorphological sequence;
2. plant species through a woodland, perhaps at different heights;
3. shop types along a road; this situation is implicit in such assertions as 'High-value-goods shops tend to cluster' or that 'Banks (solicitors, estate agents) tend to congregate whereas newsagents tend to occur at regular intervals'.

It is often worthwhile to select one category and treat the remainder as homogeneous and this reduced measurement can be analysed as a dichotomy.

A sequence of like objects or equivalent attribute states is termed a *run.* A run is either bounded by an unlike object or unbounded. We shall introduce the straightforward background to the most common and useful runs tests. Once the principles are understood, modifications to established procedures can be made to cover particular situations.

3.1.1 Number of runs tests, two attributes

Consider a sequence of n elements of two types, n_1 of type 1, and n_2 of type 2, where $n_1 + n_2 = n$. Suppose there are 8 elements of type 1 and 7 elements of type 2, arranged in a sequence (fig. 3.1).

Type 1 ○

Type 2 ●

3.1 Arrangement of objects of two types to give the maximum number of runs.

We should imagine intuitively that such a regular alternation of objects was unusual, just as we should consider the separation shown in fig. 3.2 to be unlikely. In less clear cases we may

○○○○○○○○●●●●●●●

3.2 Arrangement of objects of two types to give the minimum number of runs.

vary in our opinion as to the degree of clustering or mixing that is evident in a pattern. In such circumstances we may wish to have some straightforward statistical rule for helping us to decide how likely is any particular arrangement of n objects of types n_1 and n_2.

One clear starting point is that with n objects there are $n!$ different arrangements, and, when these objects are of n_1 of one type and n_2 of the other type, then for each of the $n!$ configurations the n_1 objects of type 1 can be rearranged in $n_1!$ ways without affecting the overall distinguishable pattern. The same is true for the n_2 objects; these can be rearranged in $n_2!$ ways without noticeable effect on the arrangement. Thus there are $n!/n_1!n_2!$ distinguishably different arrangements. Even in this relatively small example there are $15!/8!7! = 6435$ different arrangements. Each of these arrangements consists of a number of runs : thus there are 15 runs in fig. 3.1 and 2 runs in fig. 3.2. We can identify each arrangement with a number of runs and use the frequency of each of the possible number of runs as the numerator of our probability fraction. The denominator is the total number of distinguishable arrangements. In order to determine the frequency of r runs from $n_1 + n_2$ elements we need a counting rule as direct enumeration rapidly becomes unwieldy.

Counting rule 4

The number of distinguishable ways of distributing n like objects into r distinguishable cells with no cells empty is $\binom{n-1}{r-1}$.

The proof of this rule can be approached in more than one way; the following explanation is related to a diagram below (fig. 3.3) which may clarify the occurrence of the terms $n-1$ and $r-1$.

Suppose the n objects are arranged in a line. There are $n-1$ spaces. To divide these n objects into r cells requires $r+1$ cell walls. The two end walls are necessary and fixed, so there remain $r+1-2=r-1$ cell walls to be distributed amongst the $n-1$ spaces between the objects. This can be done in $\binom{n-1}{r-1}$ ways. In the case of 5 objects and 3 cells or boxes we have the following illustration where the cell walls are shown by solid circles and the objects by open circles (fig. 3.3).

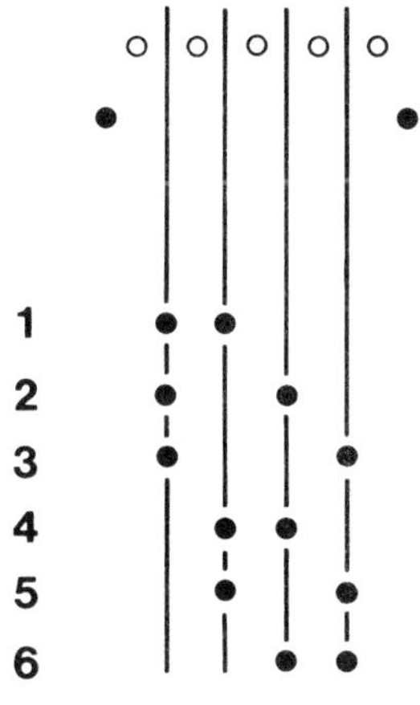

3.3 Counting rule 4.

If we regard the number of runs of the objects as the number of cells we can see that a sequence of r_1 runs of objects of type 1 can be obtained in

$$\binom{n_1-1}{r_1-1}$$

different ways. Let us suppose that we use objects of a different type, type 2, to delimit the runs of type 1 objects by acting as cell walls, then in the same fashion we may argue that the r_2 runs of the type 2 objects can arise in

$$\binom{n_2-1}{r_2-1}$$

different ways. Consequently the total number of distinguishable arrangements is given by the product of these two arrangements.

$$\binom{n_1-1}{r_1-1}\binom{n_2-1}{r_2-1}$$

As the two types of object are mutually delimiting, and each forms the cell wall for the other, the runs of type 1 must alternate with those of type 2. If $r_1=r_2$

then the sequence could begin with either a run of type 1 or of type 2 and the product must be doubled to give the frequency. The number of runs, r_1, cannot be more than 1 more or 1 less than the number of runs of r_2. If this is the case and $r_1 = r_2 \pm 1$, then we cannot simply double the frequency because either r_1 or r_2 must be even for their sum to be odd. Thus we must have

$$\binom{n_1 - 1}{r_1}\binom{n_2 - 1}{r_2 - 1} + \binom{n_1 - 1}{r_1 - 1}\binom{n_2 - 1}{r_2}$$

as the frequency of r_1 and r_2 runs when r_1 and r_2 are not equal. We can combine these two expressions to give the probability distribution of the total number of runs, r, of $n_1 + n_2 = n$ objects. To make arithmetic a little easier equation 3.1.1 uses r rather than r_1 and r_2, which were used to make understanding easier. The general probability formula is given as equation 3.1.1 and values for $n_1 = 8$, $n_2 = 7$ as fig. 3.4, some of which are illustrated in fig. 3.5.

$$P(r) = \begin{cases} \dfrac{2\binom{n_1 - 1}{r/2 - 1}\binom{n_2 - 1}{r/2 - 1}}{\binom{n}{n_1}} & \text{for } r_1 = r_2 = r/2 \\[2ex] \dfrac{\binom{n_1 - 1}{(r-1)/2}\binom{n_2 - 1}{(r-3)/2} + \binom{n_1 + 1}{(r-2)/2}\binom{n_2 - 1}{(r-1)/2}}{\binom{n}{n_1}} & \text{for } \begin{matrix} r_1 = r_2 + 1 \\ r_2 = r_1 + 1 \end{matrix} \end{cases} \qquad (3.1.1)$$

The diagrams shown in fig. 3.5 emphasize the task of visual discrimination that is implied by runs of objects with two attribute-states even for comparatively small numbers. Some numerical procedure is useful in this discrimination. The general null hypothesis is of randomness in the incidence of the attribute-states and the general alternative of non-randomness is accepted if there are too few runs or too many runs. Thus the rejection region is defined in terms of r when r corresponds to a number of runs that is extreme for n_1 and n_2. When both extremes are considered simultaneously the test is two-tailed, and when either too few runs or too many runs is considered separately the test is one-tailed. If r is large then there are too many runs to accept H_0 and H_0 is rejected in favour of H_1 that the pattern is too regular to be random and gives too many runs. If r is small then H_0 is rejected in favour of H_1 that there is clustering which gives too few runs. When either of these alternatives is acceptable the rejection region is defined on both tails. The probability distribution is often asymmetric and a considerable increase in sensitivity can result from confining the rejection region to one tail.

For the case $n_1 = 8$, $n_2 = 7$, with $\alpha = 0{\cdot}05$, H_0 is rejected in favour of H_1 that the pattern is too regular to be random if $r \geqslant 13$, and H_0 is rejected in favour of H_1 that the pattern is too clustered to be random if $r \leqslant 5$. The distribution is almost symmetric and in this case the two-tailed rejection region leads to rejection of H_0 in favour of the general alternative for the same values of r (fig. 3.6).

Number of runs r	*Frequency*	*Point probability*	*Cumulative probability*	$[1 - \Phi(Z)]$	Z
15	1	0·0001554	0·0001554	0·0002174	3·5180
14	14	0·0021756	0·0023310	0·0014436	2·9795
13	63	0·0097902	0·0121212	0·0073233	2·4411
12	252	0·0391608	0·0512820	0·0285464	1·9026
11	525	0·0815851	0·1328671	0·0862680	1·3641
10	1050	0·1631702	0·2960373	0·2045155	0·8256
9	1225	0·1903652	0·4864025	0·3869796	0·2872
8	1400	0·2175602	0·7039627	0·5992089	−0·2513
7	945	0·1468531	0·8508158	0·7851777	−0·7898
6	630	0·0979021	0·9487179	0·9079605	−1·3283
5	231	0·0358974	0·9846153	0·9690283	−1·8667
4	84	0·0130536	0·9976689	0·9919182	−2·4052
3	13	0·0020202	0·9996891	0·9983784	−2·9437
2	2	0·0003108	0·9999999	0·9997513	−3·4821
	6435				

$\bar{r} = 8{\cdot}4667$

$\sqrt{\text{Var}\, r} = 1{\cdot}8571$

At $\alpha : 0{\cdot}05$ we would make a different decision for $r \geqslant 12$ using the normal approximation, but the same lower-tailed decision for $r \leqslant 5$. What is the difference at $\alpha : 0{\cdot}05$ for a two-tailed test?

3.4 Probability of the number of runs in fig. 3.1, showing *CDF* and normal approximation.

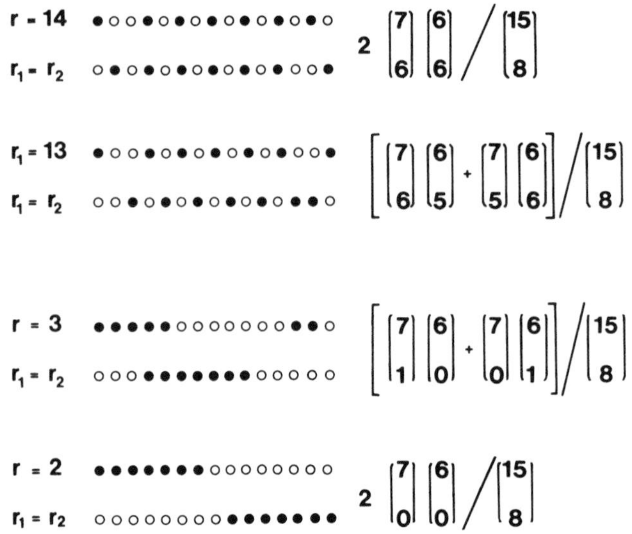

r - 14 ●○○●○●○●○●○●○●○

r_1 - r_2 ○●○●○●○●○●○●○○●

$$2\binom{7}{6}\binom{6}{6}\Big/\binom{15}{8}$$

r_1 = 13 ●○○●○●○●○●○●○○●

r_1 = r_2 ○○●○●○●○●○●○●●○

$$\left[\binom{7}{6}\binom{6}{5}+\binom{7}{5}\binom{6}{6}\right]\Big/\binom{15}{8}$$

r = 3 ●●●●●○○○○○○○●●○

r_1 = r_2 ○○○●●●●●●●○○○○○

$$\left[\binom{7}{1}\binom{6}{0}+\binom{7}{0}\binom{6}{1}\right]\Big/\binom{15}{8}$$

r = 2 ●●●●●●●○○○○○○○○

r_1 = r_2 ○○○○○○○○●●●●●●●

$$2\binom{7}{0}\binom{6}{0}\Big/\binom{15}{8}$$

3.5 Arrangements of runs of $n_1 = 8$, $n_2 = 7$ objects to compare with the frequencies in fig. 3.4.

However the values listed in table 10 are only appropriate for one-tailed tests. The authors of that table, Swed and Eisenhart, do give values for the two-tailed case. Entering table 10 with $n_1 = 8$, $n_2 = 7$ we see that we reject H_0 for $r \leqslant 4$ in favour of H_1 that there is evidence of clustering.

Tables for $P(r)$ are available for all cases where $n_1 \leqslant n_2 \leqslant 20$ and are given as table 10 in appendix 2. In those cases where the test is chosen as the appropriate test of randomness and for which tabulated values of the exact probabilities are not available then the statistic

$$Z = \frac{r - (\text{mean})}{\sqrt{\text{variance}}} \tag{3.1.2}$$

is a standard normal deviate and can be referred to tables of the normal probability distribution (table 1 in appendix 2). The mean and variance of r are given as (Gibbons 1971)

$$\text{mean} = 1 + \frac{2n_1 n_2}{n_1 + n_2} \tag{3.1.3}$$

$$\text{variance} = \frac{2n_1 n_2 (2n_1 n_2 - n_1 - n_2)}{(n_1 + n_2)^2 (n_1 + n_2 - 1)} \tag{3.1.4}$$

In the present example we have

$$\begin{aligned} \text{mean} &= 1 + 2(8)(7)/(8 + 7) \\ &= 8{\cdot}4667 \end{aligned}$$

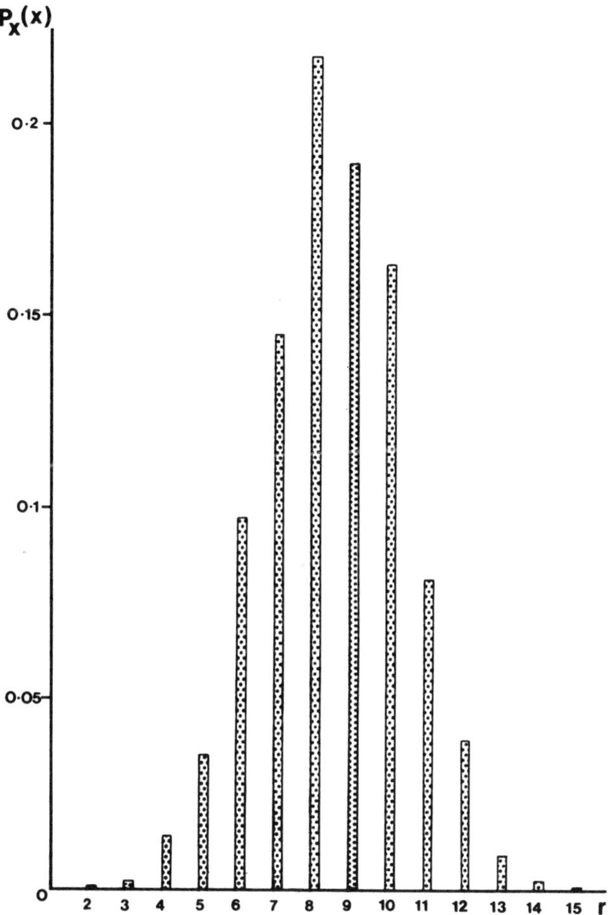

3.6 Probability distribution of r in fig. 3.4 to illustrate asymmetry.

and

$$\text{variance} = \frac{2(8)(7)[2(8)(7) - 8 - 7]}{(8+7)^2\,(8+7-1)}$$

$$= 3{\cdot}4489 \quad \sqrt{3{\cdot}4489} = 1{\cdot}8571$$

Thus for $r = 15$ we have

$$Z = \frac{15 - 8{\cdot}4667}{1{\cdot}8571}$$

$$= 3{\cdot}5180$$

The full results are tabulated in fig. 3.4 and the same decision is made on the same rejection regions in the one- and two-tailed cases.

3.1.2 Number of runs tests, one attribute

If we are interested only in the number of runs of, say, 1's, but we do not care whether $r_1 = r_2$ or whether $r_1 = r_2 \pm 1$, then the r_1 runs can still be selected in $\binom{n_1 - 1}{r_1 - 1}$ ways. However, we can now allow these runs to occupy the one unbounded space in front of and the one unbounded space after the 2's as well as the $n_2 - 1$ spaces between the 2's. There are $[(n_2 - 1) + 2] = n_2 + 1$ such spaces available for the r_1 runs, and of these $n_2 + 1$ spaces the required r_1 of them can be selected in $\binom{n_2 + 1}{r_1}$ ways. Thus the probability that there will be exactly r_1 runs of 1's when there are n_1 ones and n_2 twos is

$$P(r_1) = \frac{\binom{n_1 - 1}{r_1 - 1}\binom{n_2 + 1}{r_1}}{\binom{n}{n_1}}$$

Now $\binom{n_1 - 1}{r_1 - 1}$ is equivalent to $\binom{n_1 - 1}{n_1 - r_1}$. This identity can be verified readily by expanding the binomial coefficients. Thus $P(r_1)$ can be expressed as

$$P(r_1) = \frac{\binom{n_1 - 1}{n_1 - r_1}\binom{n_2 + 1}{r_1}}{\binom{n_1 + n_2}{n_1}} \qquad (3.1.5)$$

We note that the sum of the upper-level numbers in the numerator is the upper-level term of the denominator and the lower-level numbers in the numerator sum to give the lower-level term in the denominator. This is the hypergeometric probability distribution. Let us put this in the familiar diagrammatic form (fig. 3.7). We can see that we are finding the point probability of using only r_1 of the $n_2 + 1$ spaces available for runs of type 1 attribute-states. The list of point probabilities illustrates the difference in probability which results from this increase in flexibility for the arrangement.

Example 1

Let us suppose that two conflicting suppositions arise in a discussion about the incidence of premise occupation in large cities. One supposition is that retail and non-retail uses tend to be segregated and show clustering along the sequence. The other is that there will be as much intermixing and alternation as possible. In order to provide evidence in support of these notions the shops and other premises along the north side of Market Street, Manchester, in 1967 were classified by type of use. The sequence shown in fig. 3.8 was observed.

A simple test of the suppositions is to put

H_0 : the arrangement of retail, non-retail is random
H_1 : the retail uses are clustered

with α : 0·05 for a one-tailed test of too few runs. The details are put into two

$(n_1 - r_1)$	$(r_1 - 1)$	$n_1 - 1$
r_1	$(n_2 + 1 - r_1)$	$n_2 + 1$
n_1	n_2	n

0	7	7
8	0	8
8	7	15

6	1	7
2	6	8
8	7	15

$$P(r_1 = 8) = \binom{7}{0}\binom{8}{8}\Big/\binom{15}{8} = 1/6435 = 0{\cdot}0001554$$
$$P(r_1 = 7) = \binom{7}{1}\binom{8}{7}\Big/\binom{15}{8} = 56/6435 = 0{\cdot}0087024$$
$$P(r_1 = 6) = \binom{7}{2}\binom{8}{6}\Big/\binom{15}{8} = 588/6435 = 0{\cdot}0913753$$
$$P(r_1 = 5) = \binom{7}{3}\binom{8}{5}\Big/\binom{15}{8} = 1960/6435 = 0{\cdot}3045843$$
$$P(r_1 = 4) = \binom{7}{4}\binom{8}{4}\Big/\binom{15}{8} = 2450/6435 = 0{\cdot}3807304$$
$$P(r_1 = 3) = \binom{7}{5}\binom{8}{3}\Big/\binom{15}{8} = 1176/6435 = 0{\cdot}1827506$$
$$P(r_1 = 2) = \binom{7}{6}\binom{8}{2}\Big/\binom{15}{8} = 196/6435 = 0{\cdot}0304584$$
$$P(r_1 = 1) = \binom{7}{7}\binom{8}{1}\Big/\binom{15}{8} = 8/6435 = 0{\cdot}0012432$$

3.7 Probability distribution for r treated as a hypergeometric random variable.

C B A A B B A D A B A D C C B C C F C A E A B E A C C

A = n_1 = 8
B = n_2 = 6
C = n_3 = 8
D = n_4 = 2
E = n_5 = 2
F = n_6 = 1

N = 27

3.8 Occurrence of shop type in Market Street, Manchester, 1967.

categories, R for retail and S for non-retail, as in fig. 3.9. There could be as many as 27 runs or as few as 2 runs and we wish to know whether the actual number $r = 13$ lies in the rejection region or not.

The point probability of $r = 13$ is

$$P(r = 13) = \left[\binom{13}{6}\binom{12}{5} + \binom{13}{5}\binom{12}{6}\right]\Big/\binom{27}{14}$$
$$= 0{\cdot}127$$

which lies outside the rejection region; consequently the cumulative probability

S R R R R R R S R R R S S S R S S S S R S R R S R S S
r_2 r_1 r_2 r_1 r_2 r_1 r_2 r_1 r_2 r_1 r_2 r_1 r_2

n_1 = R = retail
n_2 = S = non-retail; includes personal services, commerce, entertainment

$n_1 = 14$; $n_2 = 13$; $n = 27$
$r_1 = 6$; $r_2 = 7$; $r = 13$

3.9 Data in fig. 3.8 reduced to two types, R=retail, S=not retail.

must lie outside the rejection region and we see no reason to reject H_0. For our particular alternative to randomness, that is of segregation along the line, the cumulative point probabilities for $r \leqslant 13$ are appropriate. For complete intermixing we should accumulate the point probabilities for $r \geqslant 13$. A two-tailed test is appropriate when either of these alternatives is acceptable.

Each of the $\binom{n_1 + n_2}{n_1}$ distinguishable arrangements of R, S, is assumed to be equally likely. The null hypothesis implies that the probability of an R or an S remains the same over the whole sequence of premises. In other words it implies that there is no increase or decrease in likelihood of a retail use in one sense of the line sequence.

Such a hypothesis is valid and interesting, but it may well be felt that it is insufficiently subtle to discriminate between the intricate land uses actually observed. When we have such a many-category classification of shop type (fig. 3.8) we can choose one occupancy sequence and test an alternative to randomness in the same way that we did for the rather coarse class of all retail. Depending upon our hypothesis we could use the total number of runs or just runs of one type.

3.1.3 Number of paired attributes

It is possible to test for pairings of types by a straightforward extension of the marginal probability for runs of one type given in equation 3.1.5. For instance in fig. 3.8 we may imagine that the incidence of food shops is likely to show a pairing with non-food retail shops rather than with the non-retail premises of categories C, D, E or F.

In this particular case there are 5 such pairs and 6 would have been the maximum because only 6 B's were recorded. The question is whether this is an unusually high number of pairs given that there are 6 B's, 8 A's, and 13 S's which are neither A nor B. Of course in other circumstances we could well have tested for other paired attribute-states.

Let us call the pairs, r, then in this case $r = 5$. In any particular case where r is the number of n_1, n_2 pairs, r can take any value from 0 to the minimum (n_1, n_2). In this case min $(n_1, n_2) = 6$. The number of pairs, r, is the test statistic and we need the frequency distribution of r from 0 to r max. The form of the argument is related closely to the earlier argument used for the runs formulae in articles 3.1.1 and 3.1.2.

Suppose there are n objects altogether and we distinguish n_1 of one kind, n_2 of a second kind and leave $n-(n_1+n_2)$ which we treat as one kind. Unlike the previous arguments we ask now that r of the (n_1, n_2) objects are adjacent and this clearly restricts both kinds of object when it restricts one kind. The question to be resolved is in how many ways can we combine exactly r (n_1, n_2)'s and allow the remaining objects to be arranged in any way which does not increase the $(n_1 n_2)$ pairs.

First we are asking that we take r of (n_1, n_2) objects. This can be done in $\binom{n_2}{r}$ ways with respect to the n_2 elements. It leaves $n-n_2$ objects. But because r involves the same number of n_1 elements it also leaves n_1-r of the n_1 objects. Secondly, we must not allow any further $(n_1 n_2)$ pairs, thus we need to know how many ways there are in which we can take n_1-r objects from $n-n_2$ objects. This is clearly $\binom{n-n_2}{n_1-r}$ and the upper number excludes the (n_1, n_2) pairs being increased as required. Thirdly, for each of the $\binom{n_2}{r}$ ways of arranging the r n_2's there are $\binom{n-n_2}{n_1-r}$ ways of arranging the remaining n_1-r objects. Their product is the numerator of our fraction. Finally the denominator is simply the number of ways we can arrange n_1 of the n objects, and this is $\binom{n}{n_1}$. The probability distribution we want is given by

$$P(r) = \frac{\binom{n_2}{r}\binom{n-n_2}{n_1-r}}{\binom{n}{n_1}} \qquad (3.1.6)$$

Again we see this is the probability distribution of a hypergeometric random variable. The nature of the point probability is clarified if we put this in tabular form (fig. 3.10).

n_1-r	$n-n_1-n_2+r$	$n-n_2$
r	n_2-r	n_2
n_1	$n-n_1$	n

3	18	21
5	1	6
8	19	27

$$P(r=6) = \binom{6}{6}\binom{21}{2}\Big/\binom{27}{8} = 0{\cdot}0000946$$

$$P(r=5) = \binom{6}{5}\binom{21}{3}\Big/\binom{27}{8} = 0{\cdot}0035945 \quad 0{\cdot}0036891$$

$$P(r=4) = \binom{6}{4}\binom{21}{4}\Big/\binom{27}{8} = 0{\cdot}0404378 \quad 0{\cdot}0441269$$

$$P(r=3) = \binom{6}{3}\binom{21}{5}\Big/\binom{27}{8} = 0{\cdot}1833181 \quad 0{\cdot}2274450$$

$$P(r=2) = \binom{6}{2}\binom{21}{6}\Big/\binom{27}{8} = 0{\cdot}3666363 \quad 0{\cdot}5940813$$

$P(r=1)$	$= \binom{6}{1}\binom{21}{7}\Big/\binom{27}{8}$	$= 0{\cdot}3142597$	$0{\cdot}9083410$
$P(r=0)$	$= \binom{6}{0}\binom{21}{8}\Big/\binom{27}{8}$	$= 0{\cdot}0916591$	$1{\cdot}0000001$

3.10 Probability distribution of incidence of paired types in fig. 3.8.

From fig. 3.10 we can see that the point probability for the market street data is $P(r=5) = 0{\cdot}0035945$ and $r = 5$ lies within the rejection region for H_0 at $\alpha : 0{\cdot}05$. The complete probability distribution is tabulated for convenience. It is asymmetric and we note that a lack of pairs, $r = 0$, would not lead to rejecting H_0 in favour of an H_1 that asserted that n_1 and n_2 uses were antithetic. However if we wished to determine the truth of an assertion that n_1 and n_2 uses tended to occupy adjacent sites then any value of $r \geqslant 4$ would lead to accepting that alternative to randomness.

Other more involved assertions can be made from events in a sequence such as shop type in a street. As soon as the geographer wishes

1. to compare events linked both ways along a line, that is links $n_1 n_2 n_1$ counted as two pairs;
2. to allow interrupted links of the form $n_1 z\, z\, n_2 n_1\, z\, n_1$ when the z are interruptions of a particular size;
3. to allow links across adjacent transects such as

$$n_1 n_2\ z\ z\ n_1\ z\ n_1\ z\ z\ n_2$$
$$z\ n_1\ n_1\ n_2\ z\ z\ z\ n_1\ z\ z$$

4. to consider triples or higher links

then the combinatorial problems become very much more involved. The joint probability distributions have to be derived for particular cases or an approximating continuous distribution has to be shown to be appropriate.

Runs tests can be kept comparatively simple and yet be extended to other situations by assuming that more information is contained in the numerals than is the case for nominal measurement.

3.2 Propositions of randomness (2)

Properties: location (position in a sequence), one other
Measurement scale: ordinal, continuous
Statistical procedure: Edgington's test for trend; Noether's test for cyclic change.

In some cases the line represents the object and observations on one or more attributes of the object are made at intervals along the line. These observations may yield measurements on a continuous or on an ordinal scale, as for example measurements showing certain physical and chemical properties of water samples taken at various stations along a river. Such values could be reduced to classes by dichotomizing them at some quantile such as the median and then the methods of articles 3.1.1 and 3.1.2 could be applied to test for randomness in the sequence of runs above and below that value. A considerable amount of the information in the measurements is suppressed by this many-to-one relation. This information could well yield some pattern in the line-ordering of the attribute. In the case of categoric information along a line there is no implication of systematic change with distance, only of pattern-type, and equivalent patterns could arise in various distance sequences. In this part we are considering test statistics sensitive to systematic variation of attribute with position in a sequence. This means that we are dropping the previous assumption that the probability of a particular attribute-level remained constant along the line.

3.2.1 Edgington's number of runs up and down test

The general null hypothesis is that the sequence of values is random and this is rejected if there are too few or if there are too many runs of increasing or of decreasing values. The alternative is then one of an increasing or decreasing sequence of values.

Each observation magnitude could be compared with the next in sequence. If the next value is larger, then a run up begins, if it is smaller, then a run down begins. Thus we define a run up as an unbroken sequence of increasing values and a run down as an unbroken sequence of decreasing values. If the values are arranged at random in the sequence we should not expect large runs up or down, as a large run up *or* down would indicate some sort of trend in the sequence, and a large number of runs up *and* down could well indicate some periodicity. Whereas in the case of a quantile such as the median only one value is used in the comparison, in the present case each value is used. A run up or a run down entails a sequence of *increasingly* extreme values, and the probability of each such extreme value will depend on the form of the distribution of the magnitudes. However, a general, distribution-free approach can be made by enumerating the sorts of effect a new value can have on a set of existing values. We shall not follow the proof through completely but shall indicate the logic that gives rise to the test statistic and the probability distribution. A more complete treatment is found in Edgington (1961) and in Gibbons (1971).

If we have three numbers, $n_1 < n_2 < n_3$, there can only be a run of length 1 or of length 2. Of the 3! arrangements of these numbers only 2 of them (1, 2, 3) (3, 2, 1) result in an arrangement where the subscripts are increasing or decreasing. The remaining four give rise to 2 runs of length one (1, 3, 2) (2, 1, 3) (2, 3, 1) (3, 1, 2). A fourth value, n_4, can be put in one of four places in each of the six arrangements of the previous three values. Consider the way this insertion affects cases (1, 2, 3) and (2, 1, 3); for ease of visualizing the effect the outcomes are put diagrammatically (fig. 3.11).

In general each extra observation must

1. split an existing run
2. lengthen an existing run
3. introduce a new run of length one.

As an existing run can be split to give two runs of equal or unequal lengths there are four mutually exclusive and exhaustive ways that a new observation can affect the incidence of run lengths. A recursive equation defining the relationship between a new observation and the incidence of run lengths enables us to calculate

Numerals and runs up +, down −	Number of runs of length r_i		
	r_1	r_2	r_3
1 + 2 + 3	0	1	0
+ + + 4	0	0	1
+ − 4 −	1	1	0
+ 4 − +	3	0	0
4 − + +	1	1	0
2 − 1 + 3	2	0	0
− + + 4	1	1	0
− + 4 −	3	0	0
+ 4 − +	3	0	0
4 − − +	1	1	0

3.11 Effect of a new measurement on three existing measurements.

the frequency with which a sequence of n observations will have exactly r_1 runs of length 1, r_2 runs of length 2 to r_{n-1} runs of length $n-1$. These frequencies are the numerator of the probability fraction whose denominator is $n!$.

Tables of the exact probabilities of at least r runs of a specified length are given for $n \leqslant 14$ in Owen (1962). Since the total number of runs is related to the number of runs of length r_i, the total number of runs may be used as a test for randomness. The hypothesis of randomness is rejected as the number of runs becomes too small. Exact cumulative probabilities of this statistic for $n \leqslant 25$ are given in Edgington (1961) and reproduced here as table 11 in appendix 2. For values outside the range of this table we can refer Z, a standard normal deviate, to tables of the normal probability function. We have

$$\text{mean runs} = \frac{(2n-1)}{3} \tag{3.2.1}$$

$$\text{variance runs} = \frac{(16n-29)}{90} \tag{3.2.2}$$

and

$$Z = \frac{r - (\text{mean})}{\sqrt{\text{variance}}} \tag{3.2.3}$$

Example

The control of water pollution is one of the principal tasks of a river authority. The Trent river system presented substantial problems of pollution control and the regular and well-documented reports of that authority provide an interesting example of the use of this test. Sources of pollution occur in all sections of the river: these sources are both direct effluent discharge and confluent tributaries whose levels of pollution vary. All sources of pollution are controlled by legislation. Although the eventual aim is to reduce pollution to negligible levels, an earlier aim could well be to eliminate any systematic changes in pollution increases. We may expect the effectiveness of the legislation controls to vary. If these variations are random along the river then the earlier aim is met. If the sources of polluting discharge are distributed systematically along the river – and this could occur if remedial controls were operating inefficiently – then there would be too few runs up and down.

The level of pollution can be measured in a number of ways. Two of the more reliable scales are biochemical oxygen demand, *BOD*, and ammoniacal nitrogen, *AN*. The values observed on these scales for the River Trent at 21 sampling points are given in fig. 3.12, and a map showing these values by class is given in fig. 3.13.

Number	*Station name*	Flow *MGD* 1965/7	*BOD* 1965/7	*BOD* 1959/61	Ammonia as *N* 1965/7	Ammonia as *N* 1959/61
1	Milton	10	4	5	0·6	1·1
2	Hanley		4	5	0·8	0·7
3	Stoke-on-Trent		9	12	3·1	5·6
4	Hanford Bridge	40	15	25	5·6	11·1
5	Stone		10	11	6·6	9·7
6	Great Haywood	60	12	13	5·4	6·6
7	Handsacre		9	9	2·2	2·9
8	Yoxall		7	9	1·8	2·4
9	Wychnor	150	7	9	1·4	1·6
10	Walton-on-Trent	390	14	15	6·7	7·7
11	Burton-on-Trent		20	15	4·6	7·3
12	Willington		13	14	3·8	5·6
13	Swarkestone		13	15	2·9	4·8
14	Shardlow	505	13	16	3·0	3·6
15	Sawley	700	10	13	2·4	2·7
17	Nottingham	800	11	11	2·2	2·4
18	Gunthorpe		9	12	2·4	2·9
19	Kelham	865	11	13	2·2	2·5
20	Dunham	890	9	11	1·7	2·3
21	Gainsborough		8	11	1·5	1·8
23	Keadby	980	8	13	1·5	1·4

Source: Trent River Authority, *Fourth annual report* (1969).

3.12 Mean pollution-level measurements at sample stations along the River Trent in 1959–61 and 1965–7.

The values are given in their natural order, downstream, and using the mean *BOD* values for 1959/61 we get the following runs of signs of first differences.

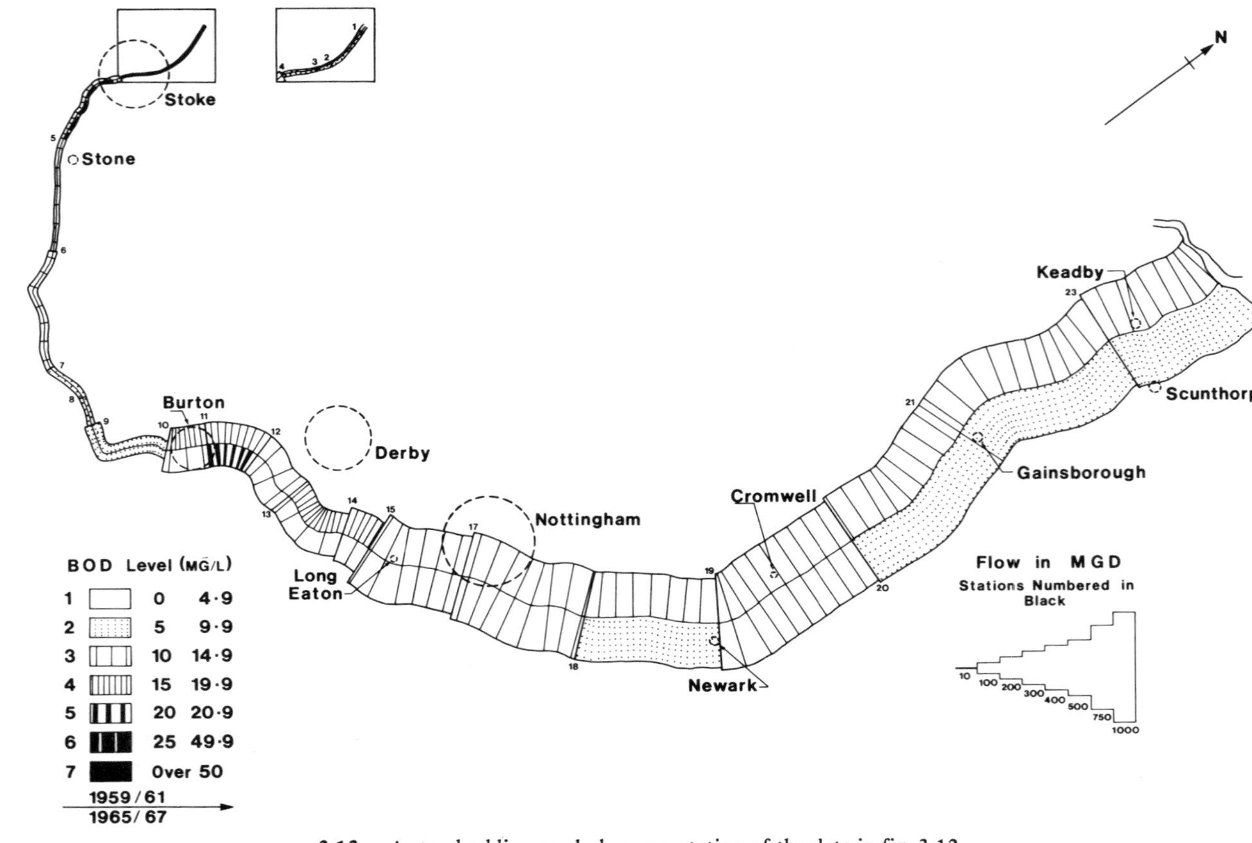

3.13 A standard line-symbol representation of the data in fig. 3.12.

$$0 + + - + - 0\ 0 + 0 - + + - - + + - 0 +$$

The zeros result from equal adjacent values and we can treat them in one of the following ways.

1. We can say that they provide no evidence and so ignore them by reducing n by the number of zeros.
2. We can argue that, on a continuous measurement scale, zeros imply insufficient precision in measurement and proceed as if measurement had been fine enough to discriminate between adjacent values. There are two cases:
   a) give the zeros the sign that increases the number of runs, r, to a maximum;
   b) give the zeros the sign of the adjacent, say lower, run except in the case where the first difference is zero, thus the number of runs remains the same and n is not reduced.

Case 2(b) gives a conservative test because it reduces the cases in which H_0 is accepted. In this particular situation we may suppose the river authority would prefer to risk rejecting the true null hypothesis that variations in pollution levels is random, than accepting that sequential variation in pollution is attained when in fact the null hypothesis is false. In practical terms the cost of a type I error is likely to be less than the cost of a type II error.

With $n = 21$, the exact probability that $r = 11$ is 0·1202 under the assumptions of the null hypothesis; consequently we accept H_0. Had we chosen case 2, then $r = 14$ with exact probability 0·6707 which emphasizes the change in rejection likelihood. This increase to $r = 14$ occurs by making the first zero a (−), the second a (+), the third a (−) and the fourth and fifth zeros do not affect the number of runs of signs as they do not split a run. As one purpose of this test is to trace the sources of non-randomness that have operated, the use of this conservative approach is advisable. Once these sources are corrected we may expect the sequence of results to become random and the levels diminish.

If there is a decrease in pollution level downstream as volume increases then the runs test will be less sensitive than a rank correlation test, because it uses less information. However, if the trend is periodic or U-shaped it will be more efficient. In the present example we may suppose that there are less likely to be sources of pollution when flow is low in the headstreams, to increase sharply in the middle reaches and decline below Shardlow as volume doubles.

3.2.2 Noether's test for cyclic change

Very often the sequence may be expected to have an intrinsic periodicity. The previous test certainly is sensitive to a trend, monotonic or periodic, as the alternative to randomness, but a test introduced by Noether (1956) is designed specifically for the hypothesis of randomness against an alternative of a cyclic trend. It is based upon the binomial distribution and tests whether the hypothesized probability, p, is acceptable against the alternative that the probability is p'.

Consider a sequence of measurements made on a continuous scale. If there is no trend of any sort then the sequence of numbers will be random. Replacing these measurements by ranks will yield a corresponding random sequence of ranks. The test proposed by Noether is directed at a cyclical pattern and for which he suggests that before replacing the measures by ranks the values are split into groups of three and then ranks (1, 2, 3) are given for each group.

For three numbers there are $3! = 6$ different arrangements in which they can occur: (i) 1, 2, 3 (ii) 1, 3, 2 (iii) 2, 1, 3 (iv) 2, 3, 1 (v) 3, 1, 2 (vi) 3, 2, 1. Only two of these (i), (vi) increase or decrease and thus the probability of a monotonic sequence is $2/6 = 1/3$ if the values are random. Thus the frequency of monotonic sequences under the null hypothesis is a binomial random variable with $p = 1/3$. It is plain that if there is periodicity in the data then more than 1/3 of the triples will be monotonic. Let us call the number of triples, n, each monotonic triple a run, r. Then the number of runs from n triples has a probability given by the binomial.

$$\binom{n}{r} p^r q^{n-r} \qquad \text{for } p = 1/3,\ q = 2/3$$

Example

The sequence of 21 values of ammoniacal nitrogen at the River Trent recording stations in 1959/61 (fig. 3.12) reduces to 7 sets of triples, of which 5 are monotonic. There were no tied values. Tied values only matter if they are adjacent and in the same triple; in such a case omit one of the values and make the triples up progressively down the sequence.

1·1, 0·7, 5·6; 11·1, 9·7, 6·6; 2·9, 2·4, 1·6; 7·7, 7·3, 5·6; 4·8, 3·6, 2·7;

2·4, 2·9, 2·5; 2·3, 1·8, 1·4

The null hypothesis is that the pollution levels downstream are random and the alternative is that the pollution levels are periodic downstream. In this particular case we have got 5 monotonic sequences out of 7 when the expectation of each set is $p = 1/3$. We reject the null hypothesis on the basis of how likely such a result is. Most tables give values for $p = 0{\cdot}33$, and so we shall calculate the exact distribution of the test statistic using 1/3, 2/3 as for this n the task is minor. The results are calculated as shown in fig. 3.14.

	Point probability	*Cumulative probability*
$P(r=7) = \binom{7}{7}\frac{1^7}{3}\frac{2^0}{3} = 1 \cdot \frac{1}{3^7} \cdot 1$	= 0·0004572	0·0004572
$P(r=6) = \binom{7}{6}\frac{1^6}{3}\frac{2^1}{3} = 7 \cdot \frac{1}{729} \cdot \frac{2}{3}$	= 0·0064015	0·0068587
$P(r=5) = \binom{7}{5}\frac{1^5}{3}\frac{2^2}{3} = 21 \cdot \frac{1}{243} \cdot \frac{4}{9}$	= 0·0384090	0·0452677
$P(r=4) = \binom{7}{4}\frac{1^4}{3}\frac{2^3}{3} = 35 \cdot \frac{1}{81} \cdot \frac{8}{27}$	= 0·1280300	0·1732977
$P(r=3) = \binom{7}{3}\frac{1^3}{3}\frac{2^4}{3} = 35 \cdot \frac{1}{27} \cdot \frac{16}{81}$	= 0·2560600	0·4293577
$P(r=2) = \binom{7}{2}\frac{1^2}{3}\frac{2^5}{3} = 21 \cdot \frac{1}{9} \cdot \frac{32}{243}$	= 0·3072699	0·7366276

$$P(r=1) = \binom{7}{1}\frac{1^1}{3}\frac{2^6}{3} = 7 \cdot \frac{1}{3} \cdot \frac{64}{729} \qquad = 0{\cdot}2048466 \quad 0{\cdot}9414742$$

$$P(r=0) = \binom{7}{0}\frac{1^0}{3}\frac{2^7}{3} = 1 \cdot 1 \cdot \frac{128}{2187} \qquad = 0{\cdot}0585277 \quad 1{\cdot}0000019$$

3.14 Exact probability distribution of a binomial random variable with $p=\frac{1}{3}$ for $r \leqslant 7$.

The cumulative probability $r \geqslant 5$ for $n = 7$ is 0·0452677, and we reject the hypothesis of randomness at our standard α : 0·05.

By specifying particular alternatives to randomness we have introduced tests that on rejection of the null hypothesis imply association between position in the sequence and magnitude of the measured attribute. These tests form a natural link to ideas and to tests of dependence. We have covered a variety of such tests already, based on chi-square and log-likelihood procedures. Such tests can be adapted to a great many situations of line maps. A number of important tests remain, and three of these are introduced in this section: Cox and Stuart, Spearman/Daniel/Hotelling-Pabst, and Kendall. Again they can be applied to many maps that do not use line symbols, but because they are suited to cases of position in a linear sequence I have chosen to put them in this section.

Two extensions to the methods of this part need mentioning. The first takes sequential events as a Poisson random variable. The second introduces the idea of a *markov process* in which events are assumed to be dependent upon adjacent values. This opens up a wider field of model than we have covered. Such models are implicit in the transitional probabilities of the standardized contingency tables. The interested reader should consult the following references: Bailey 1964; Bartholomew 1973.

3.3 Propositions of dependence

Property: two attributes; one can be location in a sequence
Measurement scale: continuous, ordinal
Statistical procedure: Cox and Stuart, Spearman's rho, Kendall's tau

Rejection of the null hypothesis in Edgington's test and in Noether's test implied some sort of trend in the measurements with respect to position in the sequence, and to this extent they can be seen as much to be tests of dependence as the tests we shall include under this present heading. Indeed the first test to be discussed is very similar to Noether's test and also is based upon the binomial distribution. This is Cox and Stuart's test for trend which can be used with one or two data sets and so provides a clear analogy and comparison with the better-known Spearman and Kendall tests of relationship. However these latter tests provide a measure of the association between the two variables and no such measure is ascribed to the Cox and Stuart test.

3.3.1 Cox and Stuart's test for trend

Cox and Stuart's test is an adaptation of the sign test and is applicable in those cases where we wish to test for a systematic increase or decrease in magnitude of an attribute according to position in a sequence. The sequence can be time or distance and it is generated by the order of observation of the random variable, $X_1, X_2, \ldots, X_n$. The idea of the test can be illustrated clearly by a diagram (fig. 3.15).

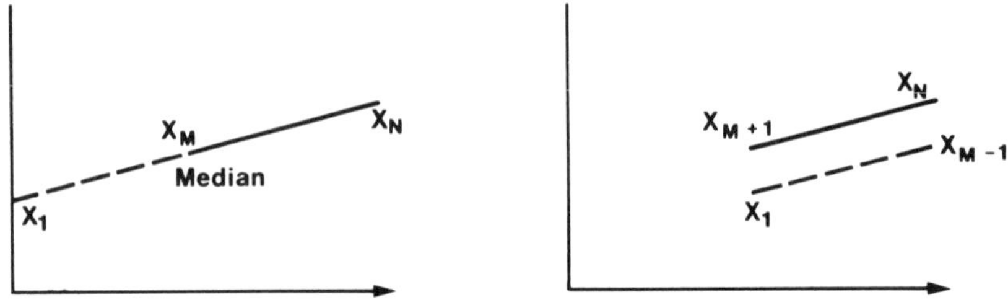

3.15 A graphical interpretation of Cox and Stuart's test.

In the simplest case, Y, the value of the attribute is a linear function of the position, X_i, of the attribute in the sequence. If we bisect these values and slide the first half beneath the remainder we get the situation in the right-hand diagram of fig. 3.15. In an exact linear function all the values in the first half of the sequence are smaller than the corresponding values in the second half of the sequence. If the Y_i, that is the measures of the attribute, are random with respect to the sequence $(X_1, X_2, \ldots, X_n)$, then the differences between the two halves of the graph are equally likely to be positive as to be negative. Thus, if the signs of these differences are recorded, plusses and minuses should occur randomly along the sequence. The general null hypothesis is of randomness in the sequence of values and the alternative can be: (1) two-tailed for either an upward or a downward trend, (2) one-tailed for an upward trend, (3) one-tailed for a downward trend in one sense of the sequence. The test statistic is simply the number of plusses in the sequence of n observations. Tied values yield zero differences and are omitted, thus reducing n and the plusses without impairing the test.

Example 1

This test is readily used to test for trend in the ammoniacal nitrogen measure of pollution level in the Trent for 1959/61. We are interested in the order of these observations because we may suppose that the rapid increase in volume downstream will dilute the pollutants and lead to a decrease in pollution level downstream from source. We are using a one-tailed rejection region, and if our alternative is true the downstream values, X_d, will be smaller than the upstream values, X_u, thus the difference, $X_d - X_u$ will be negative. The data is recorded in fig. 3.16. There are 21 values, thus the value of the median observation is $X_{Md} = (n + 1)/2$. This value has no pair and is omitted. For n even, clearly we take the $X_{n/2}$ value and use all n values in the pairs. In the present example we match X_i, X_{Md+i} and

5·6 − 1·1	+	
4·8 − 0·7	+	
3·6 − 5·6	−	
2·7 − 11·1	−	$n = 10$
2·4 − 9·7	−	$i = + = 3$
2·9 − 6·6	−	
2·5 − 2·9	−	
2·3 − 2·4	−	
1·8 − 1·6	+	
1·4 − 7·7	−	

3.16 Cox and Stuart's test results for the data in fig. 3.12, case 1.

use the sign of the differences, sign $(X_{Md+i} - X_i)$ as shown in fig. 3.16. There are $n = 10$ pairs and $r = 3$ plusses. The number of plusses is the test statistic, it is a binomial random variable and can be calculated from

$$\binom{n}{i} p^i (1-p)^{n-i} \quad \text{for } i \text{ plusses} \tag{3.3.1}$$

The point probability is

$$\binom{10}{3} \tfrac{1}{2}^3 \tfrac{1}{2}^7 = 120 \cdot \tfrac{1}{2}^{3+7} = 0{\cdot}1171875$$

and the cumulative probability for $i \leqslant 3$ is 0·1718750. H_0 is accepted customarily on such grounds. This can be inferred using table 3a in appendix 2. As $i \leqslant 3$ is equivalent to $r \geqslant 4$, the probability of $i \leqslant 3 = 1 - Pr(r \geqslant 4) = 1 - 0{\cdot}82813 = 0{\cdot}17187$ as calculated.

The efficiency of this test, which is considerable with respect to the normal regression and distribution-free regression tests with a normal population, is increased by removing the central third of the data and finding the differences between pairs of the first and final thirds of the values.

Example 2
This is best illustrated with the same data (fig. 3.17)

2·7 − 1·1	+	
2·4 − 0·7	+	
2·9 − 5·6	−	$n = 7$
2·5 − 11·1	−	$i = + = 2$
2·3 − 9·7	−	
1·8 − 6·6	−	
1·4 − 2·9	−	

3.17 Cox and Stuart's test results for the data in fig. 3.12, case 2.

which shows for $n = 7, i \leqslant 2, p = 1 - p = 0{\cdot}5$, the probability is 0·2265625.

This test seems a particularly useful and quick test for many situations where measurements are continuous, but whose form is unknown, especially where site to site variation is uncontrollable and when emphasis is on the direction of differences at sites. These differences may be between years, as for example with ground-water flow, sediment load or rate of stream discharge, or at one sampling period where the difference is sequential down the line as with the pollution. The pattern may be periodic over a year and difference comparisons taken at equivalent times in different years to see whether the rates of, say, ground-water flow are changing. Graphs of such measurements are more useful when accompanied by the results of such a test, particularly if bands of likelihood are superimposed (Noether 1956).

3.3.2 Rank correlation tests

In the previous test, Cox and Stuart, we argued that if a set of observations on an attribute taken in some sequence yield values of that attribute which are distributed systematically with respect to that sequence, then we say there is a trend. When we say the values are random along the sequence we are saying that the sequence and the values are independent. In many circumstances when we have a two-dimensional random variable (X, Y) whose values are dependent we wish to measure that dependence numerically. Measures of the degree of association between two attributes, X and Y, are termed *correlation coefficients.*

The usual measure is known as *Pearson's product moment* coefficient of correlation and is denoted by $r_{X,Y}$ and is expressed as a quotient whose numerator is the covariance of (X, Y) and whose denominator is the product of their standard deviations. This fraction was introduced in article 1.8.2 and was used in the regression equation in article 2.5.4. Thus we have

$$r_{XY} = \frac{\frac{1}{n}\Sigma\,(X - E(X))(Y - E(Y))}{\left[\frac{1}{n}\Sigma\,(X - E(X))^2 \frac{1}{n}\Sigma\,(Y - E(Y))^2\right]^{\frac{1}{2}}} \qquad (3.3.2)$$

This measure, r_{XY}, plays a very important part in descriptive statistics and it has certain properties that have become accepted as desirable in measures of association.

1. It can take values only in the range from -1 to $+1$. Note how we restricted the cross-product ratio to the same interval.
2. When an increase (decrease) in X corresponds to an increase (decrease) in Y for (X_i, Y_i) the coefficient is positive. Conversely when an increase (decrease) in X corresponds to a decrease (increase) in Y the coefficient is negative.
3. A value near zero indicates that the values of Y are random with respect to the values of X. That is, for a given value of X, the value of Y is as likely to be large as it is to be small.
   In this case (X, Y) are said to be uncorrelated.

If X and Y are independent, then $r = 0$; this is an immediate consequence of the fact that $E(X, Y) = E(X)E(Y)$ for independent events and thus the numerator r is zero. However the converse is not true and (XY) may have $r = 0$ and not be independent, thus when $r = 0$ the proper statement is that the variables (X, Y) are *uncorrelated.*

This coefficient is quite properly used in a wide variety of circumstances and there is no constraint in the formula which precludes its application with data of unrestricted measurement scales. The same is true for other descriptive parameters such as $E(X)$ and $Var(X)$, which can be used legitimately under very wide conditions. However, r is a random variable with a distribution function which depends on the form of the distribution function of the two-dimensional random variable (X, Y). Tests of significance of particular values of r do have stringent assumptions and the reliability of the test is only certain when these assumptions are met. The correlation coefficients introduced next imitate properties 1, 2 and 3 of r given above, but they are functions of the ranks of the observations and have distribution functions that are independent of the form of the two-dimensional random variable (X, Y). In the case of nominal measurements the *phi coefficient* presented in the previous section is analogous to r for contingency tables.

3.3.2.1 Spearman's rho, ρ

Consider a two-dimensional random variable (X, Y) such that a measurement has been made on X and Y on a continuous measurement scale and that there are no tied values among the X measures nor among the Y measures. Each of the X values is replaced by its rank with rank 1 given to the smallest value and rank n to the largest. Each Y value is given its rank similarly. Arrange the X values in rank order with the rank of the corresponding Y in the same sequence. If the Y values are independent of the X values, the distribution of Y ranks will be random with respect to the ranks of the paired X values. If the null hypothesis of independence is correct then each of the $n!$ arrangements of Y ranks is equally likely to have been the observed sequence. The sum of the squared differences between each X rank and its paired Y rank is the basis of the rank correlation coefficient test statistic. The statistic

$$D = \sum_i (R_{Xi} - R_{Yi})^2 \qquad (3.3.3)$$

is known as the Hotelling-Pabst test statistic, and can be referred to appropriate tables directly.

Spearman's rho, ρ, gives identical probabilities because it is a monotone function of D. The basic formula for ρ, which precedes D, is

$$\rho = \frac{\sum_i \left(R_{Xi} - \frac{n+1}{2}\right)\left(R_{Yi} - \frac{n+1}{2}\right)}{n(n^2 - 1)/12} \qquad (3.3.4)$$

and shows that rho uses the same fraction as Pearson's r, with the covariance of the values replaced by the covariance of their ranks, because the expected value of their ranks is $\frac{n+1}{2}$ as before. The denominator is derived from the product for X and Y of the variance of the first n integers which are the ranks. This equation reduces to the more familiar and computationally easier form of

$$\rho = 1 - \frac{6D}{n(n^2 - 1)} \quad (3.3.5)$$

with D defined in equation (3.3.3).

These two equations reveal the relationship between the parametric correlation coefficient and the randomization distribution of the sum, D, which we argued for initially. D and ρ give the same decision, but ρ is symmetric about zero in the interval $(-1, +1)$. Tied values cause this distribution to be inexact. If the number of tied values in X or in Y is moderate only, then the discrepancy appears to be slight.

In general, the null hypothesis is that the two variables X and Y are independent, although if H_0 is accepted the correct statement is that they are uncorrelated. H_0 is rejected in favour of

1. the two-sided alternative of an association of increasing ranks of X with increasing or decreasing ranks of Y;
2. the one-sided alternative of increasing ranks of X with increasing ranks of Y in which case there is a positive correlation;
3. the one-sided alternative of increasing ranks of X with decreasing ranks of Y in which case there is a negative correlation.

Example

Using the data presented in fig. 3.12 for the Cox and Stuart test we put X as the position in sequence downstream and Y as the measure of AN; the values of Y are replaced by ranks as in fig. 3.18.

R_{Xi}	Y_i	R_{Yi}	$(R_{Xi} - R_{Yi})^2$
1	1·1	2	1
2	0·7	1	1
3	5·6	15·5	156·25
4	11·1	21	289
5	9·7	20	225
6	6·6	17	121
7	2·9	11·5	20·25
8	2·4	7·5	2·25
9	1·6	4	25
10	7·7	19	81
11	7·3	18	49
12	5·6	15·5	12·25
13	4·8	14	1
14	3·6	13	1
15	2·7	10	25
16	2·4	7·5	72·25
17	2·9	11·5	30·25
18	2·5	9	81
19	2·3	6	169
20	1·8	5	225
21	1·4	3	324

$$D = 1911{\cdot}50$$

$$\begin{aligned} \rho &= 1 - \frac{6.1911{\cdot}50}{21(21^2-1)} \\ &= 1 - 1{\cdot}3003 \\ &= -0{\cdot}3003 \end{aligned}$$

3.18 Calculations for Spearman's coefficient of rank correlation, ρ.

The 0·95 quantile of ρ for $n = 21$ is $\pm\, 0{\cdot}3688$ (see table 12 in appendix 2). The null hypothesis is accepted and we state that pollution level is not correlated with distance from the source. There is no evidence on these attributes for any systematic increase or decrease in pollution level downstream. When Spearman's rho is used with a naturally increasing sequence, such as position in time or along a distance line, it is termed *Daniel's test* for trend.

3.3.2.2 Kendall's tau, τ

This test is practically equivalent to the previous test in the sense that the two random variables, ρ and τ, are very highly correlated under common conditions of observed measurements. However they are not mathematically equivalent. Spearman's rho is used more widely and probably is better known. Perhaps this preference reflects its similarity to Pearson's coefficient and to its use of squared differences which gives greater weight to large discrepancies. Spearman's rho is also easier to calculate. However the fact that exact tables are available for τ for a much larger number of cases and that it is a more versatile test make it preferable in many circumstances. The test and its variations are discussed thoroughly in Kendall (1948).

The assumptions for the test are the same as for Spearman's rho. We have a two-dimensional random variable (X, Y) with measurements made without ties on X and on Y. If the values are independent the pairs of ranks, $(R_{Xi}\; R_{Yi})$, will be random. However, if there is a tendency for positive (negative) correlation the number of increasing R_{Xi} with increasing (decreasing) R_{Yi} will tend to be larger than if they are uncorrelated. A pair of observations, say (X_1, Y_1), (X_2, Y_2) are said to be *concordant* if when X_2 is larger than X_1, then Y_2 is larger than Y_1. If X_2 is larger than X_1, but Y_2 is smaller than Y_1, the pairs are said to be *discordant*.

The n, (X, Y) observations can be paired in $\binom{n}{2}$ different ways. Of these N_C are concordant and $N_D = \binom{n}{2} - N_C$ are discordant. The test statistic, τ, uses the fraction

$$\tau = \frac{N_C - N_D}{\binom{n}{2}} = \frac{N_C - N_D}{n(n-1)/2} \qquad (3.3.6)$$

It is plain that when $N_C = n(n-1)/2$, $\tau = +1$, and when $N_D = n(n-1)/2$, $r = -1$ and when $N_C = N_D$, $\tau = 0$.

The enumeration of the concordant and discordant pairs is the only part of the test that seems to cause confusion. The confusion is reduced if the values of one of the variables are put in ascending order with the appropriate pair. In fact ranks need not be used, but replacing the values by their ranks within X and within Y makes the enumeration more obvious.

Example
Repeating the data used for Spearman's rho we have fig. 3.19.

R_X	R_Y	N_c	N_D
1	2	19	1
2	1	19	0
3	15·5	5	12
4	21	0	17
5	20	0	16
6	17	2	13
7	11·5	5	8
8	7·5	8	4
9	4	11	1
10	19	0	11
11	18	0	10
12	15·5	0	9
13	14	0	8
14	13	0	7
15	10	1	5
16	7·5	2	3
17	11·5	0	4
18	9	0	3
19	6	0	2
20	5	0	1
21	3	0	0
		72	135

$$\tau = \frac{72 - 135}{21(20)/2}$$

$$= -0{\cdot}300$$

$N_C + N_D \neq n(n-1)/2$ because of the tied values.

3.19 Calculations for Kendall's coefficient of rank correlation, τ, illustrating the use of concordant and discordant values.

To calculate N_C we have a counting problem to that of the Mann-Whitney test. For each (X_i, Y_i) we have to count all the other pairs that are larger in X and in Y than the X_i, Y_ith pair. We put the X_i in ascending order, in our case we use rank order. Thus all values X_i are less than all values X_{i+1}, but the Y_i are not all in ascending order and so we count the number of Y values that are larger than the Y_i value and lower down the list than the ith value. Thus (X_1, Y_1) has values

(R_{X1}, R_{Y2}), the next pair (X_2, Y_2) has values (R_{X2}, R_{Y1}) and is *discordant* because X has increased, but Y has not. For (X_3, Y_3) we have $(R_{X3}, R_{Y15\cdot5})$ and this is concordant with $(X_1\, Y_1)$. Indeed all subsequent (X_i, Y_i) are concordant with (X_1, Y_1). Now consider (X_8, Y_8) with values $(R_{X8}, R_{Y7\cdot5})$ then the following (X_i, Y_i) are concordant: (X_{10}, Y_{10}), (X_{11}, Y_{11}), (X_{12}, Y_{12}), (X_{13}, Y_{13}), (X_{14}, Y_{14}), (X_{15}, Y_{15}), (X_{17}, Y_{17}), (X_{18}, Y_{18}), but (X_{16}, Y_{16}) is not because its values are $(R_{X16}, R_{Y7\cdot5})$ and is equal to the eighth pair for Y. The discordant values are counted in a similar way. As expected the decision is the same for $\tau = -0{\cdot}3$ and we accept H_0 that order and pollution measure are uncorrelated.

Kendall devised a test known as Kendall's test of concordance, usually denoted by W, which tests for dependence in a multi-dimensional random variable $(X_1, \ldots, X_k)$. This test assesses the degree of similarity between the k sets of ranks on n objects each with k ranked attributes, and provides an estimate of the best composite ranking based on the sum of ranks over the k attributes for each object. It is equivalent to Friedman's test statistic as it is a simple function of it and will not be discussed further.

The notion of correlation in the bivariate case can be extended to the multivariate case. In those cases where we wish to isolate the association between one pair of variables while suppressing the influence of the remaining variables, the appropriate technique is known as partial correlation. Kendall's tau can be adapted to provide a measure of partial association for ranked data. The technique is discussed in Kendall (1948), together with his concordance measure, W. The reader who wishes to explore the ideas and methods of multivariate and partial correlation should do so by becoming familiar with the theory of Pearson's coefficient and its assumptions.

3.4 Propositions of equality of location and identity of distribution

Properties: two, of which location often one
Measurement scale: nominal, ordinal, continuous
Statistical procedure: Wald-Wolfowitz, Westenberg-Mood, Wilcoxon signed rank, Kolmogorov-Smirnov

These tests all require two sets of values. These are usually two samples for which we wish to establish whether the measurements are taken from identical populations of values or, in a slightly more restricted comparison, whether they have a common measure of statistical location. A second situation arises when one set of measurements of some attribute is compared with the values that are expected under a set of precise conditions; for example the population of measurements may be assumed to be normal, or to be Poisson or some other form. Tests for comparisons involving an empirical distribution and an expected distribution are known as goodness-of-fit tests. The two best-known are Kolmogorov-Smirnov tests and chi-square. Two data sets for lines arise as measurements of a common attribute of two objects, each of which is represented as a line, such as land use along a road, or a characteristic of a river. Alternatively the line is an artifice for sampling purposes, and the transect-lines are considered able to characterize some property such as hill-slope, and individual readings of slope are made along the two or more transects. There are no changes in line symbolism implied by these propositions corresponding to the changes in point symbolism associated with increases in information.

We have seen already how closely related certain tests of randomness are to tests for independence; their similarity rests in the conditions and in the alternative chosen if the null hypothesis is rejected. We shall now adopt a runs test for randomness in dichotomies to a test for a null hypothesis of identity of two distributions of measurements. This test is known as the Wald-Wolfowitz test.

3.4.1 Wald-Wolfowitz number of runs test

The logic of this test is the same as when it is used as a one-sample test against randomness. In the present case the values come from an X population and from a Y population of measurements or of attribute classes. There are n_1 X values and n_2 Y values. These two sets of values are combined into one set of $n_1 + n_2$ values and the combined set is ordered, a run is defined as a sequence of consecutive X values, or as a sequence of consecutive Y values. The total number of runs of X's and Y's is the test statistic. The null hypothesis is that the two samples are drawn from the same population or from identical populations of measurements. If the null hypothesis is true then each of the $\binom{n_1+n_2}{n_1}$ ordered sequences of values is equally likely, with probability $1\Big/\binom{n_1+n_2}{n_1}$. If the populations are not the same then the values from one population are expected to occur at one part of the ordered sequence of the combined sample. The randomization distribution of number of runs is the same as in the one-sample test.

Example 1

In a study of an exposed section of glacial deposits it was argued that two different horizons had differences in glacial history which would be evident in the mean angularity of the particles. Eleven samples from one horizon, X, and 14 from the other, Y, were measured and combined to give the following sequence.

$$\overline{Y\,Y}\,\underline{X\,X\,X\,X\,X}\,\overline{Y\,Y\,Y}\,\underline{X\,X\,X\,X}\,\overline{Y\,Y\,Y\,Y\,Y}\,\underline{X\,X}\,\overline{Y\,Y\,Y\,Y}$$

$$n_1 + n_2 = 11 + 14 = 25$$

$$r_1 + r_2 = 3 + 4 = 7$$

The 0·95 quantile for $n_1 = 11$, $n_2 = 14$ is 8 runs (see table 10 in appendix 2), so for values of $r \leqslant 8$, H_0 is rejected and we conclude that there is a difference in angularity.

Example 2

Using the information on farm income increments of fig. 2.75 (article 2.7.1) we have the following sequence:

$$Y\,X\,Y\,X\,X\,Y\,X\,X\,X\,Y\,Y\,X\,Y\,Y\,Y\,Y\,Y\,Y\,Y$$

$$n_1 + n_2 = 7 + 12 = 19$$

$$r_1 + r_2 = 4 + 5 = 9$$

The 0·95 quantile is $r \leqslant 6$ and thus we would accept H_0, whereas with the Wilcoxon rank-sum test we rejected in favour of a difference in distribution with the X population being larger than the Y with the same rejection region.

The difference in decision reflects the difference in sensitivity of the two tests. Wilcoxon's test uses more information and offers greater discrimination. Runs tests generally have comparatively little power, but they are useful where values cannot be measured on more restricted scales.

3.4.2 The matched-pairs sign test

The binomial distribution has been the basis of some of the simplest and most powerful tests we have considered. The one-sample sign test was used as the device for making clear some notions of hypotheses and hypothesis testing in article 1.9. An adaptation of this test to matched observations in two sequences in time or along a line is useful, easily applied and powerful. You will have noted that a frequent response to a situation that yields a two-dimensional random variable is to reduce the pairs of values to one value and then treat this as a random variable.

In the present case the two-dimensional random variable $(X\ Y)$ is reduced to difference scores. These are differences in two sequences, in time or along a line, in which the observations are paired in some logical way. For each (X_i, Y_i) we form the difference $(X_i - Y_i)$ and replace this difference by its sign; putting a + if $Y_i > X_i$, a − if $X_i > Y_i$ and a zero if $X_i = Y_i$. This implies that our measurements cannot be nominal because the differences must be able to be ordered.

The null hypothesis is that the two populations are identical and thus have identical medians. The probability that a value will be greater than this median equals the probability that it will be less than it. Then, in forming the differences, the probability that an X is greater than its paired Y is equal to the probability that an X is less than its paired Y. Thus the test is based on the binomial distribution with $p = \frac{1}{2}$. The alternative hypotheses are:

1. two-tailed such that the two populations are different;
2. one-tailed such that the X population yields larger values than the Y population, in which case there will be a greater number of plusses than expected;
3. one-tailed such that the X population yields smaller values than the Y population, in which case X_i will tend to be less than Y_i and result in a preponderance of minus differences.

Example

The 21 sites along the River Trent have pollution-level measurements for different years. If there had been no change in pollution-level values between say, 1959–61 and 1965–7 then we should expect the X value, 1959–61, to be larger than the Y value, 1965–7, as often as it was smaller, and also that the median difference score is zero. If there had been some improvement, as hoped, then we would expect the Y_i to be smaller than the X_i sufficiently often to reject the null hypothesis. Thus we put

$$H_0 : P(X_i > Y_i) \leqslant P(X_i < Y_i) = 0{\cdot}5$$
$$H_1 : P(X_i > Y_i) > P(X_i < Y_i)$$

From fig. 3.12 we find for ammoniacal nitrogen that we get the following series of difference signs.

− + − − − − − − − − − − − − − − − − − − +

Referring the number of plusses, 2, with $n = 21$ to the binomial tables we find that the two-tailed likelihood is 0·001; consequently the one-tailed probability which we require is half this, giving 0·0005. The only reasonable inference is to accept H_1, that the median of the difference scores is not zero.

A number of factors may affect changes in the level of pollution along the river: such things as confluent streams, pollutant sources, channel configuration and protected reaches of the river. These conditions remain common to the observations at each site; thus, although these may all affect the value of the pollution measure, it is argued that the direction of the difference in these values can be attributed to legislation controlling the nature and the quantity of the effluent.

The form of the statement of H_1 reflects that the test concerns the median of the difference-score population and *not* the difference between the medians of the two sequences, X and Y. The difference-score population may well be non-zero when the combined X and Y measures have a common median. This is proved by Gibbons (1971) and can be illustrated by the data in the above example.

The median of the combined population is 2·6; if we dichotomize the 42 values on this median and treat the X, 1959–61, sequence as sample 1 and the Y, 1965–7, sequence as sample 2 we can form the 2 × 2 contingency table in fig. 3.20.

	Pollution score about median		
	$<2{\cdot}6$	$>2{\cdot}6$	
Sample 1	9	12	21
Sample 2	12	9	21
	21	21	42

3.20 Westenberg-Mood median test table for data in fig. 3.12.

The null hypothesis is that equal proportions of the two populations of measurements from which samples 1 and 2 are taken lie below their common median. This is clearly a hypergeometric random variable and we accept the null hypothesis of no difference at any reasonable level of significance.

In this form, Fisher's exact test is known as the *Westenberg-Mood median test.* It can be extended to divide the combined sample on any quantile that is specified in advance.

3.4.3 Wilcoxon signed-rank test

The ordinary sign test can be used with nominal measures such as success, failure.

The sign test applied to difference scores required ordinal measurement because the differences had to satisfy one of the relations >, <, =, but any information about the size of those differences was ignored. A closely related test, the Wilcoxon signed-rank test, does take these magnitudes into consideration.

In comparison with the sign test, Wilcoxon's test is more powerful because it uses more of the available information. Like the sign test, the Wilcoxon test is applied to a random sample of n observations, $X_1, X_2, \ldots, X_n$, from a continuous population with median M, but the additional assumption is made that the population of measurements is symmetric. Using the same null hypothesis as in the sign test, namely that the hypothesized median, Md, equals the population median, M, the assumption of symmetry entails that the differences, $d_i = X_i - Md$, are distributed symmetrically about zero. Whereas in the sign test we had just that the numbers of plusses and minuses were equal, in the present case we have that the positive and negative differences of equal magnitude are equal in number; or, more exactly, that they have the same probability of occurrence.

These differences are arranged in order of their absolute magnitude, that is their sign is disregarded. Each difference is given its appropriate rank, with rank 1, R_1, given to the smallest difference and rank n, R_n, to the largest. The signs of the original differences are now attached to these ranks. Because of the symmetry of the continuous population, the sum of the ranks of the positive differences, R^+, should equal the sum of the ranks of the negative differences, R^-, if Md is the true median. We can use R^+ or R^- as the test statistic because under the null hypothesis they are equal. Now we can imagine we have n magnitudes arranged in order but the sign attached to each magnitude, and therefore its rank, is as likely to be positive as negative. Thus there are 2^n possible assignments of signs to ranks and each has probability $\frac{1}{2}^n$. For each of these assignments we can calculate the sum of positive (or negative) ranks, and by multiplying the number of times each sum occurs by the probability $\frac{1}{2}^n$ we find the probability of any particular sum. By cumulating these probabilities over all sums for positive (or negative) ranks of the difference scores we have the null distribution.

The idea can be illustrated very easily by considering four measurements to which we give ranks 1, 2, 3, 4. The probability that any number is given a plus sign is $\frac{1}{2}$ and thus the probability for all four numbers is $\frac{1}{2}^4 = \frac{1}{16}$.

Assume in the first case that none of these is positive and so the sum of positive ranks, $R^+ = 0$. Now $R^+ = 0$ can arise in only one way, $(-R_1, -R_2, -R_3, -R_4)$. Similarly $R^+ = 1$ can also occur in only one way, $(+R_1, -R_2, -R_3, -R_4)$, and so too with $R^+ = 2$ as $(-R_1, +R_2, -R_3, -R_4)$. Each of these has probability $\frac{1}{16}$ of occurring. Now $R^+ = 3$ can occur in two ways, $(+R_1, +R_2, -R_3, -R_4)$ and $(-R_1, -R_2, +R_3, -R_4)$ and $R^+ = 3$ has probability $\frac{2}{16}$. The results for $n = 4$ are given and the arrangements are shown symbolically. In the same table, fig. 3.21, a few arrangements for $n = 10$ are also shown.

Systematic enumeration soon becomes tedious. Values of R^+ are tabulated for all values $n = 5(1)20$ in table 14 in appendix 2, while for larger values a standard normal deviate, Z, is available using

$$\text{mean} = \frac{n(n+1)}{4} \tag{3.4.1}$$

Sum of positive ranks	Ranks in the sequence with positive values for sequence of $n = 4$ numbers	Number of ways, r_{n_1} that R_i^+ can arise for given n					
R_i^+	$n = 4$	r_4	$P(R^+ \leqslant R_i^+)$	R_i^+	$n = 10$	r_{10}	$P(R^+ \leqslant R_i^+)$
0	(0)	1	0·0625	0	(0)	1	0·000977
1	(1)	1	0·1250	1	(1)	1	0·001953
2	(2)	1	0·1875	2	(2)	1	0·002930
3	(1,2) (3)	2	0·3125	3	(1,2) (3)	2	0·004883
4	(1,3) (4)	2	0·4375	4	(1,3) (4)	2	0·006836
5	(1,4) (5)	2	0·5625	5	(1,4) (2,3) (5)	3	0·009766
6	(1,2,3) (2,4)	2	0·6875	6	(1,5) (2,4) (1,2,3) (6)	4	0·013672
7	(1,2,4) (3,4)	2	0·8125	.	.	.	.
8	(1,3,4)	1	0·8800	.	.	.	.
9	(2,3,4)	1	0·9375	.	.	.	.
R^+ max	(1,2,3,4)	1	1·0000	55	(1,2, . . .,10)	1	1·0000
$R^+_{max} = \frac{n(n+1)}{2}$		$16 = 2^n$				$1024 = 2^n$	

3.21 Probability calculations for the events implied by the Wilcoxon signed-rank test for $n=4$, full, $n=10$, partial.

$$\text{variance} = \frac{n(n+1)(2n+1)}{24} \tag{3.4.2}$$

with R^+ to give

$$Z = \frac{R^+ - \text{Mean}}{\sqrt{\text{variance}}} \tag{3.4.3}$$

which can be referred to table 1 in appendix 2.

Example

As Wilcoxon's test can be used for matched pairs the previous data, fig. 3.12, is used to illustrate the arithmetic procedure and format, fig. 3.22. There are only two positive differences, X_2, X_{21}, and each is given rank 1·5 because they are equally the two smallest differences and their sum is 3. Substituting in equation 3.4.3 we have

$$Z = \frac{3 - 115{\cdot}5}{\sqrt{827{\cdot}75}} = 3{\cdot}9102$$

and reject H_0 that $Md = M = 0$ and accept H_1 that the population of difference measures is symmetric with median not equal to zero, or H_1' that the population is asymmetric.

Wilcoxon's test can be used as a test for symmetry if the only assumption that is made is that the random sample is taken from a continuous distribution.

When the test is used under the assumption of continuity we can extend the procedure to make an inference concerning the value of the median difference at a given confidence level of its accuracy. Let us suppose we wish to know what range of values the median difference in the previous example, fig. 3.22, could have taken so that the sum of ranks, R^+, would be just greater than the $\alpha/2 = 0{\cdot}05$ significance level and so that the sum of ranks, R^-, would also be just greater than the $\alpha/2 = 0{\cdot}05$ significance level. By taking $\alpha/2$ we are keeping a rejection region of $\alpha = 0{\cdot}05$ on the two-tailed alternative and find what are the least values of the test statistic that lie in that rejection region.

One way of doing this is to take each of the possible average differences in turn and sum for R^+ or R^- until we come to that difference for which R^+ is just significant and then to that difference for which R^- is just significant at a specified, common, α level. This would give a confidence interval for the median difference. Forming these average differences means that each of the n numerals is compared with every other numeral and with itself. There are $\binom{n}{2}$ ways of selecting two numbers from n numerals and for n numerals there are n self-comparisons. Thus there are $\binom{n}{2} + n$ comparisons altogether. This simplifies to

$$\begin{aligned}\binom{n}{2} + n &= \frac{n!}{(n-2)!\,2!} + n = \frac{n(n-1)}{2} + \frac{2n}{2} \\ &= \frac{n(n+1)}{2}\end{aligned} \tag{3.4.4}$$

1959–61 X	1·1	0·7	5·6	11·1	9·7	6·6	2·9	2·4	1·6	7·7	7·3	5·6	4·8	3·6	2·7	2·4	2·9	2·5	2·3	1·8	1·4
1965–67 Y	0·6	0·8	3·1	5·6	6·6	5·4	2·2	1·8	1·4	6·7	4·6	3·8	2·9	3·0	2·4	2·2	2·4	2·2	1·7	1·5	1·5
$d_i = (Y) - (X)$	−0·5	+0·1	−2·5	−5·5	−3·1	−1·2	−0·7	−0·6	−0·2	−1·0	−2·7	−1·8	−1·8	−0·6	−0·3	−0·2	−0·5	−0·3	−0·6	−0·3	+0·1
$R(d_i)$	8·5	1·5	18	21	20	15	13	11	3·5	14	19	16·5	16·5	11	6	3·5	8·5	6	11	6	1·5
$R+ = 3 =$		1·5																			1·5

3.22 Calculating the Wilcoxon signed-rank test statistic for the data of fig. 3.12.

possible averages $(d_i + d_j)/2$ for all i, j including $i = j$. With $n = 21$, there are 231 comparisons. To make the counting task straightforward we can form these $n(n + 1)/2$ averages graphically in the following manner.

Arrange the signed differences in order along a line from smallest to largest. Bisect this range. From fig. 3.23 we see that the values range from +0·1 to −5·5 giving the median of the range as −2·7. This line is shown as *B*. The values 0·1 and −5·5 are joined to some point along *B* forming an isosceles triangle *ABC*. Each of the values of the differences are shown along *AC* and from each of these points a line is drawn parallel to *AB* and *CB*. Each of these intersections gives one

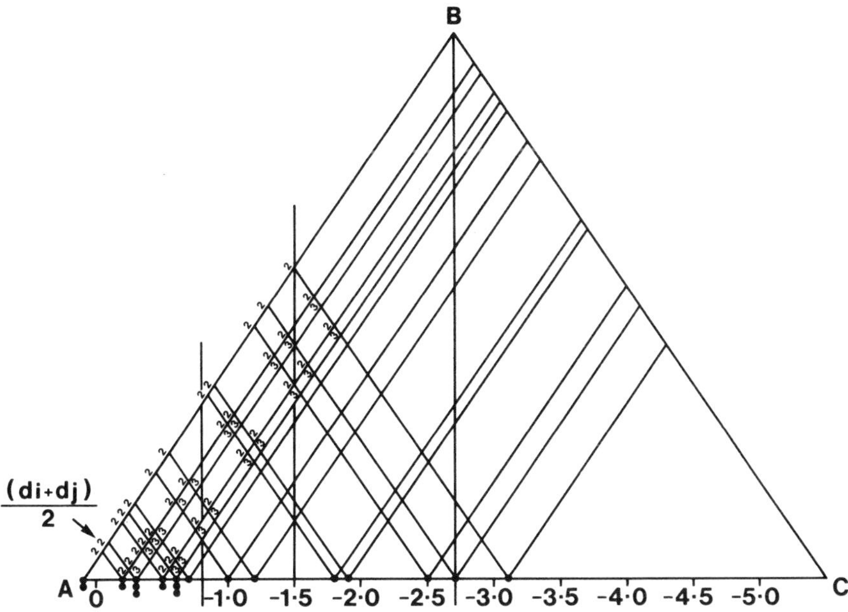

3.23 Confidence-interval for the median difference in the Wilcoxon signed-rank test.

of the averages $(d_i + d_j)/2$; thus the one shown on the diagram is for $d_i = +0{\cdot}1$, $d_j = -0{\cdot}2$, $[0{\cdot}1 + (-0{\cdot}2)]/2 = 0{\cdot}05$. Where there are tied difference values, each intersection involving those lines is weighted by the number of tied values as indicated in fig. 3.23. For $n = 21$, R^+ or $R^- = 67$ corresponds to an $\alpha = 0{\cdot}0479$, and so if we cumulate the number of points as we go along the interval +0·1 to −5·5 until we reach the value that gives a sum of 67 or less we shall have defined the interval for the value of the median difference at a nominal confidence level of α : 0·05. If the sum is 68 the exact $\alpha = 0{\cdot}0516$. We can see that the interval for the median difference in this case is $-0{\cdot}8 \leqslant Md \leqslant -1{\cdot}5$.

These two sign tests have potentially great usefulness in geography and their form is related particularly closely to the form of propositions contained in line maps. The assumptions of continuity at *M* in the ordinary sign test and over the whole population for Wilcoxon's test are often met. Where ties occur the sign test accommodates this by reducing n accordingly. By entailing similarity of conditions for the paired values rather than for all values, these tests are applicable to many situations in geomorphology where measurements on, say, discharge, sediment load, ground-water flow or precipitation are collected throughout the year for many years. In such cases comparisons of differences at the same site at the same period between years may well be more relevant than comparisons of differences between adjacent months or similar periods. There need be no violation of population form, because no assumption is made about its form in the sign test and *need not* be made in Wilcoxon's. This ability of using the ordinary sign test for dichotomies makes its application in land-use transect studies feasible, and its power should permit assertions of considerable strength to be made.

3.4.4 Kolmogorov-Smirnov test statistics: Smirnov's two-sample test

To conclude this part we shall discuss Kolmogorov-Smirnov statistics. This is a class of test statistic which compares two distribution functions in terms of the largest vertical difference between their graphs. Statistics that compare an empirical distribution function with an hypothesized distribution function such as the normal or the Poisson are called Kolmogorov-type statistics. Those that compare two empirical distribution functions are called Smirnov-type statistics. The distinction is often ignored and tests that use this largest-deviation approach are called Kolmogorov-Smirnov tests. They are important because

1. they use an approach not shared by the other statistics discussed;
2. they are appropriate to a number of combinations of proposition and measurement scale;
3. as goodness-of-fit statistics they provide an alternative to chi-square.

We begin by discussing the two-sample test as this is a natural extension to tests of the proposition of equality and identity.

3.4.4.1 Smirnov test

This test is an alternative to the other two-sample tests we have discussed so far for the hypothesis that the two distribution functions are identical. These tests are sensitive in particular to differences in location and this is reflected in the alternative to the null hypothesis of, for example, the Wilcoxon test or the Westenberg-Mood median test. The Smirnov test is sensitive to *any difference* in the two populations. Thus the general null hypothesis is one of identity, but the alternatives are not specific and the non-identity on rejection of H_0 may be of location, dispersion or skewness.

Suppose we have two random variables X and Y with $X_1, X_2, \ldots, X_m$ measurements on X and $Y_1, Y_2, \ldots, Y_n$ measurements on Y and that these measurements are made on a continuous scale. We wish to test against general alternatives whether these two samples come from populations of measurements that have identical distributions, or equivalently from the same population of measurements. Under the assumptions of such a null hypothesis we can assume that we have a sample of $m + n$ measurements from this common population of measurements and then have allotted randomly m events as the X sample and the remaining n

events as the Y sample. Such an allocation can be made in $\binom{m+n}{m}$ ways, each equally likely with probability $1\Big/\binom{m+n}{m}$.

Each of these arrangements can be ordered from smallest to largest value giving a result such as $X_1 < X_2 < Y_1 < \ldots < X_m < Y_n$. The two most extreme arrangements are $X_1 < X_2 < \ldots < X_m < Y_1 < Y_2 < \ldots < Y_n$ and $Y_1 < Y_2 < \ldots < Y_n < X_1 < X_2 < \ldots < X_m$. For each such arrangement we can construct a graph so that each time an X occurs we move one step along the X axis and each time a Y occurs we move one step up the Y axis. There is one path for each of the $\binom{m+n}{m}$ combinations. Each intersection of the m, n coordinates is involved in at least one such path, some of the intersections are involved in a great many of the paths and they all finish at $(X_m Y_n)$ so the total number of paths is found at that intersection. The diagonal from $(0, 0)$ to (m, n) is the most equitable, least extreme arrangement of the $m + n$ X, Y values under the assumption of continuous measurements. The vertical distance between this diagonal and any intersection of any vertical path can be measured. We want to know how many paths would have intersections at least as extreme as a particular path. The denominator of our probability fraction is $\binom{m+n}{m}$, the total number of paths, and the numerator is the number of times the intersections, that are at least as extreme as the most extreme intersection of the actual path, would be involved in a path.

A diagram helps to fix the problem for $m = 4$, $n = 3$ (fig. 3.24).

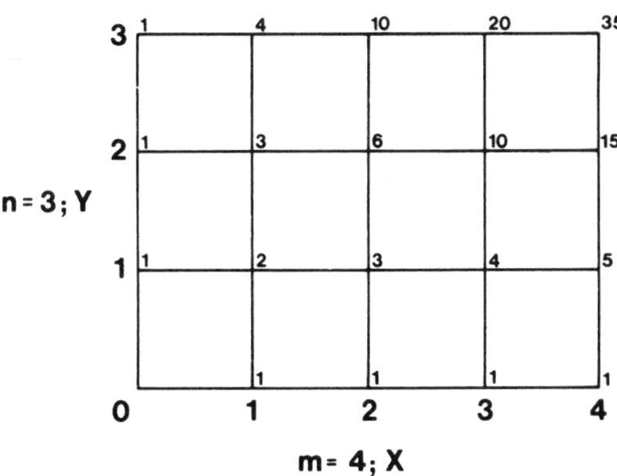

3.24 Possible paths in the Smirnov two-sample empirical distribution function statistic for m=4, n=3.

The m X values are shown on the horizontal axis and the n Y values on the vertical axis. The number at each coordinate shows the number of paths up to that intersection which include that intersection. Thus (1, 1) is involved in paths beginning

X, Y and Y, X and no others. Intersection (3, 2) involves the 3 smallest X values and the 2 smallest Y values; there are 10 such paths up to that point made up from the 6 up to intersection (2, 2) and the 4 up to intersection (3, 1). You will have noticed that the number for any intersection (X_i, Y_j) is the sum of the intersections $(X_{i-1}, Y_j) + (X_i, Y_{j-1})$. The following diagram shows the sequence of $(X_i Y_j)$ values for the first few intersections.

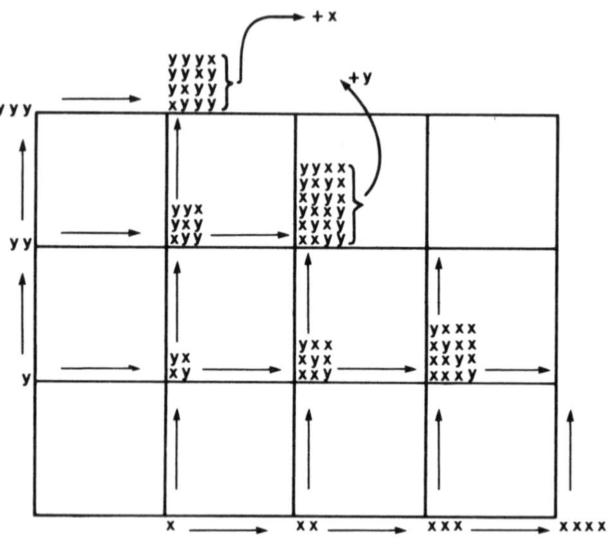

3.25 Forming the 35 (X, Y) arrangements for $m=4$, $n=3$ in fig. 3.24.

In this case of $m = 4$, $n = 3$ there are 35 combinations and these are listed in fig. 3.26.

$\binom{n+m}{m}$ arrangements	$D = (Y - X)$ twelfths						
$YYXXYXX$	4	8	5	2	6	3	0
$YXYXYXX$	4	1	5	2	6	3	0
$XYYXYXX$	−3	1	5	2	6	3	0
$YXXYYXX$	4	1	−2	2	6	3	0
$XYXYYXX$	−3	1	−2	2	6	3	0
$XXYYYXX$	−3	−6	−2	2	6	3	0
$YYYXXXX$	4	8	12	9	6	3	0
$YYXYXXX$	4	8	5	9	6	3	0
$YXYYXXX$	4	1	5	9	6	3	0
$XYYYXXX$	−3	1	5	9	6	3	0
$YYXXXYX$	4	8	5	2	−1	3	0
$YXYXXYX$	4	1	5	2	−1	3	0

X Y Y X X Y X	−3	1	5	2	−1	3	0
Y X X Y X Y X	4	1	−2	2	−1	3	0
X Y X Y X Y X	−3	1	−2	2	−1	3	0
X X Y Y X Y X	−3	−6	−2	2	−1	3	0
Y X X X Y Y X	4	1	−2	−5	−1	3	0
X Y X X Y Y X	−3	1	−2	−5	−1	3	0
X X Y X Y Y X	−3	−6	−2	−5	−1	3	0
X X X Y Y Y X	−3	−6	−9	−5	−1	3	0
Y Y X X X X Y	4	8	5	2	−1	−4	0
Y X Y X X X Y	4	1	5	2	−1	−4	0
X Y Y X X X Y	−3	1	5	2	−1	−4	0
Y X X Y X X Y	4	1	−2	2	−1	−4	0
X Y X Y X X Y	−3	1	−2	2	−1	−4	0
X X Y Y X X Y	−3	−6	−2	2	−1	−4	0
Y X X X Y X Y	4	−1	−2	−5	−1	−4	0
X Y X X Y X Y	−3	1	−2	−5	−1	−4	0
X X Y X Y X Y	−3	−6	−2	−5	−1	−4	0
X X X Y Y X Y	−3	−6	−9	−5	−1	−4	0
Y X X X X Y Y	4	1	−2	−5	−8	−4	0
X Y X X X Y Y	−3	1	−2	−5	−8	−4	0
X X Y X X Y Y	−3	−6	−2	−5	−8	−4	0
X X X Y X Y Y	−3	−6	−9	−5	−8	−4	0
X X X X Y Y Y	−3	−6	−9	−12	−8	−4	0

3.26 The 35 arrangements of fig. 3.24 showing the step-wise deviations.

Each sequence increases to the right, the inequality signs are omitted. Thus the sequence *X Y X X Y X Y* represents the order of values $X < Y < X < X < Y < X < Y$. This sequence is shown by a heavy line on fig. 3.27(a). We want to know how likely is a path as extreme or more extreme than that path.

Unfortunately we cannot count these paths directly from this graph (fig. 3.25), and we have to construct a separate graph (fig. 3.27) in which only intersections less extreme than the intersection in the observed path are available. From the three diagrams in fig. 3.27 it is easy to see what the statistic implies.

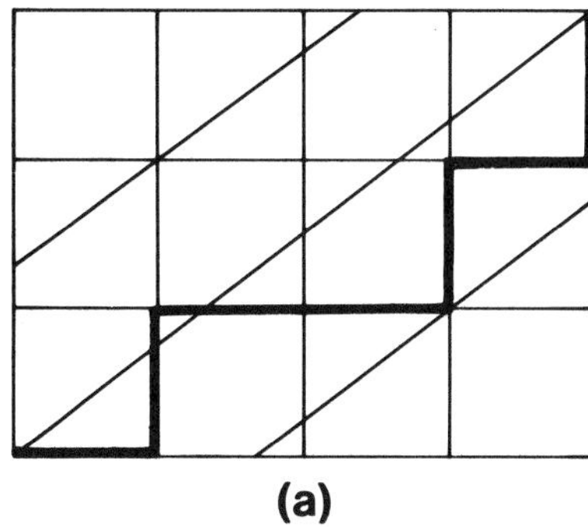

3.27(a) Fixing a rejection region in Smirnov's test.

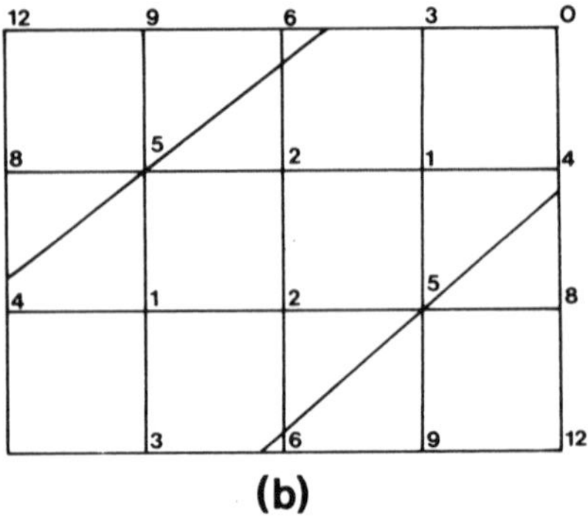

3.27(b) Rejection region in relation to step-wise deviations in Smirnov's test.

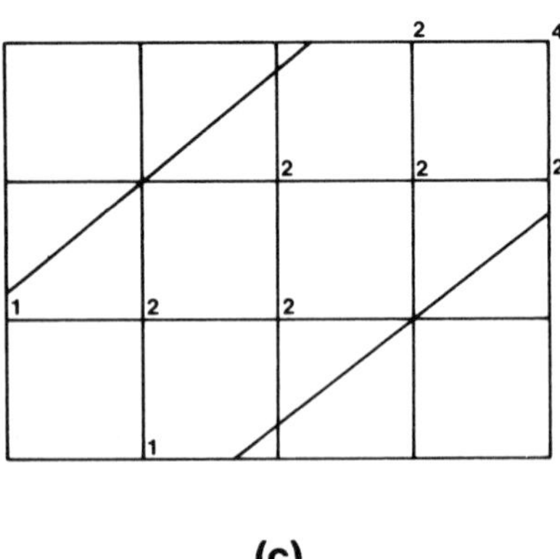

3.27(c) Possible arrangements and their paths within a given rejection region in Smirnov's test.

1. Diagram (a) shows the greatest difference to be at intersection (3, 1) where the proportions differ by 5/12. Lines parallel to the main diagonal at this difference show admissable paths.
2. Diagram (b) superimposes these bounds on a graph of deviations at each intersection. They pass through the 5/12 intersections and leave intersections with deviations <5, i.e. 1, 2, 3, 4.
3. Diagram (c) shows the number of possible paths which avoid intersections giving differences as great as 5. There are only 4 such paths
   - (i) *Y X X Y X Y X*
   - (ii) *X Y X Y X Y X*
   - (iii) *Y X X Y X X Y*
   - (iv) *X Y X Y X X Y*

These are seen readily from the table of differences (fig. 3.26), or inferred from building them up from diagram (c). Thus there are (35—4) = 31 paths that involve intersections yielding deviations at least this extreme. The probability fraction is (35—4)/35 = 31/35 = 0·885714 and tells us exactly how likely such a result is. No result is sufficiently unlikely to give a value in the critical region for $\alpha : 0{\cdot}05$ on a two-tailed test for $m = 4$, $n = 3$. This is seen readily from diagram (c), in which 2 out of the 35 arrangements give this most extreme value and 2/35 = 0·057143. And only a result this extreme falls in the one-tailed rejection region at $\alpha : 0{\cdot}05$.

Example
Using the figures for farm income increment from fig. 2.75, we have the following arrangement (fig. 3.28).

$Y_1 < Y_2 < Y_3 < Y_4 < Y_5 < Y_6 < Y_7 < X_1 < Y_8 < Y_9 < X_2 < X_3 < X_4 < Y_{10} < X_5 < X_6 < Y_{11} < X_m < Y_n$

$n = 12$
$m = 7$

3.28 Smirnov's test applied to the data of fig. 2.75.

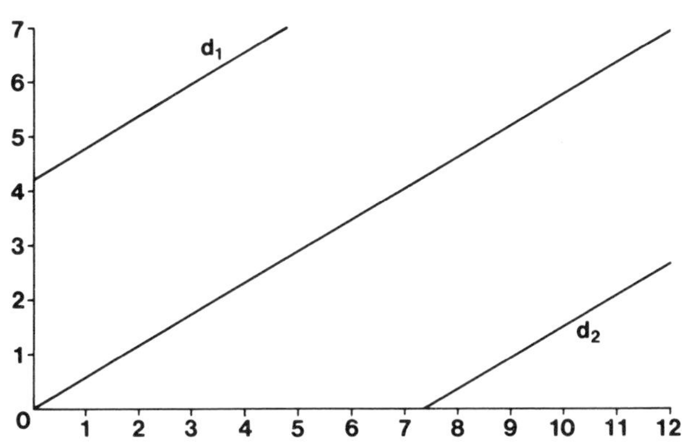

3.29 Total possible arrangements for the data in fig. 2.75.

For $X_m = 7$, $Y_n = 12$, $m + n = 19$, there are $\binom{19}{12} = 50388$ arrangements or paths, and we wish to determine how many of these arrangements are at least as extreme as this particular arrangement. To emphasize the similarity between this situation and the illustrative material used earlier this farm income data is shown diagrammatically in figs. 3.29 and 3.30. The first diagram shows the accumulation of available paths, with the main diagonal representing the most equitable path under continuity and the area of the graph containing paths at least as extreme as the actual path — these are the lines d_1 and d_2. The second diagram shows the actual path together with the total number of possible paths no more extreme than it. There are 47,856 such paths, consequently the probability of getting a path as extreme as or more extreme than the actual path is (50388—47856)/50388 =

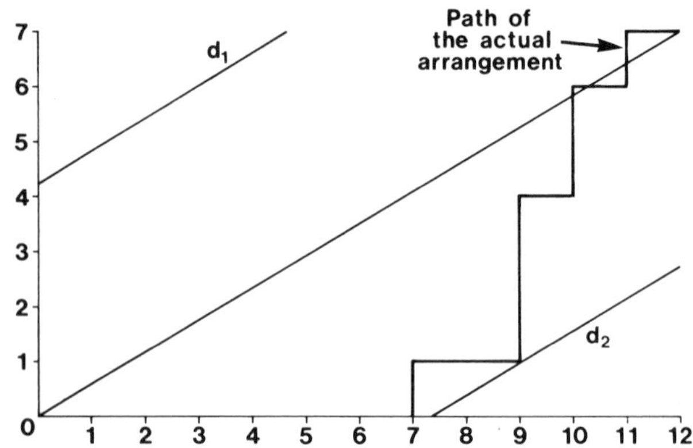

3.30 Total possible arrangements for a given rejection region for the data in fig. 2.75.

2532/50388 = 0·05025. On this basis the null hypothesis would be accepted at a nominal $\alpha : 0{\cdot}05$, against the two-sided alternative. Against the one-sided alternative that the X population is larger than the Y population the number of paths available at $\alpha : 0{\cdot}05$ is 49522, thus the exact probability of a sequence as extreme or more extreme than the sequence obtained is (50388−49522)/50388 = 0·01719, and H_1 is accepted.

Tables of exact probabilities are available for two samples, n and m where $n = m \leqslant 40$ and for two samples of unequal size such that $n < m \leqslant 20$. See table 15 in appendix 2 where an approximation formula for larger sample sizes is also given. To use these tables the sample values can be set out as in fig. 3.31.

The $n + m$ values are arranged in ascending order; the m X values are put in one column in their appropriate rank, and the n Y values are put in the second column in their correct position in the combined sequence. The number of X values up to a particular rank is expressed as a fraction of the total number, m, of the X values. Thus if the smallest value is a Y then the first X fraction is $^0/m$ ths of the X values. This continues through the ranks until the first X value is reached giving $^1/m$. The same applies to the Y values, with n as the denominator. The greatest difference is found simply by subtracting each of n+m fractions. The easiest way is to use nm as the denominator of the differences and as the denominator of the X and Y values and then finally to simplify the fraction for use with exact tables. In some tables the fraction is calculated from an asymptotic distribution and a decimal fraction is used. In the present example the greatest difference is at $\frac{1}{7} - \frac{9}{12}$ or $\frac{12}{84} - \frac{63}{84} = \frac{51}{84} = 0{\cdot}607$, for which we have established the exact probability.

3.5 Goodness-of-fit statistics

In section 1 we made a distinction between classical parametric statistical tests and distribution-free tests on the basis that the classical tests are derived under the assumption that the population of measurements has a normal distribution,

$S_1(x)$	$-\ S_2(Y)$	$D_{n,m} = \dfrac{S_1(x) - S_2(y)}{n.m}$
0	$\frac{1}{12}$	$-\frac{7}{84}$
0	$\frac{2}{12}$	$-\frac{14}{84}$
0	$\frac{3}{12}$	$-\frac{21}{84}$
0	$\frac{4}{12}$	$-\frac{28}{84}$
0	$\frac{5}{12}$	$-\frac{35}{84}$
0	$\frac{6}{12}$	$-\frac{42}{84}$
0	$\frac{7}{12}$	$-\frac{49}{84}$
$\frac{1}{7}$	$\frac{7}{12}$	$-\frac{37}{84}$
$\frac{1}{7}$	$\frac{8}{12}$	$-\frac{44}{84}$
$\frac{1}{7}$	$\frac{9}{12}$	$-\frac{51}{84} \rightarrow 0 \cdot 607$
$\frac{2}{7}$	$\frac{9}{12}$	$-\frac{39}{84}$
$\frac{3}{7}$	$\frac{9}{12}$	$-\frac{27}{84}$
$\frac{4}{7}$	$\frac{9}{12}$	$-\frac{15}{84}$
$\frac{4}{7}$	$\frac{10}{12}$	$-\frac{12}{84}$
$\frac{5}{7}$	$\frac{10}{12}$	$-\frac{10}{84}$
$\frac{6}{7}$	$\frac{10}{12}$	$+\frac{2}{84}$
$\frac{6}{7}$	$\frac{11}{12}$	$-\frac{5}{84}$
$\frac{7}{7}$	$\frac{11}{12}$	$+\frac{7}{84}$
$\frac{7}{7}$	$\frac{12}{12}$	$\frac{0}{84}$

3.31 Standard tabulation of values for calculating Smirnov's test statistic.

in contrast to distribution-free test statistics which make no assumptions about the form of the population of measurements. When it is unreasonable to assume a normal distribution, inferences based on the classical test statistics are inexact and distribution-free tests are preferable. It would be contrary to the general spirit of statistical inference if the judgement as to the form of the population of measurements were left quite arbitrary. Tests which are designed to check whether an observed distribution is compatible with the set of values expected if the population were a normal distribution or of any other specified form are known as *goodness-of-fit* tests. The two best-known tests are the chi-square test and the Kolmogorov test. Both tests are appropriate to a null hypothesis that the observed distribution comes from a distribution function which is compatible with a specified distribution function. The acceptability of the null hypothesis is judged on the similarity of the observed values and the values expected for the specified distribution. The basic difference between the two tests is that the chi-square test compares the differences between the two frequency distributions and the Kolmogorov test compares the differences between the two cumulative frequency distributions. In both cases the alternative to the null hypothesis is that the form of the population of measurements from which the sample is taken is not of the specified form. It is customary for goodness-of-fit tests to be used in the expectation that H_0 will be accepted.

3.5.1 Kolmogorov's test, Lilliefor's modification

The idea of the test is very similar to that of the two-sample Smirnov test considered already; the principal difference is that in the Kolmogorov test one of the distributions is an infinite population of measurements. Although two distributions are being compared as one of these is a population, the Kolmogorov test is known as a one-sample test. The sample is a random sample, $X_1, X_2, \ldots, X_n$ taken from a continuous population of measurements. This population has an unknown *cdf*, $F_X(X)$, and the observed values are a random sample from this unknown population. These observed values can be put as an empirical cumulative distribution, $S_n(X)$, and this empirical distribution is a consistent estimator of the unknown distribution, $F_X(X)$. As the sample size, n, increases, $S_n(X)$ provides an increasingly close approximation to $F_X(X)$. If we suppose in our null hypothesis that the population of measurements $F_X(X)$ has some specified form $F_O(X)$, say a normal distribution, then the discrepancy between $S_n(X)$ and $F_O(X)$ can be used to judge the acceptability of H_0, when $F_O(X)$ is the specified distribution function. In fact the test statistic is the largest difference between the empirical *cdf*, $S_n(X)$, and the hypothesized *cdf*, $F_O(X)$.

The task of deriving the probability distribution of this difference is substantial and the interested reader should consult a more advanced text (Gibbons 1971). Some indication of the essence of the test is gained by analogy with the Smirnov test in which two empirical distribution functions are compared. The *cdf* of the hypothesized population can be envisaged as a continuous cumulative frequency graph, and the *cdf* of the observed values is a discrete cumulative frequency graph whose step-size is a multiple of $1/n$ and so is related to sample size (fig. 3.33a). As the nX values are a random sample from the population, $F_X(X)$, any n values are equally likely to have arisen. For each selection the largest differences have to be calculated and their distribution is the probability distribution of the test statistic. Rejection of H_0 implies that the obtained largest difference between $S_n(X)$ and $F_O(X)$ is sufficiently great to suppose that $S_n(X)$ is a sample from a population with a different *cdf*, in other words $F_X(X) \neq F_O(X)$.

Example 1

Suppose we have a distribution of objects as in fig. 3.32 and we wish to determine whether their incidence is uniform with distance from the centre. These objects occur along the unit radius as it describes the circle to give the pattern shown in fig. 3.33(b). The null hypothesis is that the distribution function, $F_X(X)$, of the distance measurements is the *uniform distribution*, $F_U(X)$. That is we are supposing that the distance of each object from the centre is taken from a population of measurements that is indistinguishable from the uniform distribution. If H_0 is rejected we accept the alternative hypothesis that for at least one object the distance measurement is not from the uniform distribution. Thus we put:

$$\begin{aligned} H_0 : F_X(X) &= F_U(X) \text{ for all } X \\ H_1 : F_X(X) &\neq F_U(X) \text{ for at least one } X \end{aligned}$$

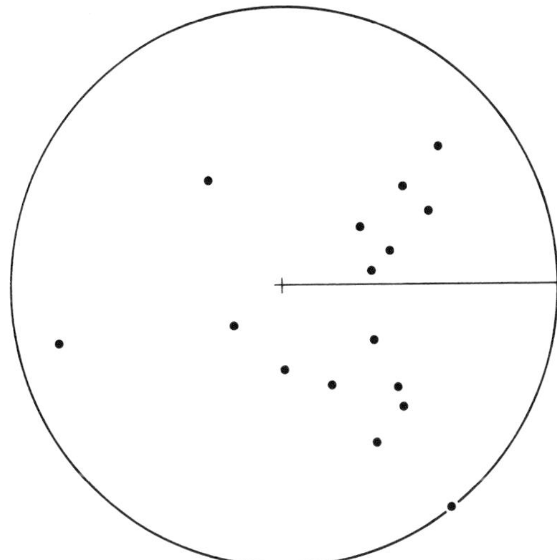

3.32 Location of homes of 16 visitors to a National Park.

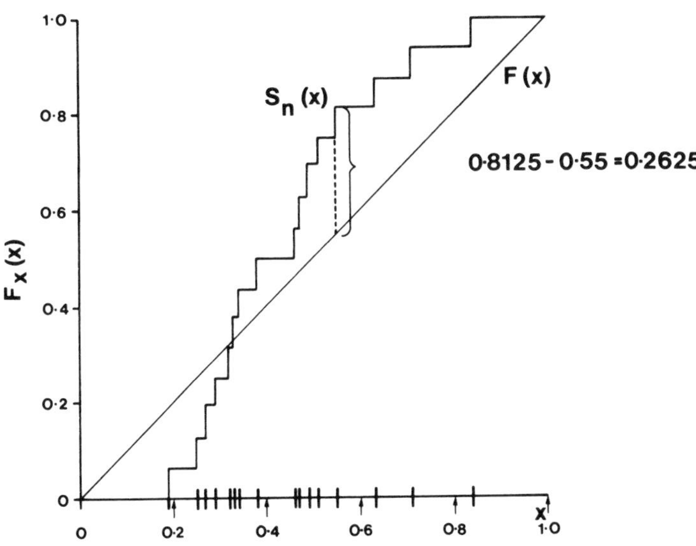

3.33(a) Graphical illustration of the Kolomogorov one-sample test applied to data of fig. 3.32.

3.33(b) Arrangement of symbols in fig. 3.32 along the unit radius.

The hypothesized distribution, $F_U(X)$, is shown in fig. **3.33(a)** and the empirical distribution, $S_n(X)$, which represents the unknown distribution function of the X, is superimposed as a stepped cumulative frequency. In this case n=16 and, under H_0, it is expected that the sixteen objects would occur uniformly along the unit radius at $1/n$=1/16 increments. The greatest discrepancy occurs at the thirteenth ranked object when $S_n(X) = 0{\cdot}55$ and the expected distance under H_0 is 0·8125, giving a greatest difference regardless of sign of 0·2625. The details are tabulated as fig. 3.34.

$S_n(X)$	$F_U(X)$
0·19	0·0625
0·25	0·1250
0·27	0·1875
0·29	0·2500
0·32	0·3125
0·33	0·3750
0·34	0·4375
0·38	0·5000
0·46	0·5625
0·47	0·6250
0·49	0·6875
0·51	0·7500
0·55	0·8125
0·63	0·8750
0·71	0·9375
0·84	1·0000

$S_n(X)$ represents $F_X(X)$, the unknown distribution function of the X, and is compared with the values expected if $F_X(X) = F_U(X)$.

3.34 Standard format for calculating the value of the one-sample Kolmogorov test applied to the data of fig. 3.33.

For n=16 a difference of 0·258 defines α : 0·80 and 0·295 defines α : 0·90, as is seen by entering table 16 in appendix 2 with n=16. By linear interpolation a value of 0·2625 defines an $\hat{\alpha}$: 0·812 from

$$\hat{\alpha} = 0{\cdot}80 + (0{\cdot}9 - 0{\cdot}8)\,\frac{(0{\cdot}2625 - 0{\cdot}2580)}{(0{\cdot}295 - 0{\cdot}258)}$$

$$= 0{\cdot}8 + 0{\cdot}1\,\frac{0{\cdot}0045}{0{\cdot}037}$$

$$= 0{\cdot}8 + 0{\cdot}012$$

Consequently we accept the null hypothesis and conclude that the actual distribution function of distances is insufficiently different from the uniform distribution to be discerned. We do not assert that it is the uniform distribution. Of course there is no need to graph the functions; it is customary to tabulate the values as in fig. 3.34.

This test can be used with other hypothesized distribution functions if they are specified completely. To illustrate the application of the Kolmogorov test with a hypothesized normal distribution we shall use the percentage of votes polled by the Conservative and Labour parties in the British general election of 1966 for various random samples of constituencies.

Example 2

A random sample of constituency results is taken sequentially for sample sizes n_1=5, n_2=10, n_3=15, and the random variable, X, is the percentage of total votes polled for the Labour party. The hypotheses are:

H_0 : The random samples have a normal distribution with *specified* mean and variance.

H_1 : The samples have a distribution function which is non-normal.

The mean and variance specified is in fact the population mean and variance for the Labour party in 1966, namely with $\mu = 49{\cdot}3$, $\sigma = 16{\cdot}152$. For ease of reference the three samples, n_1, n_2, n_3, are distinguished in fig. 3.35, although the constituencies in one sample are included in the larger sample. Also the results are arranged in rank order. As we are hypothesizing that the unknown population

X_i	Z_i	$F_N(X)=P(Z)$	$S_n(X)$	$D=\lvert F_N(X)-S_n(X)\rvert$
27·9	−1·3249	0·0926	0·2	0·1074
37·6	−0·7244	0·2358	0·4	0·1642
44·4	−0·3034	0·3821	0·6	[0·2179]
55·2	+0·3653	0·6425	0·8	0·1575
65·9	+1·0277	0·8480	1·0	0·1520

(a) Sample size, $n = 5$. $\alpha : 0{\cdot}05, D \geqslant 0{\cdot}563$

X_i	Z_i	$F_N(X)=P(Z)$	$S_n(X)$	$D=\lvert F_N(X)-S_n(X)\rvert$
27·9	−1·3249	0·0926	0·1	0·0074
35·4	−0·8606	0·1949	0·2	0·0051
37·6	−0·7244	0·2358	0·3	0·0642
44·4	−0·3034	0·3821	0·4	0·0179
44·4	−0·3034	0·3821	0·5	0·0179
45·1	−0·2600	0·3974	0·6	[0·2026]
50·5	+0·0743	0·5295	0·7	0·1705
55·2	+0·3653	0·6425	0·8	0·1575
59·5	+0·6191	0·7321	0·9	0·1679
65·9	+1·0277	0·8480	1·0	0·1520

(b) Sample size, $n = 10$. $\alpha : 0{\cdot}05, D \geqslant 0{\cdot}409$

X_i	Z_i	$F_N(X)=P(Z)$	$S_n(X)$	$D=\lvert F_N(X)-S_n(X)\rvert$
27·9	−1·3249	0·0926	0·0667	0·0259
35·4	−0·8606	0·1949	0·1333	0·0616
36·9	−0·7677	0·2212	0·2000	0·0212
37·6	−0·7244	0·2358	0·2667	0·0309
44·4	−0·3034	0·3821	0·3333	0·0488
44·4	−0·3034	0·3821	0·4000	0·0179
45·1	−0·2600	0·3974	0·4667	0·0693
50·5	+0·0743	0·5295	0·5333	0·0038
52·0	+0·1672	0·5664	0·6000	0·0336
55·2	+0·3653	0·6425	0·6667	0·0242
59·2	+0·6129	0·7301	0·7333	0·0032
59·5	+0·6191	0·7321	0·8000	0·0679
65·9	+1·0277	0·8480	0·8667	0·0187
73·3	+1·4859	0·9313	0·9333	0·0020
74·9	+1·5849	0·9435	1·0000	0·0565

(c) Sample size, $n = 15$. $\alpha : 0{\cdot}05, D \geqslant 0{\cdot}338$

The z_i are calculated from the population $\mu = 49{\cdot}3$, $\sigma = 16{\cdot}152$.

3.35 Random samples, taken sequentially, of British Parliamentary constituency results for the Labour Party, 1966, at sample sizes n=5, n=10, n=15.

is a normal distribution with given μ, σ, the obtained percentage values must be converted to standard deviates, z_i, by

$$z_i = (X_i - \mu)/\tau$$

These are shown in column 2. The cumulative probability of these Z values is found from table 1 in appendix 2, and recorded in column 3. For example, $x_1 = 27{\cdot}9$ and gives $z_1 = (27{\cdot}9 - 49{\cdot}3)/16{\cdot}152 = -1{\cdot}3249$; the sign means it lies in the lower half of the normal distribution. The value for $+1{\cdot}32$ is 0·9066 and for $+1{\cdot}33$ is 0·9082, thus the value for $+1{\cdot}325$ is taken to be half this interval, that is 0·9074. Consequently the value for $-1{\cdot}325$ is $1 - 0{\cdot}9074 = 0{\cdot}0926$, which means that 9·26 per cent of the area under the normal distribution function lies below this value and 90·74 per cent lies above it. For x_4 in n_1 we have $x = 55{\cdot}2$, $z = +0{\cdot}3653$, and the cumulative probability is found directly from table 1 in appendix 2 as 0·6425; that is 64·25 per cent of the area lies below this Z value and 35·75 per cent lies above it. In column 4, the empirical *cdf* is given as the sum of successive increments. Each step is $1/n$, where n is the sample size. Thus for n=5, the steps are 0·2, for n_2 they are 0·1, and for n_3 they are 0·0667.

The test statistic is the largest difference, regardless of sign, between values in columns 3 and 4, that is between $F_N(X)$ and $S_n(X)$. These differences are shown in column 5. As anticipated, the differences diminish as n increases; the empirical *cdf* becomes a better approximation to the unknown population $F_X(X)$ which is hypothesized to be indistinguishable from a normal distribution, $F_N(X)$ with $\mu = 49{\cdot}3$, $\sigma = 16{\cdot}152$. In all three samples H_0 is accepted because the test statistic

D is considerably less than the value of D under the null hypothesis for a rejection region at $\alpha : 0{\cdot}05$ (table 1 in appendix 2). We infer that the distribution of percentage votes polled by the Labour party is not noticeably different from a normal distribution with the parameters specified.

Example 3

For the same sequential sample of constituencies a second random variable, Y, was recorded as the percentage vote gained by the Conservative party. Keeping the same hypotheses and using the same specified parameters, $\mu = 49{\cdot}3$, $\sigma = 16{\cdot}152$, we get the values shown in fig. 3.36.

Y_i	Z_i	$F_N(Y)=P(Z)$	$S_n(Y)$	$D=\|F_N(Y)-S_n(Y)\|$
34·1	−0·9411	0·1733	0·2	0·0267
37·0	−0·7615	0·2230	0·4	0·1770
43·4	−0·3653	0·3575	0·6	0·2425
44·6	−0·2910	0·3855	0·8	**0·4145**
56·5	+0·4458	0·6721	1·0	0·3279
(a) Sample size, $n = 5$. $\alpha : 0{\cdot}05, D \geqslant 0{\cdot}563$				
34·1	−0·9411	0·1733	0·1	0·0733
35·1	−0·8791	0·1897	0·2	0·0103
37·0	−0·7615	0·2230	0·3	0·0770
40·5	−0·5448	0·2929	0·4	0·1071
43·4	−0·3653	0·3575	0·5	0·1425
43·4	−0·3653	0·3575	0·6	0·2425
43·5	−0·3591	0·3598	0·7	0·3402
44·6	−0·2910	0·3855	0·8	**0·4145**
48·2	−0·0681	0·5271	0·9	0·3729
56·5	+0·4458	0·6721	1·0	0·3279
(b) Sample size, $n = 10$, $\alpha : 0{\cdot}05, D \geqslant 0{\cdot}409$				

The z_i are calculated from the Labour $\mu = 49{\cdot}3$, $\sigma = 16{\cdot}152$. Using Conservative $\mu = 40{\cdot}865$, $\sigma = 11{\cdot}589$ the results for sample (a), $n = 5$ are as below.

Y_i	Z_i	$F_N(Y)$	$S_N(Y)$	D
34·1	−0·5837	0·2797	0·2	0·0797
37·0	−0·3335	0·3696	0·4	0·0304
43·4	+0·2187	0·5868	0·6	0·0132
44·6	+0·3223	0·6262	0·8	**0·1738**
56·5	+1·3491	0·9113	1·0	0·0887

3.36 Random samples, taken sequentially, of British Parliamentary constituency results for the Conservative party, 1966, at sample sizes n=5, n=10.

For n_1, n_2 the greatest difference between $S_n(Y)$ and $F_N(Y)$ is $0{\cdot}8 - 0{\cdot}3855 = 0{\cdot}4145$ which is significant at $\alpha : 0{\cdot}05$ for $n = 10$. Thus we reject H_0 that $F_Y(Y)$ is indistinguishable from a normal distribution with $\mu = 49{\cdot}3$, $\sigma = 16{\cdot}152$. However we accept H_0 that $F_Y(Y)$ is indistinguishable from a distribution function $N(40{\cdot}865, 11{\cdot}589)$ for $n = 5$, $n = 10$.

When $F_X(X)$ is specified fully then the Kolmogorov test is appropriate in this form; when the parameters are estimated from the observed values the Kolmogorov test needs modifying to test H_0 that $F_X(X)$ is a standardized normal distribution function such that $N(0, 1)$ is $F_N(X)$ and *Lilliefor's* test is appropriate. The reader should consult Conover (1971), Gibbons (1971) and Lilliefors (1967). However, in such situations many readers may prefer to use the chi-square test.

3.5.2 Chi-square test

The intention is the same as in the previous case, but the procedure is different. The similarity lies in the null hypothesis. A sample of size n of observed measurements is taken from a population of measurements with a particular but unknown distribution function $F(X)$. The null hypothesis is that this distribution function, $F(X)$, has a specified form, $F_O(X)$, or that it is indistinguishable from it. The alternative, if H_0 is rejected, is that the population, $F(X)$, is not of the form specified, $F_O(X)$. The difference lies in the fact that it is the frequency distribution rather than the cumulative frequency distribution which is used in calculating the test statistic. In order to get a frequency distribution, the values must be allocated to categories and the number in each class counted. Clearly this procedure can be used with all forms of measurement, but for the more restricted scales any grouping leads to some loss of information.

If H_0 is correct, then it is possible to determine the probability that an observed value will fall in any specified category and these probabilities, multiplied by n, give the expected frequency of observations in each class under the null hypothesis. A comparison of the observed and expected frequencies for each class is the basis of the test statistic, X^2, which is given as

$$X^2 = \sum_{i=1}^{n} \frac{(O_i - E_i)^2}{E_i} \tag{3.5.1}$$

and X^2 is distributed approximately as a chi-square random variable.

It is plain from our earlier discussion of chi-square random variables that if H_0 is true then the observed frequency equals the expected frequency. This follows because $E_i = np_i = O_i$ if $p_i = p_{oi}$ when p_{oi} is the probability under H_0 for $F_0(X)$ and p_i is the probability in the actual distribution $F(X)$ that a value will lie in the ith class. Again we have noted earlier that X^2 has a complicated distribution which is approximated by χ^2 under conditions stated in article 2.3.2.

We make the same distinction that we made for the Kolmogorov test between testing for a completely specified distribution and testing for a distribution in which some parameters are estimated from the data.

Example 1

Using Labour party percentage votes in the 1966 general election for a random sample of size $n = 100$ we group the values into ten classes each of the same-sized interval. The ten classes and their common boundary-values are shown in columns 1 and 2 in fig. 3.37.

1	2	3	4	5	6	7	8
Class	*Boundary value*	*Observed frequency*	z_j	p_j	p_i	np_i	X^2
i	j				$p_i = p_j - p_{j-1}$		
1		0			0·00649	0·649	0·649
	10		−2·4331	0·00649			
2		2			0·02831	2·831	0·2439
	20		−1·8140	0·03480			
3		8			0·08130	8·130	0·0021
	30		−1·1949	0·1161			
4		13			0·16620	16·620	0·7885
	40		−0·5758	0·2823			
5		23			0·23490	23·490	0·0102
	50		+0·0433	0·5172			
6		26			0·22900	22·900	0·4197
	60		+0·6625	0·7462			
7		16			0·15380	15·380	0·0250
	70		+1·2816	0·9000			
8		9			0·0713	7·13	0·4904
	80		+1·9007	0·9713			
9		3			0·02283	2·283	0·2252
	90		+2·5198	0·99413			
10		0			0·00587	0·587	0·587
		$n = 100$				100·000	3·4410

Classes $= k = 10$

Degrees of freedom $= k - 1 = 10 - 1 = 9$

at α : 0·05, $\chi^2_9 = 16{\cdot}92$

3.37 Chi-square as a goodness-of-fit test statistic under H_0 of a normal population of measurements for grouped data and population parameters.

The third column shows the observed frequencies. In column 4 the boundary-values of column 2 are expressed as standard deviates with respect to $\mu = 49{\cdot}3$, $\sigma = 16{\cdot}152$, thus the boundary-value of classes $i = 1$ and 2 is $j = 10$, for which $z_j = (10 - 49{\cdot}3)/16{\cdot}152 = -2{\cdot}4331$. These are expressed as probabilities in column 5; that is the area of the normal probability function below that z value. For example $z = +2{\cdot}4331$ is given the value 0·99251 from table 1 in appendix 2; consequently $z = -2{\cdot}4331$ has probability $1 - 0{\cdot}99251 = 0{\cdot}00649$. In column 6 the probabilities for each class interval are given and they represent the differences between the values up to successive ordinates, z_j. Thus the probability of a value occurring in the interval 40—50 per cent is the probability up to 50, 0·5172, minus the probability up to 40, 0·2823, giving $0{\cdot}5172 - 0{\cdot}2823 = 0{\cdot}2349$.

In column 7, these class-interval-probabilities are multiplied by n to give the frequencies expected in a sample of size n. In column 8 the contributions to X^2 are recorded by substituting the values in columns 3 and 7 in to equation (3.5.1). The total X^2 is 3·4410 and is referred to χ^2 with k—1 degrees of freedom, k of course being the number of classes. The null hypothesis is accepted as $X^2 =$ 3·4410 is less than $\chi^2 = 16{\cdot}92$ with 9 degrees of freedom at α : 0·05. We accept that the sample is drawn from a population of measurements whose form is not distinguishable from that of a normal distribution with the specified parameters.

An analogous procedure applies to other distribution functions, discrete and continuous. It is common to find that the parameters of the supposed, hypothesized distribution function are unknown and have to be estimated from the observed values. The use of these estimates implies that the probabilities associated with the classes are no longer constant but are themselves random variables. It is then quite remarkable that the χ^2 distribution still provides an approximation to the test statistic X^2 with reduced degrees of freedom. The degrees of freedom are reduced by one for each parameter, r, estimated so long as $r < k$.

If we had used $\bar{x}$, Var(X) in the previous example we should have lost 2 more degrees of freedom giving 10—1—2 = 7. Of course the z_j and p_i would also be different and you may care to treat this calculation as an exercise.

As a final illustration of a goodness-of-fit statistic we shall apply chi-square to the data given in fig. 2.15.

Example 2

Does the random variable, $X =$ the number of student homes in specified unit areas in London have a Poisson distribution? We estimated the parameter, λ, from the observations and used this to calculate the probability of $X = 0, 1, 2, 3, 4$. We used these probabilities to calculate the frequencies expected for each X. All these values are tabulated in fig. 2.15 (p.77). In the final column the contribution to X^2 is given for each x, and the total X^2 is 0·576. There are $k = 5$ classes, we lose 1 degree of freedom as usual in such a table, but in addition we lose a further degree of freedom because we estimated one parameter, λ. Consequently we refer 0·576 to χ^2 with 3 degrees of freedom at α : 0·05. We accept the hypothesis that the population distribution $F(X)$ is indistinguishable from $F_O(X)$ where $F_O(X)$ is the Poisson distribution with λ as estimated.

3.6 Propositions and angular measurements

At the beginning of the book we recognized that direction was an important characteristic of location in the plane or on the sphere, but we remarked that direction was neglected in its use as a measure of location. The assertions made about the locational relations of objects in terms of distance can also be made in terms of direction. The propositions and the hypotheses are the same, but the procedures used to develop test statistics need to be modified because of the differences in geometry between the unbounded real line and the closed circle. The importance of considerations of periodicity have been recognized for many years, and under the stimulus of problems in geology and biology an increasing number of statistical tests applicable to hypotheses on the circle have become available and can be used by the geographer to great advantage. The most comprehensive statement of these procedures is combined with a rigorous theoretical justification of them in the recent work by Mardia (1972). To conclude this introductory text we shall consider the real line to be wrapped to form the unit circle and illustrate two distribution-free tests for a proposition of goodness-of-fit and a proposition of equality in two distributions of angular measurements. These seem to arise regularly in undergraduate work and they relate easily to the tests and ideas developed for the line. By joining the ends of the real line we preclude the possibility of a unique ordering of measurements, because we can vary the starting point or zero direction. The distribution-free test statistics, like their linear counterparts, use order properties as the basis for their probability distribution; consequently we require our test statistics to be invariant under rotation, so that the choice of zero direction is unimportant.

3.6.1 Kuiper's test, a Kolmogorov analogue

Consider the situation depicted in fig. 3.32 (p.221) for which we found that the location measures of the objects were indistinguishable from a uniform distribution in terms of distance from the origin. Suppose we wish to test whether the objects' directions are from a uniform distribution.

We have three distributions to consider as before. First the unknown distribution function of the population of angular measurements, $F_X(X)$, secondly the empirical distribution function, $S_n(X)$, which provides a consistent estimator of $F_X(X)$, and thirdly the hypothesized distribution function, $F_O(X)$. We shall suppose $F_O(X)$ to have some particular form, such as the uniform distribution, $F_U(X)$. The null hypothesis is again that $F_X(X) = F_O(X)$ and the alternative if H_0 is rejected is that $F_X(X) \neq F_O(X)$.

The test we use, Kuiper's test, is invariant under rotation of the circle and it does not matter where we put our zero direction for measuring the angles. This invariance property is not possessed by the unmodified Kolmogorov test statistic applied to the circle.

θ_i	$S_n(X) = \theta_i/360$	$F_U(X)$	$S_n(X) - F_U(X)$
31	0·086	0·0625	0·0235
42	0·117	0·1250	−0·0080
45	0·125	0·1875	−0·0625
53	0·147	0·2500	−0·1030
59	0·164	0·3125	−0·1485
64	0·178	0·3750	−0·1970
89	0·247	0·4375	−0·1905
142	0·394	0·5000	−0·1060
167	0·464	0·5625	−0·0985
235	0·653	0·6250	−0·0280
319	0·886	0·6875	0·1985
321	0·892	0·7500	0·1420
323	0·897	0·8125	0·0845
334	0·928	0·8750	0·0530
342	0·950	0·9375	0·0125
350	0·972	1·0000	0·0280

$$
\begin{aligned}
D &= D^{+} - D^{-} + 1/n \\
D &= 0{\cdot}1985 - (-0{\cdot}1970) + 0{\cdot}0625 \\
&= 0{\cdot}3955 + 0{\cdot}0625 \\
&= 0{\cdot}4580 \\
D* &= \sqrt{n}\, D \\
&= \sqrt{16}\ 0{\cdot}4580 \\
&= \underline{1{\cdot}832}
\end{aligned}
$$

$\alpha : 0{\cdot}05\ D^{*}_{16} \geqslant 1{\cdot}66$

reject $H_0\ F(X) = F_U(X)$

accept $H_1\ F(X) \neq F_U(X)$

3.38 Calculating Kuiper's test of goodness-of-fit for angular measurements using data from fig. 3.33.

We put 0° as the X axis. Each object's position is measured, clockwise in the example, from the origin as degrees θ_i. The null hypothesis is that $F_X(X) = F_U(X)$ and again we have to provide two distribution functions. The first is the empirical distribution function of the θ_i, which are converted to the unit interval by dividing θ_i by 360° to give $S_n(X)$. The expected cumulative distribution is given by cumulating the reciprocal of the sample size, that is $1/n$. These details are given as the first three columns of fig. 3.38 whose similarity to fig. 3.34 is plain. The final column shows the largest positive difference, D^{+}, and the largest negative difference, D^{-}, between the two distributions. The test statistic, D^{*}, is given as

$$D^{*} = \sqrt{n}\, D \qquad (3.6.1)$$

with

$$D = D^{+} - (D^{-}) + 1/n \qquad (3.6.2)$$

which for this example is

$$
\begin{aligned}
D &= 0{\cdot}1985 - (-0{\cdot}1970) + 0{\cdot}0625 \\
&= 0{\cdot}4580
\end{aligned}
$$

so that

$$
\begin{aligned}
D^{*} &= \sqrt{16}\ (0{\cdot}4580) \\
&= 4.\ 0{\cdot}4580 \\
&= 1{\cdot}832
\end{aligned}
$$

From table 17 in appendix 2 we see that at $\alpha : 0{\cdot}05$, $D* = 1{\cdot}66$ and values greater than or equal to this value lie in the rejection region. Thus we reject H_0 and infer that the population distribution, $F_X(X)$, is distinguishably different from the uniform distribution, $F_U(X)$, of angular measurements. The details in fig. 3.38 are shown graphically in fig. 3.39 and this emphasizes the similarity of the Kuiper test and Kolmogorov's. Neither D^{+} nor D^{-}, that is the Kolmogorov analogues, is invariant under rotation; proof that D^{*} is, can be found in Mardia and related references. Other tests of uniformity and goodness-of-fit exist, but Kuiper's test seems preferable for small sample sizes.

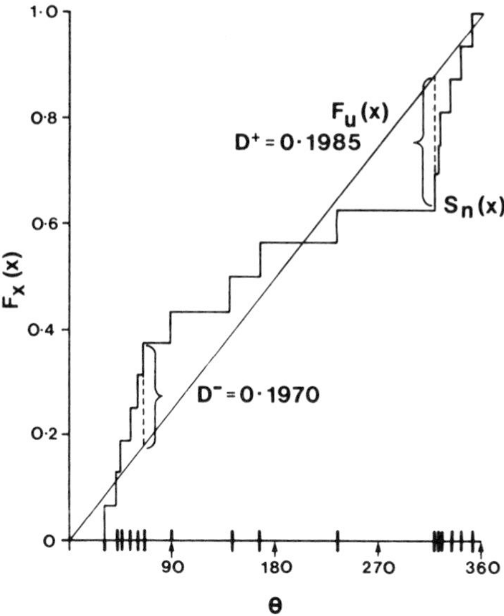

3.39 Graphical illustration of Kuiper's test as a Kolmogorov analogue.

3.6.2 Uniform scores test and propositions of identity or equality

It is often of interest to determine whether two samples of angular measurements can be considered justifiably to have come from the same population or from identical populations of measurements. Such situations are common in studies of hill-form, which use slope angles along transects, or in studies of fluvial or glacial deposition, in which distinctive deposits are sampled and the orientation of contained particles are recorded as rose diagrams, which are used as evidence for an assertion about equality of the populations of measurements. Similarly, observations on angles to the normal plane are made for specified deposits to determine whether there is evidence for different populations of orientations in separate horizons. Similar sorts of problem arise in human geography in terms of the directional incidence of different classes of user of, say, recreational facilities. The empirical evidence of diffusion of artefacts in historical geography can often be discussed in terms of two populations of angles. Indeed the general diffusion of an innovation implicitly assumes a uniform distribution of probability for direction in setting up transition probabilities. Differences between an observed and an expected incidence of towns in central place studies can be discussed in terms of angular differences as well as distance discrepancies. We shall complete the discussion of angular distribution-free tests by illustrating the uniform scores test with data from an analysis of the macrofabric of a till.

Example

Consider that we have two samples of angular measurements from a till. Each sample is taken from a separate site with the random variable $\theta_{11}, \theta_{12}, \ldots, \theta_{1n_1}$ at site one, and the random variable $\theta_{21}, \theta_{22}, \ldots, \theta_{2n_2}$ from site two. These measurements are taken randomly from two continuous populations of measurements with distribution functions $F_1(\theta)$ and $F_2(\theta)$ respectively.

There are $n = n_1 + n_2$ measurements altogether and the observed values are shown diagrammatically in fig. 3.40.

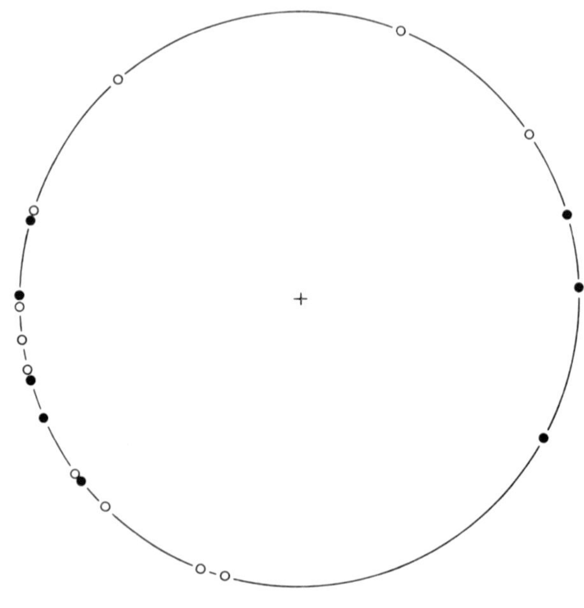

● = n_1 = 8 measurements from F_1 (θ) site 1

○ = n_2 = 11 measurements from F_2 (θ) site 2

3.40 Angular measurements of the macrofabric of a till.

We are testing the following hypotheses.

$$H_0 : F_1(\theta) = F_2(\theta)$$
$$H_1 : F_1(\theta) \neq F_2(\theta)$$

just as we did in the Wilcoxon test in section 2. The difference is that we are dealing with angular measurements and we need a test statistic that is invariant under choice of zero direction and whether measurements are ordered clockwise or anti-clockwise.

The observations are combined and treated as one sample of size n, each value is replaced by its rank, r_i, and given the value at i/n degrees of the circle. Now if the two samples are from different populations we expect this difference to be reflected in a separation of the n_1 and n_2 ranks, such that one of the samples has larger values than the other. In the linear case for the analogous test we interpreted a difference in distribution to be a difference in a location measure such as the mean or the median. However, in the circular case the equivalent measure of location is the *resultant* of the unit vectors from the origin to the position on the circumference of the unit circle (fig. 3.41).

Measurement on the unit circle.
$\theta_1, \theta_2, \ldots, \theta_n$ are measured from the x-axis OX in the anti-clockwise direction.
θ_i also represent the unit vector $\bar{O}\bar{P}_i$ making angle θ_i with the x-axis, where P_i is the point θ_i on the unit circle, $i = 1, \ldots, n$. The cartesian co-ordinates of P_i are $P_i = (\cos\theta_i, \sin\theta_i)$.
The mean direction $\bar{x}_0$ of θ_i is defined as the direction of the resultant of the unit vectors $\bar{O}\bar{P}_1, \ldots, \bar{O}\bar{P}_n$.
Thus the centre of gravity of the P_i, is $(\bar{C}, \bar{S})$, where

$$\bar{C} = \frac{1}{n}\sum_{i=1}^{n}\cos\theta_i,\ \bar{S} = \frac{1}{n}\sum_{i=1}^{n}\sin\theta_i$$

If $\bar{R} = (\bar{C}^2 + \bar{S}^2)^{\frac{1}{2}}$

then $\bar{C} = \bar{R}\cos\bar{X}_0, \bar{S} = \bar{R}\sin\bar{X}_0$ and

$R = n\,\bar{R}$ is the length of the resultant and $\bar{X}_0$ is the solution of the equations for $\bar{C}, \bar{S}$.

3.41 The unit circle and the resultant.

That the customary mean applied to the angles will be misleading is clearly illustrated by the extreme case used by Mardia of two values 359°, 1° whose mean is $\frac{1}{2}(359 + 1) = 180°$. The coordinates of each point, P_i on the unit circle are $(\cos\theta_i, \sin\theta_i)$ and the centre of gravity of the P_i is $(\bar{C}, \bar{S})$ where

$$\bar{C} = \frac{1}{n}\sum_i \cos\theta_i \tag{3.6.3}$$

$$\bar{S} = \frac{1}{n}\sum_i \sin\theta_i \tag{3.6.4}$$

We calculate the resultant of sample 1, R_1, and the resultant of sample 2, R_2, and use the resultant of the smaller sample as the basis of the test statistic. The procedure is illustrated in fig. 3.42.

The resultant of the smaller sample is found by converting the ranks of the sample 1 values to angular measurements as in column 3. Thus $3 \times 360/19 = 56{\cdot}842105$ which is 56° and 0·842105(60) minutes of arc, which give 56° 51′. These eight uniform-scores are expressed as cosines and sines, whose column sums, C_1, S_1, are substituted in

$$R_1^2 = C_1^2 + S_1^2 \tag{3.6.5}$$

to give

$$R_1^2 = -0{\cdot}4969^2 + 0{\cdot}3046^2$$
$$= 0{\cdot}33969$$

θ_i	Rank θ_i r_i	$\beta_i = 360\ r_i/n$	Cos β_i	Sin β_i
22	1			
56	2			
* 74	3	56° 51′	0·5469	0·8373
* 89	4	75° 47′	0·2456	0·9694
*120	5	94° 44′	−0·0822	0·9966
196	6			
201	7			
224	8			
*231	9	170° 32′	−0·9864	0·1645
232	10			
*246	11	208° 25′	−0·8795	−0·4759
*254	12	227° 22′	−0·6773	−0·7357
256	13			
262	14			
269	15			
*271	16	303° 9′	0·5469	−0·8373
*286	17	322° 6′	0·7891	−0·6143
287	18			
320	19			
Sum			−0·4969	+0·304

$$R_1^2 = C_1^2 + S_1^2 = 0{\cdot}24691 + 0{\cdot}092781 = \underline{0{\cdot}33969}$$

The measurements from site 1 are asterisked.

3.42 Calculating the resultant for the uniform scores test for angular measurements.

as the test statistic to be referred to table 18 in appendix 2. From this table we see that the 0·10 value for $n = 19$, $n_1 = 8$ is 11·12 and we accept H_0 that the two populations of measurements are identical.

Procedures for dealing with directional measurements are an important component of the geographer's statistical awareness and the reader is urged to consult Mardia (1972) and the references cited in that text. Many substantial problems can be tackled with greater sensitivity and, more importantly, new facets of traditional problems are revealed and subtler analysis is possible.

All the tests can be extended to other situations and there are many more tests available for various particular situations than can be covered in an introductory text. A tailor-made test is, in general, more powerful than a standard, less restricted test. However, these standard tests provide the logical basis for most of the tests that have been developed, as is illustrated by the rationale of the two tests introduced for angular measurements. The tests covered in this text certainly encompass the principal ideas and logic of constructing statistical tests and are the standard procedures to be used in the situations met commonly in the maps geographers draw. The organization of these standard tests by map type, by information available and used, and by proposition intended, should make it easier for you to assess the possibilities of given maps and make it easier to determine the requirements for particular problems. Of course the problems are the most important part of your geographic development, but the problems are more useful if you can provide answers to them, if you can know the restrictions on these answers and if you can judge how the problems can be put into a more answerable form. The problems of greatest interest are often compound and need to be put in the context of a model. It is easier to ask interesting but difficult questions than to answer them analytically. This is plain from the log-likelihood statistical model for comparatively simple maps, or the use of the compound Poisson model for point pattern analysis. This gap between problem forming and problem solving needs to be acknowledged and reduced. This process can only be aided by an appreciation of statistical inference. The modest elements introduced in this text can be used to build up a familiarity with problem-solving procedures and the more difficult problems can be pared down progressively. The importance of statistical testing is as much to be found in the thought processes engendered as in the techniques commanded. Indeed, the less obtrusive the procedures the more effective is their use likely to be. You should now be able to move onto more advanced texts.

Exercises

3.1 The following data refer to the incidence of shop and other premises, classified by type, along part of Deansgate, Manchester, in 1967.

F A B A C A B G C G G F B A C C B A A C G

where (i) *C*, *F*, *G*, represent commerce, business, personal services
(ii) *A*, *B*, represent retail trade with
A as non-food retail and
B as food retail.

Put the following ideas into appropriate null hypothesis terms and judge their acceptability:

(a) the sequence of retail, non-retail shops is random;
(b) food shops tend to be paired with non-food retail uses.

3.2 A wet day was defined as one with $\geqslant 0{\cdot}01$ inches of rain and a dry day as one which is not wet. After the 50 consecutive days of records given below the geography student asserted that the incidence of wet and dry days was not random. What justification is there for this assertion?

D W W W W D D W W W D W D D W D D D D D D D W D D D D D
W D W W W W D W W W D D D D D D D D D D D D

Assuming you will use an approximation what is the implicit distribution of the random variable?

3.3 Is there any evidence for periodicity in the *BOD* 1965/7 data and the *BOD* 1959/61 data in fig. 3.12? Use at least two test statistics and calculate the exact probabilities for Noether's test.

3.4 For the data given in fig. (i) accompanying exercise 1.29

a) use Cox and Stuart's test to determine whether there is any evidence for a monotonic trend in time;

b) apply Spearman's rho and Kendall's tau to the same data.

What similarity is there between the map (ii) in that exercise and Kendall's notion of concordance?

3.5 The following data were recorded for a small river in the sub-tropics. Is there any justification for the belief that total dissolved solids is uncorrelated with discharge?

X *Discharge* *(cusecs)*	Y *Total dissolved solids* *(ppm)*
30	58
17·8	31
152·3	29
83·6	28
109·6	34
322·1	38
217·2	36
83·0	25
177·0	32
54	39
42	41
44	42
45	31
483	28
54	34
73	32
59	36
328	30
101	28
260	33
416	32
178	29

a) calculate rho, tau.

b) Pearson's product moment coefficient is

$$r = -0{\cdot}293$$

and

$$Y = 36{\cdot}23 - 0{\cdot}0154\,X$$

c) Graph X, Y and construct the regression line.

3.6 Compare the sensitivity of the Wald-Wolfowitz test with the tests you have used already with the data of exercises 2.31 and 2.32.

3.7 Before a motorway was constructed 64 fields on each side of its proposed route were classified as having the same or different land-use: 52 fields had the same use, and 12 had different uses. After the construction of the motorway, $\frac{1}{4}$ of those that had been the same were different, and $\frac{1}{4}$ of those that had been different were the same. The information can be put in a 2 X 2 table when it refers to nominal measurements.

Land use			
Before construction	*After construction*		
	Same	*Different*	
Same	39 ^{0}	13 $^{-}$	52
Different	3 $^{+}$	9 ^{0}	12
	42	22	64

In this form it is known as the McNemar test for change, but it is a variant of the sign test, for $H_0 : P(+) = P(-)$ when *same* to *different* is (−) and *different* to *same* is (+) and the *no change* categories count as ties.

Test H_0 : The field land-use pattern is unaffected by the motorway, using

(i) sign test with $n = (13 + 3) = 16, p = \frac{1}{2}$

(ii) $T = (SD - DS)^2/(SD + DS)$ and refer to χ^2 with 1 *df.*

3.8 Use the *BOD* values in fig. 3.12 in the matched-pairs sign test. Compare the result with the ammoniacal nitrogen data.

3.9 To find out whether, in a map analysis course, preparation can raise performance level as judged by the score on a multiple-choice examination, 6 out of 13 students were selected at random and permitted to see the questions beforehand. The other 7 were not allowed to see the questions until the examination. All 13 were given the test with the following results.

Y, preparation : 55, 63, 73, 74, 77, 84
X, no preparation : 42, 50, 56, 58, 64, 67, 86

H_0 : The two sets of measurements come from identical populations.

a) Judge the acceptability of the assertion.
b) Construct graphs equivalent to figs. 3.24 and 3.27.
c) How many arrangements are there?
d) What other tests could you use? Do they lead to the same judgement; are they less sensitive?

3.10 Compare the sensitivity of Smirnov's two-sample test with the tests used in exercises 2.31, 2.32, 2.34 and 2.35.

3.11 The following data are a random sample of the percentage votes polled by the Labour party, X, and the Conservative party, Y, in the British general election of 1966.

X	Y
54·4	45·6
53·0	36·1
46·5	53·5
70·6	29·4
59·2	40·8
60·9	36·3
30·9	41·3
43·9	48·1
45·0	55·0
56·5	43·5
62·7	37·3
80·3	16·4
52·4	28·3
74·5	14·0
52·7	47·3

a) The random sample of measurements for the Labour party comes from a population with a mean 49·3, variance 16·152.
b) The random sample of measurements for the Conservative party comes from a population with a mean 40·865, variance 11·589.

Test these null hypotheses by using Kolmogorov's test.

3.12 Use chi-square to decide whether the data in exercise 2.10 are approximated satisfactorily by a Poisson distribution with λ estimated empirically.

3.13 Do the following figures support the assertion that the palaeo-current readings for the Middle Sands are from a different population from the readings for the Upper Sands?

Middle Sands	127°, 133°, 140°, 148°, 160°, 166°
Upper Sands	104°, 123°, 135°, 137°, 152°, 162°, 168°, 192°

Plot, compare and calculate the uniform scores test statistic.

3.14 Why is the point probability $P(r = 15)$ in fig. 3.4 the same as $P(r_1 = 8)$ in fig. 3.7?

Appendix 1: Using the standard normal probability distribution

1. Although discrete distributions are the basis of the text, considerable reference is made to two continuous distributions, χ^2 and the *normal* distribution. Chi-square is used because it provides the basis of many tests that are applicable to nominal and ordinal measurement. The standard normal distribution is the asymptotic distribution of so many discrete distributions or of functions of them that it provides a measure of likelihood when the number of measurements made is large. When tabulated values are not available for these exact probabilities recourse is taken to the normal distribution. This appendix indicates how it is used.

2. The normal distribution depends on two parameters, the mean, μ, and the variance, σ^2, and it is often denoted by $N(\mu, \sigma^2)$. The standard normal distribution refers to a normal distribution $N(0,1)$. A normal distribution is transformed to a standard normal distribution by subtracting the mean from each value and dividing the difference by the standard deviation as: $(x_i - \bar{x})/S_x$. Normal distributions are symmetric about their mean and have the following shape.

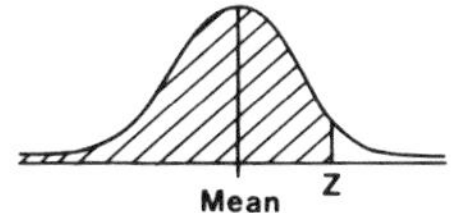

3. Suppose X is a random variable distributed as $N(0, 1)$. Then the probability that X is less than, say, 1·75 is

$$P(X \leqslant 1{\cdot}75) = 0{\cdot}9599408$$

and

$$P(X \leqslant -1{\cdot}75) = 1 - 0{\cdot}9599408 = 0{\cdot}0400592$$

because of symmetry.

And

$$\begin{aligned} P(-1{\cdot}75 \leqslant x \leqslant 1{\cdot}75) &= P(X \leqslant 1{\cdot}75) - P(X \leqslant -1{\cdot}75) \\ &= 0{\cdot}9599408 - 0{\cdot}0400592 \\ &= 0{\cdot}9198816 \end{aligned}$$

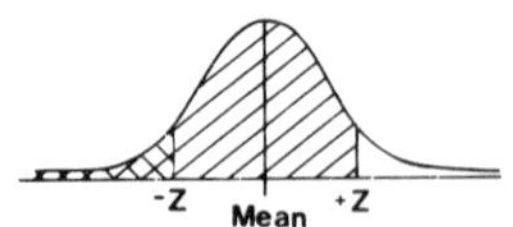

4. The same procedure applies to $+Z \neq -Z$.
   Say $+Z = 1{\cdot}75$, $-Z = 1{\cdot}0$

   then $P\,(x \leqslant 1{\cdot}75)$ and $P\,(x \geqslant -1{\cdot}0)$ is

$$
\begin{aligned}
P\,(-1{\cdot}0 \leqslant x \leqslant 1{\cdot}75) &= P\,(x \leqslant 1{\cdot}75) - P\,(x \leqslant -1{\cdot}0)\\
&= 0{\cdot}9599 - 0{\cdot}1587\\
&= 0{\cdot}8012
\end{aligned}
$$

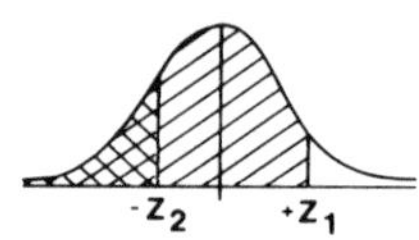

5. For x, $N\,(\mu, \sigma^2)$ then

$$P\,(X \leqslant x) = \Phi\left(\frac{x-\mu}{\sigma}\right)$$

   Say $\mu = 3$, $\sigma = 2$ then for $x = 6$

$$
\begin{aligned}
P\,(X \leqslant 6) &= \Phi\left(\frac{6-3}{2}\right) = 1{\cdot}5\\
&= 0{\cdot}933193
\end{aligned}
$$

6. Now Φ, phi, is the conventional symbol for the probability distribution of the standard normal. Often the scaling along the horizontal axis of $N\,(0,1)$ is in terms of the standard normal deviate, Z, thus

$$Z = \left(\frac{x_i - \bar{x}}{S_x}\right) \quad \text{or} \quad Z = \left(\frac{x_i - \mu}{\sigma}\right)$$

   and Z is referred to Φ to derive the probability. Because Φ is symmetric most tables give values of $\Phi\,(Z) \geqslant 0$, corresponding to $Z = 0$ to $\pm\infty$.

7. Values of Z corresponding to nominal α are important because they arise so often. Thus

   $Z = 1{\cdot}96$ corresponds to the two-tailed rejection region for $\alpha = 0{\cdot}05$; i.e. it bounds an upper-tailed region of 0·025 and a lower-tailed region of 0·025, together representing 0·05 of the total area of 1.

   $Z = 2{\cdot}58$ corresponds to the two-tailed rejection region for $\alpha = 0{\cdot}01$; i.e. it bounds an upper-tailed region of 0·005 and a lower-tailed region of 0·005, together representing 0·01 of the total area of 1.

8. In fig. 3.4 the exact cumulative probabilities are compared with the cumulative

probabilities using the normal approximation. The final column gives the standardized Z values. The probability of a value $\leqslant Z$ is found from table 1 in appendix 2, as for example $\Phi\ (Z = 3{\cdot}5180) > \Phi\ (Z_1 = 3{\cdot}5) = 0{\cdot}99977$ and $< \Phi\ (Z_2 = 3{\cdot}6) = 0{\cdot}99984$. We can estimate $\Phi\ (Z = 3{\cdot}5180)$ from these values in the following way:

$$Z_2 - Z_1 = 0{\cdot}1,\ Z - Z_1 = 0{\cdot}018$$

$$\Phi\ (Z_2) - \Phi\ (Z_1) = 0{\cdot}00007$$

then we take

$$\Phi\ (Z) \simeq \frac{(Z - Z_1)}{(Z_2 - Z_1)} \cdot [\Phi\ (Z_2) - \Phi\ (Z_1)] + \Phi\ (Z_1)$$

$$= \frac{0{\cdot}018}{0{\cdot}100} \cdot 0{\cdot}00007 + 0{\cdot}99977$$

$$= 0{\cdot}9997826$$

This value is identical with the value calculated from the exact function for the normal curve: the linear interpolation is adequate at the tails. In order to match the exact cumulative probabilities which are summed from runs_{max} to runs_{min} we must take $[1 - \Phi\ (Z)]$, and $\{1 - [1 - \Phi\ (Z)]\} = \Phi\ (Z)$ for negative Z values.

9. If we wish to estimate the likely value of the *population* mean and *population* variance from the *sample* mean, $\bar{x}$, and *sample* variance, S_x^2, many people consider it is important that the *expected value* of the sample mean and sample variance equals the population mean and population variance. The formula given on p.33 for the sample mean is

$$\bar{x} = \frac{1}{n} \Sigma x_i$$

and its expected value is μ.
The formula given on p.33 for the sample mean is

$$S_x^2 = \frac{1}{n} \Sigma (x_i - \bar{x})^2$$

and its expected value is *not* σ^2. However the expected value of

$$S_x^2 = \frac{1}{n-1} \Sigma (x_i - \bar{x})^2$$

is σ^2. The effect is greatest when n is small.

Appendix 2: Tables for distribution-free statistics

1. The normal probability distribution
2. Chi-square
3. Binomial
4. Fisher's exact test for 2 × 2 tables; a hypergeometric random variable
5. Poisson
6. Wilcoxon's rank-sum test
7. Mann-Whitney's *U* test
8. Kruskal-Wallis test
9. Friedman's test
10. Total number of runs test
11. Edgington's runs up and down test
12. Spearman's rank-order correlation test
13. Kendall's rank-order correlation test
14. Wilcoxon's signed-rank test
15. Smirnov's maximum deviation test
16. Kolmogorov's goodness-of-fit test
17. Kuiper's maximum deviation test for angular measurements
18. Mardia's uniform scores test for angular measurements

Table 1 Areas of the standardized normal distribution.
Source: J. White, A. Yeats and G. Skipworth, *Tables for statisticians* (London 1974), 18–19.

−0·09	−0·08	−0·07	−0·06	−0·05	−0·04	−0·03	−0·02	−0·01	−0·00	z
0·99997	0·99997	0·99996	0·99996	0·99996	0·99996	0·99996	0·99996	0·99995	0·99995	−3·9
0·99995	0·99995	0·99995	0·99994	0·99994	0·99994	0·99994	0·99993	0·99993	0·99993	−3·8
0·99992	0·99992	0·99992	0·99992	0·99991	0·99991	0·99990	0·99990	0·99990	0·99989	−3·7
0·99989	0·99988	0·99988	0·99987	0·99987	0·99986	0·99986	0·99985	0·99985	0·99984	−3·6
0·99983	0·99983	0·99982	0·99981	0·99981	0·99980	0·99979	0·99978	0·99978	0·99977	−3·5
0·99976	0·99975	0·99974	0·99973	0·99972	0·99971	0·99970	0·99969	0·99968	0·99966	−3·4
0·99965	0·99964	0·99962	0·99961	0·99960	0·99958	0·99957	0·99955	0·99953	0·99952	−3·3
0·99950	0·99948	0·99946	0·99944	0·99942	0·99940	0·99938	0·99936	0·99934	0·99931	−3·2
0·99929	0·99926	0·99924	0·99921	0·99918	0·99916	0·99913	0·99910	0·99906	0·99903	−3·1
0·99900	0·99896	0·99893	0·99889	0·99886	0·99882	0·99878	0·99874	0·99869	0·99865	−3·0
0·99861	0·99856	0·99851	0·99846	0·99841	0·99836	0·99831	0·99825	0·99819	0·99813	−2·9
0·99807	0·99801	0·99795	0·99788	0·99781	0·99774	0·99767	0·99760	0·99752	0·99744	−2·8
0·99736	0·99728	0·99720	0·99711	0·99702	0·99693	0·99683	0·99674	0·99664	0·99653	−2·7
0·99643	0·99632	0·99621	0·99609	0·99598	0·99585	0·99573	0·99560	0·99547	0·99534	−2·6
0·99520	0·99506	0·99492	0·99477	0·99461	0·99446	0·99430	0·99413	0·99396	0·99379	−2·5
0·99361	0·99343	0·99324	0·99305	0·99286	0·99266	0·99245	0·99224	0·99202	0·99180	−2·4
0·99158	0·99134	0·99111	0·99086	0·99061	0·99036	0·99010	0·98983	0·98956	0·98928	−2·3
0·98899	0·98870	0·98840	0·98809	0·98778	0·98745	0·98713	0·98679	0·98645	0·98610	−2·2
0·98574	0·98537	0·98500	0·98461	0·98422	0·98382	0·98341	0·98300	0·98257	0·98214	−2·1
0·98169	0·98124	0·98077	0·98030	0·97982	0·97932	0·97882	0·97831	0·97778	0·97725	−2·0
0·97670	0·97615	0·97558	0·97500	0·97441	0·97381	0·97320	0·97257	0·97193	0·97128	−1·9
0·97062	0·96995	0·96926	0·96856	0·96784	0·96712	0·96638	0·96562	0·96485	0·96407	−1·8
0·96327	0·96246	0·96164	0·96080	0·95994	0·95907	0·95818	0·95728	0·95637	0·95543	−1·7
0·95449	0·95352	0·95254	0·95154	0·95053	0·94950	0·94845	0·94738	0·94630	0·94520	−1·6
0·94408	0·94295	0·94179	0·94062	0·93943	0·93822	0·93699	0·93574	0·93448	0·93319	−1·5

0·93189	0·93056	0·92922	0·92785	0·92647	0·92507	0·92364	0·92220	0·92073	0·91924	—1·4
0·91774	0·91621	0·91466	0·91308	0·91149	0·90988	0·90824	0·90658	0·90490	0·90320	—1·3
0·90147	0·89973	0·89796	0·89617	0·89435	0·89251	0·89065	0·88877	0·88686	0·88493	—1·2
0·88298	0·88100	0·87900	0·87698	0·87493	0·87286	0·87076	0·86864	0·86650	0·86433	—1·1
0·86214	0·85993	0·85769	0·85543	0·85314	0·85083	0·84850	0·84614	0·84375	0·84134	—1·0
0·83891	0·83646	0·83398	0·83147	0·82894	0·82639	0·82381	0·82121	0·81859	0·81594	—0·9
0·81327	0·81057	0·80785	0·80511	0·80234	0·79955	0·79673	0·79389	0·79103	0·78814	—0·8
0·78524	0·78230	0·77935	0·77637	0·77337	0·77035	0·76731	0·76424	0·76115	0·75804	—0·7
0·75490	0·75175	0·74857	0·74537	0·74215	0·73891	0·73565	0·73237	0·72907	0·72575	—0·6
0·72240	0·71904	0·71566	0·71226	0·70884	0·70540	0·70194	0·69847	0·69497	0·69146	—0·5
0·68793	0·68439	0·68082	0·67724	0·67364	0·67003	0·66640	0·66276	0·65910	0·65542	—0·4
0·65173	0·64803	0·64431	0·64058	0·63683	0·63307	0·62930	0·62552	0·62172	0·61791	—0·3
0·61409	0·61026	0·60642	0·60257	0·59871	0·59483	0·59095	0·58706	0·58317	0·57926	—0·2
0·57535	0·57142	0·56750	0·56356	0·55962	0·55567	0·55172	0·54776	0·54380	0·53983	—0·1
0·53586	0·53188	0·52790	0·52392	0·51994	0·51595	0·51197	0·50798	0·50399	0·50000	—0·0

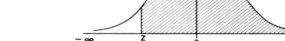

The function tabulated is $\frac{1}{\sqrt{2\pi}} \int_z^\infty e^{-x^2/2}\, dx$,

the probability that $Z > z$, where $Z \sim N(0, 1)$.

[Table 1 contd]

z	0·00	0·01	0·02	0·03	0·04	0·05	0·06	0·07	0·08	0·09
0·0	0·50000	0·49601	0·49202	0·48803	0·48405	0·48006	0·47608	0·47210	0·46812	0·46414
0·1	0·46017	0·45620	0·45224	0·44828	0·44433	0·44038	0·43644	0·43250	0·42858	0·42465
0·2	0·42074	0·41683	0·41294	0·40905	0·40517	0·40129	0·39743	0·39358	0·38974	0·38591
0·3	0·38209	0·37828	0·37448	0·37070	0·36693	0·36317	0·35942	0·35569	0·35197	0·34827
0·4	0·34458	0·34090	0·33724	0·33360	0·32997	0·32636	0·32276	0·31918	0·31561	0·31207
0·5	0·30854	0·30503	0·30153	0·29806	0·29460	0·29116	0·28774	0·28434	0·28096	0·27760
0·6	0·27425	0·27093	0·26763	0·26435	0·26109	0·25785	0·25463	0·25143	0·24825	0·24510
0·7	0·24196	0·23885	0·23576	0·23269	0·22965	0·22663	0·22363	0·22065	0·21770	0·21476
0·8	0·21186	0·20897	0·20611	0·20327	0·20045	0·19766	0·19489	0·19215	0·18943	0·18673
0·9	0·18406	0·18141	0·17879	0·17619	0·17361	0·17106	0·16853	0·16602	0·16354	0·16109
1·0	0·15866	0·15625	0·15386	0·15150	0·14917	0·14686	0·14457	0·14231	0·14007	0·13786
1·1	0·13567	0·13350	0·13136	0·12924	0·12714	0·12507	0·12302	0·12100	0·11900	0·11702
1·2	0·11507	0·11314	0·11123	0·10935	0·10749	0·10565	0·10383	0·10204	0·10027	0·09853
1·3	0·09680	0·09510	0·09342	0·09176	0·09012	0·08851	0·08692	0·08534	0·08379	0·08226
1·4	0·08076	0·07927	0·07780	0·07636	0·07493	0·07353	0·07215	0·07078	0·06944	0·06811
1·5	0·06681	0·06552	0·06426	0·06301	0·06178	0·06057	0·05938	0·05821	0·05705	0·05592
1·6	0·05480	0·05370	0·05262	0·05155	0·05050	0·04947	0·04846	0·04746	0·04648	0·04551
1·7	0·04457	0·04363	0·04272	0·04182	0·04093	0·04006	0·03920	0·03836	0·03754	0·03673
1·8	0·03593	0·03515	0·03438	0·03362	0·03288	0·03216	0·03144	0·03074	0·03005	0·02938
1·9	0·02872	0·02807	0·02743	0·02680	0·02619	0·02559	0·02500	0·02442	0·02385	0·02330
2·0	0·02275	0·02222	0·02169	0·02118	0·02068	0·02018	0·01970	0·01923	0·01876	0·01831
2·1	0·01786	0·01743	0·01700	0·01659	0·01618	0·01578	0·01539	0·01500	0·01463	0·01426
2·2	0·01390	0·01355	0·01321	0·01287	0·01255	0·01222	0·01191	0·01160	0·01130	0·01101
2·3	0·01072	0·01044	0·01017	0·00990	0·00964	0·00939	0·00914	0·00889	0·00866	0·00842
2·4	0·00820	0·00798	0·00776	0·00755	0·00734	0·00714	0·00695	0·00676	0·00657	0·00639
2·5	0·00621	0·00604	0·00587	0·00570	0·00554	0·00539	0·00523	0·00508	0·00494	0·00480
2·6	0·00466	0·00453	0·00440	0·00427	0·00415	0·00402	0·00391	0·00379	0·00368	0·00357
2·7	0·00347	0·00336	0·00326	0·00317	0·00307	0·00298	0·00289	0·00280	0·00272	0·00264
2·8	0·00256	0·00248	0·00240	0·00233	0·00226	0·00219	0·00212	0·00205	0·00199	0·00193
2·9	0·00187	0·00181	0·00175	0·00169	0·00164	0·00159	0·00154	0·00149	0·00144	0·00139

3·0	0·00135	0·00131	0·00126	0·00122	0·00118	0·00114	0·00111	0·00107	0·00104	0·00100
3·1	0·00097	0·00094	0·00090	0·00087	0·00084	0·00082	0·00079	0·00076	0·00074	0·00071
3·2	0·00069	0·00066	0·00064	0·00062	0·00060	0·00058	0·00056	0·00054	0·00052	0·00050
3·3	0·00048	0·00047	0·00045	0·00043	0·00042	0·00040	0·00039	0·00038	0·00036	0·00035
3·4	0·00034	0·00032	0·00031	0·00030	0·00029	0·00028	0·00027	0·00026	0·00025	0·00024
3·5	0·00023	0·00022	0·00022	0·00021	0·00020	0·00019	0·00019	0·00018	0·00017	0·00017
3·6	0·00016	0·00015	0·00015	0·00014	0·00014	0·00013	0·00013	0·00012	0·00012	0·00011
3·7	0·00011	0·00010	0·00010	0·00010	0·00009	0·00009	0·00008	0·00008	0·00008	0·00008
3·8	0·00007	0·00007	0·00007	0·00006	0·00006	0·00006	0·00006	0·00005	0·00005	0·00005
3·9	0·00005	0·00005	0·00004	0·00004	0·00004	0·00004	0·00004	0·00004	0·00003	0·00003

The function tabulated is $\frac{1}{\sqrt{2\pi}} \int_z^\infty e^{-x^2/2}\, dx$,

the probability that $Z > z$, where $Z \sim N(0, 1)$.

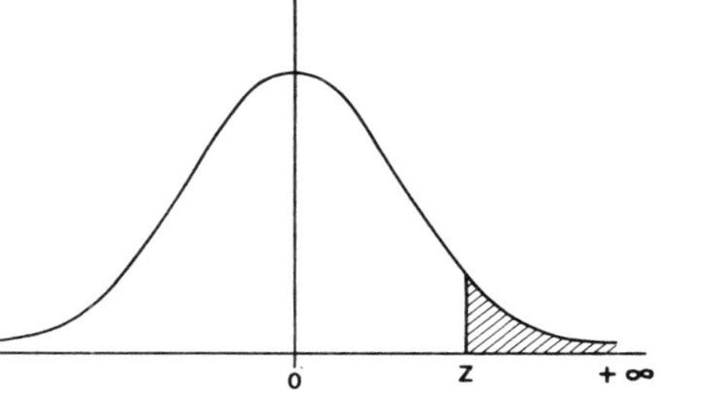

Table 2 Percentage points of the χ^2 distribution.
Source: J. White, A. Yeats and G. Skipworth, *Tables for statisticians* (London 1974), 22—3.

0·995	0·990	0·975	0·950	0·900	0·750	0·500	α / v
392704 . 10^{-10}	157088 . 10^{-9}	982069 . 10^{-9}	393214 . 10^{-8}	0·0157908	0·1015308	0·454936	1
0·0100251	0·0201007	0·0506356	0·102587	0·210721	0·575364	1·38629	2
0·0717218	0·114832	0·215795	0·351846	0·584374	1·212534	2·36597	3
0·206989	0·297109	0·484419	0·710723	1·063623	1·92256	3·35669	4
0·411742	0·554298	0·831212	1·145476	1·61031	2·67460	4·35146	5
0·675727	0·872090	1·23734	1·63538	2·20413	3·45460	5·34812	6
0·989256	1·239043	1·68987	2·16735	2·83311	4·25485	6·34581	7
1·34441	1·64650	2·17973	2·73264	3·48954	5·07064	7·34412	8
1·73493	2·08790	2·70039	3·32511	4·16816	5·89883	8·34283	9
2·15586	2·55821	3·24697	3·94030	4·86518	6·73720	9·34182	10
2·60322	3·05348	3·81575	4·57481	5·57778	7·58414	10·3410	11
3·07382	3·57057	4·40379	5·22603	6·30380	8·43842	11·3403	12
3·56503	4·10692	5·00875	5·89186	7·04150	9·29907	12·3398	13
4·07467	4·66043	5·62873	6·57063	7·78953	10·1653	13·3393	14
4·60092	5·22935	6·26214	7·26094	8·54676	11·0365	14·3389	15
5·14221	5·81221	6·90766	7·96165	9·31224	11·9122	15·3385	16
5·69722	6·40776	7·56419	8·67176	10·0852	12·7919	16·3382	17
6·26480	7·01491	8·23075	9·39046	10·8649	13·6753	17·3379	18
6·84397	7·63273	8·90652	10·1170	11·6509	14·5620	18·3377	19
7·43384	8·26040	9·59078	10·8508	12·4426	15·4518	19·3374	20
8·03365	8·89720	10·28293	11·5913	13·2396	16·3444	20·3372	21
8·64272	9·54249	10·9823	12·3380	14·0415	17·2396	21·3370	22
9·26043	10·19567	11·6886	13·0905	14·8480	18·1373	22·3369	23
9·88623	10·8564	12·4012	13·8484	15·6587	19·0373	23·3367	24

							v
10·5197	11·5240	13·1197	14·6114	16·4734	19·9393	24·3366	25
11·1602	12·1981	13·8439	15·3792	17·2919	20·8434	25·3365	26
11·8076	12·8785	14·5734	16·1514	18·1139	21·7494	26·3363	27
12·4613	13·5647	15·3079	16·9279	18·9392	22·6572	27·3362	28
13·1211	14·2565	16·0471	17·7084	19·7677	23·5666	28·3361	29
13·7867	14·9535	16·7908	18·4927	20·5992	24·4776	29·3360	30
20·7065	22·1643	24·4330	26·5093	29·0505	33·6603	39·3353	40
27·9907	29·7067	32·3574	34·7643	37·6886	42·9421	49·3349	50
35·5345	37·4849	40·4817	43·1880	46·4589	52·2938	59·3347	60
43·2752	45·4417	48·7576	51·7393	55·3289	61·6983	69·3345	70
51·1719	53·5401	57·1532	60·3915	64·2778	71·1445	79·3343	80
59·1963	61·7541	65·6466	69·1260	73·2911	80·6247	89·3342	90
67·3276	70·0649	74·2219	77·9295	82·3581	90·1332	99·3341	100

The values tabulated are $\chi_v^2(\alpha)$, where
$Pr(\chi_v^2 > \chi_v^2(\alpha)) = \alpha$, for v degrees of freedom.

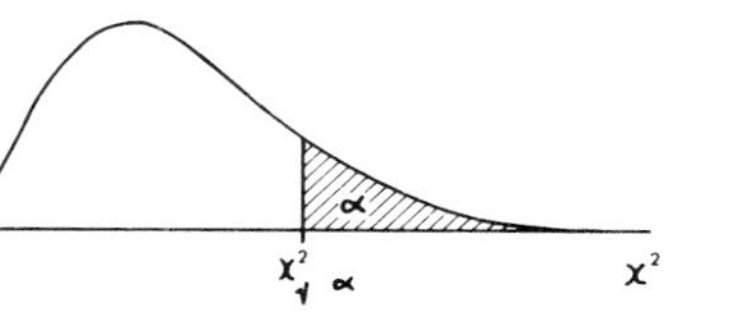

[Table 2 contd]

v \ α	0·250	0·100	0·050	0·025	0·010	0·005	0·001
1	1·32330	2·70554	3·84146	5·02389	6·63490	7·87944	10·828
2	2·77259	4·60517	5·99146	7·37776	9·21034	10·5966	13·816
3	4·10834	6·25139	7·81473	9·34840	11·3449	12·8382	16·266
4	5·38527	7·77944	9·48773	11·1433	13·2767	14·8603	18·467
5	6·62568	9·23636	11·0705	12·8325	15·0863	16·7496	20·515
6	7·84080	10·6446	12·5916	14·4494	16·8119	18·5476	22·458
7	9·03715	12·0170	14·0671	16·0128	18·4753	20·2777	24·322
8	10·2189	13·3616	15·5073	17·5345	20·0902	21·9550	26·125
9	11·3888	14·6837	16·9190	19·0228	21·6660	23·5894	27·877
10	12·5489	15·9872	18·3070	20·4832	23·2093	25·1882	29·588
11	13·7007	17·2750	19·6751	21·9200	24·7250	26·7568	31·264
12	14·8454	18·5493	21·0261	23·3367	26·2170	28·2995	32·909
13	15·9839	19·8119	22·3620	24·7356	27·6882	29·8195	34·528
14	17·1169	21·0641	23·6848	26·1189	29·1412	31·3194	36·123
15	18·2451	22·3071	24·9958	27·4884	30·5779	32·8013	37·697
16	19·3689	23·5418	26·2962	28·8454	31·9999	34·2672	39·252
17	20·4887	24·7690	27·5871	30·1910	33·4087	35·7185	40·790
18	21·6049	25·9894	28·8693	31·5264	34·8053	37·1565	42·312
19	22·7178	27·2036	30·1435	32·8523	36·1909	38·5823	43·820
20	23·8277	28·4120	31·4104	34·1696	37·5662	39·9968	45·315
21	24·9348	29·6151	32·6706	35·4789	38·9322	41·4011	46·797
22	26·0393	30·8133	33·9244	36·7807	40·2894	42·7957	48·268
23	27·1413	32·0069	35·1725	38·0756	41·6384	44·1813	49·728
24	28·2412	33·1962	36·4150	39·3641	42·9798	45·5585	51·179
25	29·3389	34·3816	37·6525	40·6465	44·3141	46·9279	52·618
26	30·4346	35·5632	38·8851	41·9232	45·6417	48·2899	54·052
27	31·5284	36·7412	40·1133	43·1945	46·9629	49·6449	55·476
28	32·6205	37·9159	41·3371	44·4608	48·2782	50·9934	56·892
29	33·7109	39·0875	42·5570	45·7223	49·5879	52·3356	58·301

30	34·7997	40·2560	43·7730	46·9792	50·8922	53·6720	59·703
40	45·6160	51·8051	55·7585	59·3417	63·6907	66·7660	73·402
50	56·3336	63·1671	67·5048	71·4202	76·1539	79·4900	86·661
60	66·9815	74·3970	79·0819	83·2977	88·3794	91·9517	99·607
70	77·5767	85·5270	90·5312	95·0232	100·425	104·215	112·317
80	88·1303	96·5782	101·879	106·629	112·329	116·321	124·839
90	98·6499	107·565	113·145	118·136	124·116	128·299	137·208
100	109·141	118·498	124·342	129·561	135·807	140·169	149·449

For $v > 30$ take $\chi_v^2(\alpha) = v\left[1 - \frac{2}{9v} + u_\alpha\sqrt{\frac{2}{9v}}\right]^3$ where u_α is such that $Pr(U > u_\alpha) = \alpha$, and $U \sim N(0,1)$.

Table 3(a) Cumulative binomial probabilities.
Source: J. White, A. Yeats and G. Skipworth, *Tables for statisticians* (London 1974), 6–9.

R has the binomial distribution with parameters n and p. The entries are the probability that r or more successes occur in n independent trials. The entries are the values of $P(R \geqslant r) = \sum_{r=0}^{n} \binom{n}{r} p^r (1-p)^{n-r}$ for p at intervals from 0·01 to 0·50.

p		0·01	0·02	0·03	0·04	0·05	0·06	0·07	0·08	0·09
$n = 2$	$r = 0$	1·00000	1·00000	1·00000	1·00000	1·00000	1·00000	1·00000	1·00000	1·00000
	1	0·01990	0·03960	0·05910	0·07840	0·09750	0·11640	0·13510	0·15360	0·17190
	2	0·00010	0·00040	0·00090	0·00160	0·00250	0·00360	0·00490	0·00640	0·00810
$n = 3$	$r = 0$	1·00000	1·00000	1·00000	1·00000	1·00000	1·00000	1·00000	1·00000	1·00000
	1	0·02970	0·05881	0·08733	0·11526	0·14263	0·16942	0·19564	0·22131	0·24643
	2	0·00030	0·00118	0·00265	0·00467	0·00725	0·01037	0·01401	0·01818	0·02284
	3		0·00001	0·00003	0·00006	0·00013	0·00022	0·00034	0·00051	0·00073
$n = 4$	$r = 0$	1·00000	1·00000	1·00000	1·00000	1·00000	1·00000	1·00000	1·00000	1·00000
	1	0·03940	0·07763	0·11471	0·15065	0·18549	0·21925	0·25195	0·28361	0·31425
	2	0·00059	0·00234	0·00519	0·00910	0·01402	0·01991	0·02673	0·03443	0·04296
	3		0·00003	0·00011	0·00025	0·00048	0·00083	0·00130	0·00193	0·00272
	4					0·00001	0·00001	0·00002	0·00004	0·00007
$n = 5$	$r = 0$	1·00000	1·00000	1·00000	1·00000	1·00000	1·00000	1·00000	1·00000	1·00000
	1	0·04901	0·09608	0·14127	0·18463	0·22622	0·26610	0·30431	0·34092	0·37597
	2	0·00098	0·00384	0·00847	0·01476	0·02259	0·03187	0·04249	0·05436	0·06738
	3	0·00001	0·00008	0·00026	0·00060	0·00116	0·00197	0·00308	0·00453	0·00634
	4				0·00001	0·00003	0·00006	0·00011	0·00019	0·00030
	5									0·00001

n	r									
$n = 6$	$r = 0$	1·00000	1·00000	1·00000	1·00000	1·00000	1·00000	1·00000	1·00000	1·00000
	1	0·05852	0·11416	0·16703	0·21724	0·26491	0·31013	0·35301	0·39364	0·43213
	2	0·00146	0·00569	0·01246	0·02155	0·03277	0·04592	0·06082	0·07729	0·09515
	3	0·00002	0·00015	0·00050	0·00117	0·00223	0·00376	0·00584	0·00851	0·01183
	4			0·00001	0·00004	0·00009	0·00018	0·00032	0·00054	0·00085
	5							0·00001	0·00002	0·00003
$n = 7$	$r = 0$	1·00000	1·00000	1·00000	1·00000	1·00000	1·00000	1·00000	1·00000	1·00000
	1	0·06793	0·13187	0·19202	0·24855	0·30166	0·35152	0·39830	0·44215	0·48324
	2	0·00203	0·00786	0·01709	0·02938	0·04438	0·06178	0·08127	0·10259	0·12548
	3	0·00003	0·00026	0·00086	0·00198	0·00376	0·00629	0·00969	0·01401	0·01933
	4		0·00001	0·00003	0·00008	0·00019	0·00039	0·00071	0·00118	0·00184
	5					0·00001	0·00001	0·00003	0·00006	0·00011
$n = 8$	$r = 0$	1·00000	1·00000	1·00000	1·00000	1·00000	1·00000	1·00000	1·00000	1·00000
	1	0·07726	0·14924	0·21626	0·27861	0·33658	0·39043	0·44042	0·48678	0·52975
	2	0·00269	0·01034	0·02234	0·03815	0·05724	0·07916	0·10347	0·12976	0·15768
	3	0·00005	0·00042	0·00135	0·00308	0·00579	0·00962	0·01470	0·02110	0·02889
	4		0·00001	0·00005	0·00016	0·00037	0·00075	0·00134	0·00220	0·00341
	5				0·00001	0·00002	0·00004	0·00008	0·00015	0·00026
	6								0·00001	0·00001
$n = 9$	$r = 0$	1·00000	1·00000	1·00000	1·00000	1·00000	1·00000	1·00000	1·00000	1·00000
	1	0·08648	0·16625	0·23977	0·30747	0·36975	0·42701	0·47959	0·52784	0·57207
	2	0·00344	0·01311	0·02816	0·04777	0·07121	0·09784	0·12705	0·15832	0·19117
	3	0·00008	0·00061	0·00198	0·00448	0·00836	0·01380	0·02091	0·02979	0·04048
	4		0·00002	0·00009	0·00027	0·00064	0·00128	0·00227	0·00372	0·00570
	5				0·00001	0·00003	0·00008	0·00017	0·00031	0·00055
	6							0·00001	0·00002	0·00004

[Table 3(a) contd]

p	0·01	0·02	0·03	0·04	0·05	0·06	0·07	0·08	0·09
$n=10$ $r=0$	1·00000	1·00000	1·00000	1·00000	1·00000	1·00000	1·00000	1·00000	1·00000
1	0·09562	0·18293	0·26258	0·33517	0·40126	0·46138	0·51602	0·56561	0·61058
2	0·00427	0·01618	0·03451	0·05815	0·08614	0·11759	0·15173	0·18788	0·22545
3	0·00011	0·00086	0·00276	0·00621	0·01150	0·01884	0·02834	0·04008	0·05404
4		0·00003	0·00015	0·00044	0·00103	0·00203	0·00358	0·00580	0·00883
5			0·00001	0·00002	0·00006	0·00015	0·00031	0·00059	0·00101
6						0·00001	0·00002	0·00004	0·00008
$n=20$ $r=0$	1·00000	1·00000	1·00000	1·00000	1·00000	1·00000	1·00000	1·00000	1·00000
1	0·18209	0·33239	0·45621	0·55800	0·64151	0·70989	0·76576	0·81131	0·84836
2	0·01686	0·05990	0·11984	0·18966	0·26416	0·33955	0·41314	0·48314	0·54840
3	0·00100	0·00707	0·02101	0·04386	0·07548	0·11497	0·16100	0·21205	0·26657
4	0·00004	0·00060	0·00267	0·00741	0·01590	0·02897	0·04713	0·07062	0·09933
5		0·00004	0·00026	0·00096	0·00257	0·00563	0·01071	0·01834	0·02904
6			0·00002	0·00010	0·00033	0·00087	0·00193	0·00380	0·00679
7				0·00001	0·00003	0·00011	0·00028	0·00064	0·00129
8						0·00001	0·00003	0·00009	0·00020
9								0·00001	0·00003
$n=50$ $r=0$	1·00000	1·00000	1·00000	1·00000	1·00000	1·00000	1·00000	1·00000	1·00000
1	0·39499	0·63583	0·78193	0·87011	0·92306	0·95467	0·97344	0·98453	0·99104
2	0·08944	0·26423	0·44472	0·59952	0·72057	0·81000	0·87351	0·91729	0·94676
3	0·01382	0·07843	0·18920	0·32329	0·45947	0·58375	0·68921	0·77403	0·83946
4	0·00160	0·01776	0·06276	0·13913	0·23959	0·35270	0·46726	0·57470	0·66966
5	0·00015	0·00321	0·01681	0·04897	0·10362	0·17940	0·27097	0·37105	0·47234
6	0·00001	0·00048	0·00374	0·01441	0·03778	0·07764	0·13505	0·20813	0·29281
7		0·00006	0·00070	0·00361	0·01179	0·02892	0·05831	0·10187	0·15963
8		0·00001	0·00011	0·00078	0·00319	0·00938	0·02201	0·04379	0·07684
9			0·00002	0·00015	0·00076	0·00267	0·00732	0·01665	0·03283

10				0·00002	0·00016	0·00067	0·00216	0·00563	0·01252
11					0·00003	0·00015	0·00057	0·00171	0·00428
12						0·00003	0·00014	0·00047	0·00132
13						0·00001	0·00003	0·00011	0·00037
14							0·00001	0·00003	0·00009
15								0·00001	0·00002
$n=100$ $r=0$	1·00000	1·00000	1·00000	1·00000	1·00000	1·00000	1·00000	1·00000	1·00000
1	0·63397	0·86738	0·95245	0·98313	0·99408	0·99795	0·99929	0·99976	0·99992
2	0·26424	0·59673	0·80538	0·91284	0·96292	0·98483	0·99399	0·99768	0·99913
3	0·07937	0·32331	0·58022	0·76786	0·88174	0·94339	0·97421	0·98873	0·99524
4	0·01837	0·14104	0·35275	0·57052	0·74216	0·85698	0·92559	0·96329	0·98270
5	0·00343	0·05083	0·18215	0·37114	0·56402	0·72322	0·83684	0·90966	0·95261
6	0·00053	0·01548	0·08084	0·21163	0·38400	0·55931	0·70858	0·82012	0·89548
7	0·00007	0·00406	0·03123	0·10639	0·23399	0·39365	0·55572	0·69684	0·80602
8	0·00001	0·00093	0·01062	0·04751	0·12796	0·25165	0·40122	0·55289	0·68721
9		0·00019	0·00322	0·01899	0·06309	0·14629	0·26603	0·40737	0·55060
10		0·00003	0·00087	0·00684	0·02819	0·07754	0·16202	0·27802	0·41249
11		0·00001	0·00021	0·00224	0·01147	0·03761	0·09078	0·17567	0·28820
12			0·00005	0·00067	0·00427	0·01675	0·04690	0·10285	0·18762
13			0·00001	0·00018	0·00146	0·00688	0·02241	0·05588	0·11384
14				0·00005	0·00046	0·00261	0·00993	0·02824	0·06445
15				0·00001	0·00014	0·00092	0·00409	0·01330	0·03410
16					0·00004	0·00030	0·00157	0·00585	0·01688
17					0·00001	0·00009	0·00056	0·00241	0·00784
18						0·00003	0·00019	0·00093	0·00342
19						0·00001	0·00006	0·00034	0·00140
20							0·00002	0·00012	0·00054
21							0·00001	0·00004	0·00020
22								0·00001	0·00007
23									0·00002
24									0·00001

[Table 3(a) contd]

p		0·10	0·15	0·20	0·25	0·30	0·35	0·40	0·45	0·50
$n = 2$	$r = 0$	1·00000	1·00000	1·00000	1·00000	1·00000	1·00000	1·00000	1·00000	1·00000
	1	0·19000	0·27750	0·36000	0·43750	0·51000	0·57750	0·64000	0·69750	0·75000
	2	0·01000	0·02250	0·04000	0·06250	0·09000	0·12250	0·16000	0·20250	0·25000
$n = 3$	$r = 0$	1·00000	1·00000	1·00000	1·00000	1·00000	1·00000	1·00000	1·00000	1·00000
	1	0·27100	0·38588	0·48800	0·57813	0·65700	0·72538	0·78400	0·83363	0·87500
	2	0·02800	0·06075	0·10400	0·15625	0·21600	0·28175	0·35200	0·42525	0·50000
	3	0·00100	0·00338	0·00800	0·01562	0·02700	0·04287	0·06400	0·09112	0·12500
$n = 4$	$r = 0$	1·00000	1·00000	1·00000	1·00000	1·00000	1·00000	1·00000	1·00000	1·00000
	1	0·34390	0·47799	0·59040	0·68359	0·75990	0·82149	0·87040	0·90849	0·93750
	2	0·05230	0·10952	0·18080	0·26172	0·34830	0·43702	0·52480	0·60902	0·68750
	3	0·00370	0·01198	0·02720	0·05078	0·08370	0·12648	0·17920	0·24148	0·31250
	4	0·00010	0·00051	0·00160	0·00391	0·00810	0·01501	0·02560	0·04101	0·06250
$n = 5$	$r = 0$	1·00000	1·00000	1·00000	1·00000	1·00000	1·00000	1·00000	1·00000	1·00000
	1	0·40951	0·55629	0·67232	0·76270	0·83193	0·88397	0·92224	0·94967	0·96875
	2	0·08146	0·16479	0·26272	0·36719	0·47178	0·57159	0·66304	0·74378	0·81250
	3	0·00856	0·02661	0·05792	0·10352	0·16308	0·23517	0·31744	0·40687	0·50000
	4	0·00046	0·00223	0·00672	0·01562	0·03078	0·05402	0·08704	0·13122	0·18750
	5	0·00001	0·00008	0·00032	0·00098	0·00243	0·00525	0·01024	0·01845	0·03125
$n = 6$	$r = 0$	1·00000	1·00000	1·00000	1·00000	1·00000	1·00000	1·00000	1·00000	1·00000
	1	0·46856	0·62285	0·73786	0·82202	0·88235	0·92458	0·95334	0·97232	0·98438
	2	0·11427	0·22352	0·34464	0·46606	0·57983	0·68092	0·76672	0·83643	0·89063
	3	0·01585	0·04734	0·09888	0·16943	0·25569	0·35291	0·45568	0·55848	0·65625
	4	0·00127	0·00589	0·01696	0·03760	0·07047	0·11742	0·17920	0·25526	0·34375
	5	0·00006	0·00040	0·00160	0·00464	0·01094	0·02232	0·04096	0·06920	0·10937
	6		0·00001	0·00006	0·00024	0·00073	0·00184	0·00410	0·00830	0·01562

n	r									
$n = 7$	$r = 0$	1·00000	1·00000	1·00000	1·00000	1·00000	1·00000	1·00000	1·00000	1·00000
	1	0·52170	0·67942	0·79028	0·86652	0·91765	0·95098	0·97201	0·98478	0·99219
	2	0·14969	0·28342	0·42328	0·55505	0·67058	0·76620	0·84137	0·89758	0·93750
	3	0·02569	0·07377	0·14803	0·24359	0·35293	0·46772	0·58010	0·68356	0·77344
	4	0·00273	0·01210	0·03334	0·07056	0·12604	0·19985	0·28975	0·39171	0·50000
	5	0·00018	0·00122	0·00467	0·01288	0·02880	0·05561	0·09626	0·15293	0·22656
	6	0·00001	0·00007	0·00037	0·00134	0·00379	0·00901	0·01884	0·03571	0·06250
	7			0·00001	0·00006	0·00022	0·00064	0·00164	0·00374	0·00781
$n = 8$	$r = 0$	1·00000	1·00000	1·00000	1·00000	1·00000	1·00000	1·00000	1·00000	1·00000
	1	0·56953	0·72751	0·83223	0·89989	0·94235	0·96814	0·98320	0·99163	0·99609
	2	0·18690	0·34282	0·49668	0·63292	0·74470	0·83087	0·89362	0·93682	0·96484
	3	0·03809	0·10521	0·20308	0·32146	0·44823	0·57219	0·68461	0·77987	0·85547
	4	0·00502	0·02135	0·05628	0·11382	0·19410	0·29360	0·40591	0·52304	0·63672
	5	0·00043	0·00285	0·01041	0·02730	0·05797	0·10609	0·17367	0·26038	0·36328
	6	0·00002	0·00024	0·00123	0·00423	0·01129	0·02532	0·04981	0·08846	0·14453
	7		0·00001	0·00008	0·00038	0·00129	0·00357	0·00852	0·01812	0·03516
	8				0·00002	0·00007	0·00023	0·00066	0·00168	0·00391
$n = 9$	$r = 0$	1·00000	1·00000	1·00000	1·00000	1·00000	1·00000	1·00000	1·00000	1·00000
	1	0·61258	0·76838	0·86578	0·92492	0·95965	0·97929	0·98992	0·99539	0·99805
	2	0·22516	0·40052	0·56379	0·69966	0·80400	0·87891	0·92946	0·96148	0·98047
	3	0·05297	0·14085	0·26180	0·39932	0·53717	0·66273	0·76821	0·85050	0·91016
	4	0·00833	0·03393	0·08564	0·16573	0·27034	0·39111	0·51739	0·63862	0·74609
	5	0·00089	0·00563	0·01958	0·04893	0·09881	0·17172	0·26657	0·37858	0·50000
	6	0·00006	0·00063	0·00307	0·00999	0·02529	0·05359	0·09935	0·16582	0·25391
	7		0·00005	0·00031	0·00134	0·00429	0·01118	0·02503	0·04977	0·08984
	8			0·00002	0·00011	0·00043	0·00140	0·00380	0·00908	0·01953
	9					0·00002	0·00008	0·00026	0·00076	0·00195
$n = 10$	$r = 0$	1·00000	1·00000	1·00000	1·00000	1·00000	1·00000	1·00000	1·00000	1·00000
	1	0·65132	0·80313	0·89263	0·94369	0·97175	0·98654	0·99395	0·99747	0·99902
	2	0·26390	0·45570	0·62419	0·75597	0·85069	0·91405	0·95364	0·97674	0·98926

[Table 3(a) contd]

p	0·10	0·15	0·20	0·25	0·30	0·35	0·40	0·45	0·50
$n = 10$ $r = 3$	0·07019	0·17980	0·32220	0·47441	0·61722	0·73839	0·83271	0·90044	0·94531
4	0·01280	0·04997	0·12087	0·22412	0·35039	0·48617	0·61772	0·73396	0·82813
5	0·00163	0·00987	0·03279	0·07813	0·15027	0·24850	0·36690	0·49560	0·62305
6	0·00015	0·00138	0·00637	0·01973	0·04735	0·09493	0·16624	0·26156	0·37695
7	0·00001	0·00013	0·00086	0·00351	0·01059	0·02602	0·05476	0·10199	0·17187
8		0·00001	0·00008	0·00042	0·00159	0·00482	0·01229	0·02739	0·05469
9				0·00003	0·00014	0·00054	0·00168	0·00450	0·01074
10					0·00001	0·00003	0·00010	0·00034	0·00098
$n = 20$ $r = 0$	1·00000	1·00000	1·00000	1·00000	1·00000	1·00000	1·00000	1·00000	1·00000
1	0·87842	0·96124	0·98847	0·99683	0·99920	0·99982	0·99996	0·99999	1·00000
2	0·60825	0·82444	0·93082	0·97569	0·99236	0·99787	0·99948	0·99989	0·99998
3	0·32307	0·59510	0·79392	0·90874	0·96452	0·98788	0·99639	0·99907	0·99980
4	0·13295	0·35227	0·58855	0·77484	0·89291	0·95562	0·98404	0·99507	0·99871
5	0·04317	0·17015	0·37035	0·58516	0·76249	0·88180	0·94905	0·98114	0·99409
6	0·01125	0·06731	0·19579	0·38283	0·58363	0·75460	0·87440	0·94467	0·97931
7	0·00239	0·02194	0·08669	0·21422	0·39199	0·58337	0·74999	0·87007	0·94234
8	0·00042	0·00592	0·03214	0·10181	0·22773	0·39897	0·58411	0·74799	0·86841
9	0·00006	0·00133	0·00998	0·04093	0·11333	0·23762	0·40440	0·58569	0·74828
10	0·00001	0·00025	0·00259	0·01386	0·04796	0·12178	0·24466	0·40864	0·58810
11		0·00004	0·00056	0·00394	0·01714	0·05317	0·12752	0·24929	0·41190
12			0·00010	0·00094	0·00514	0·01958	0·05653	0·13076	0·25172
13			0·00002	0·00018	0·00128	0·00602	0·02103	0·05803	0·13159
14				0·00003	0·00026	0·00152	0·00647	0·02141	0·05766
15					0·00004	0·00031	0·00161	0·00643	0·02069
16					0·00001	0·00005	0·00032	0·00153	0·00591
17						0·00001	0·00005	0·00028	0·00129
18							0·00001	0·00004	0·00020
19									0·00002

Table 3(b) Critical values, r, for the less frequently occurring sign for the two-tailed sign test when $p = 0{\cdot}5$.
Source: W. J. Mackinnon, 'Table for both the sign test and distribution-free confidence intervals of the median for sample sizes to 1000', *J. Am. Stat. Assoc.* 59 (1964), 937.

Largest value of r for which $P(r \text{ or fewer } +) + P(r \text{ or fewer } -) \leqslant \alpha$. Or $P(r \text{ or fewer } +) \leqslant \alpha/2$.

n	Two-tailed probability, α, for r					
	0·001	0·01	0·02	0·05	0·10	0·50
1	—	—	—	—	—	—
2	—	—	—	—	—	0
3	—	—	—	—	—	0
4	—	—	—	—	—	0
5	—	—	—	—	0	1
6	—	—	—	0	0	1
7	—	—	0	0	0	2
8	—	0	0	0	1	2
9	—	0	0	1	1	2
10	—	0	0	1	1	3
11	0	0	1	1	2	3
12	0	1	1	2	2	4
13	0	1	1	2	3	4
14	0	1	2	2	3	5
15	1	2	2	3	3	5
16	1	2	2	3	4	6
17	1	2	3	4	4	6
18	1	3	3	4	5	7
19	2	3	4	4	5	7
20	2	3	4	5	5	7
21	2	4	4	5	6	8
22	3	4	5	5	6	8
23	3	4	5	6	7	9
24	3	5	5	6	7	9
25	4	5	6	7	7	10
26	4	6	6	7	8	10
27	4	6	7	7	8	11
28	5	6	7	8	9	11
29	5	7	7	8	9	12
30	5	7	8	9	10	12
31	6	7	8	9	10	13
32	6	8	8	9	10	13
33	6	8	9	10	11	14
34	7	9	9	10	11	14
35	7	9	10	11	12	15
36	7	9	10	11	12	15
37	8	10	10	12	13	15
38	8	10	11	12	13	16
39	8	11	11	12	13	16

n	Two-tailed probability, α, for r					
	0·001	0·01	0·02	0·05	0·10	0·50
40	9	11	12	13	14	17
41	9	11	12	13	14	17
42	10	12	13	14	15	18
43	10	12	13	14	15	18
44	10	13	13	15	16	19
45	11	13	14	15	16	19
46	11	13	14	15	16	20
47	11	14	15	16	17	20
48	12	14	15	16	17	21
49	12	15	15	17	18	21
50	13	15	16	17	18	22
51	13	15	16	18	19	22
52	13	16	17	18	19	23
53	14	16	17	18	20	23
54	14	17	18	19	20	24
55	14	17	18	19	20	24
56	15	17	18	20	21	24
57	15	18	19	20	21	25
58	16	18	19	21	22	25
59	16	19	20	21	22	26
60	16	19	20	21	23	26
61	17	20	20	22	23	27
62	17	20	21	22	24	27
63	18	20	21	23	24	28
64	18	21	22	23	24	28
65	18	21	22	24	25	29
66	19	22	23	24	25	29
67	19	22	23	25	26	30
68	20	22	23	25	26	30
69	20	23	24	25	27	31
70	20	23	24	26	27	31
71	21	24	25	26	28	32
72	21	24	25	27	28	32
73	22	25	26	27	28	33
74	22	25	26	28	29	33
75	22	25	26	28	29	34
76	23	26	27	28	30	34
77	23	26	27	29	30	35

[Table 3(b) contd]

n	Two-tailed probability, α, for r						n	Two-tailed probability, α, for r					
	0·001	0·01	0·02	0·05	0·10	0·50		0·001	0·01	0·02	0·05	0·10	0·50
78	24	27	28	29	31	35	84	26	29	30	32	33	38
79	24	27	28	30	31	36	85	26	30	31	32	34	38
80	24	28	29	30	32	36	86	27	30	31	33	34	39
							87	27	31	32	33	35	39
81	25	28	29	31	32	36	88	28	31	32	34	35	40
82	25	28	30	31	33	37	89	28	31	33	34	36	40
83	26	29	30	32	33	37	90	29	32	33	35	36	41

Table 4 Critical lower-tail values for Fisher's exact test for 2 × 2 contingency tables.
Source: E. S. Pearson and H. O. Hartley (eds.), *Biometrika tables for statisticians: Vol. 1*, 3rd edn. (Cambridge 1970), 212–17.

Largest value of b, with fixed values for A, B and a, for the ratio a/A to be just significantly greater than b/B at a given nominal α and an exact probability for the 2 × 2 table.

a	$A-a$	A
b	$B-b$	B
$a+b$	$(A+B)-(a+b)$	$A+B$

With $A \geqslant B$, and $a/A \geqslant b/B$.

The table shows for given a, A, B, the value of b ($< a$) which is just significant at the given probability. The table also shows for given A, B and $r = a + b$, the exact probability under independence that b is less than or equal to the tabulated value of b.

A	B	a	Probability 0·05	0·025	0·01	0·005
$A=3$	$B=3$	3	0 ·050	—	—	—
$A=4$	$B=4$	4	0 ·014	0 ·014	—	—
	3	4	0 ·029	—	—	—
$A=5$	$B=5$	5	1 ·024	1 ·024	0 ·004	0 ·004
		4	0 ·024	0 ·024	—	—
	4	5	1 ·048	0 ·008	0 ·008	—
		4	0 ·040	—	—	—
	3	5	0 ·018	0 ·018	—	—
	2	5	0 ·048	—	—	—
$A=6$	$B=6$	6	2 ·030	1 ·008	1 ·008	0 ·001
		5	1 ·040	0 ·008	0 ·008	—
		4	0 ·030	—	—	—
	5	6	1 ·015^{+}	1 ·015^{+}	0 ·002	0 ·002
		5	0 ·013	0 ·013	—	—
		4	0 ·045^{+}	—	—	—
	4	6	1 ·033	0 ·005^{-}	0 ·005^{-}	0 ·005^{-}
		5	0 ·024	0 ·024	—	—
	3	6	0 ·012	0 ·012	—	—
		5	0 ·048	—	—	—
	2	6	0 ·036	—	—	—
$A=7$	$B=7$	7	3 ·035^{-}	2 ·010^{+}	1 ·002	1 ·002
		6	1 ·015^{-}	1 ·015^{-}	0 ·002	0 ·002
		5	0 ·010^{+}	0 ·010^{+}	—	—
		4	0 ·035^{-}	—	—	—

A	B	a	Probability 0·05	0·025	0·01	0·005
$A=7$	$B=6$	7	2 ·021	2 ·021	1 ·005^{-}	1 ·005^{-}
		6	1 ·025^{+}	0 ·004	0 ·004	0 ·004
		5	0 ·016	0 ·016	—	—
		4	0 ·049	—	—	—
	5	7	2 ·045^{+}	1 ·010^{+}	0 ·001	0 ·001
		6	1 ·045^{+}	0 ·008	0 ·008	—
		5	0 ·027	—	—	—
	4	7	1 ·024	1 ·024	0 ·003	0 ·003
		6	0 ·015^{+}	0 ·015^{+}	—	—
		5	0 ·045^{+}	—	—	—
	3	7	0 ·008	0 ·008	0 ·008	—
		6	0 ·033	—	—	—
	2	7	0 ·028	—	—	—
$A=8$	$B=8$	8	4 ·038	3 ·013	2 ·003	2 ·003
		7	2 ·020	2 ·020	1 ·005^{+}	0 ·001
		6	1 ·020	1 ·020	0 ·003	0 ·003
		5	0 ·013	0 ·013	—	—
		4	0 ·038	—	—	—
	7	8	3 ·026	2 ·007	2 ·007	1 ·001
		7	2 ·035^{-}	1 ·009	1 ·009	0 ·001
		6	1 ·032	0 ·006	0 ·006	—
		5	0 ·019	0 ·019	—	—
	6	8	2 ·015^{-}	2 ·015^{-}	1 ·003	1 ·003
		7	1 ·016	1 ·016	0 ·002	0 ·002
		6	0 ·009	0 ·009	0 ·009	—
		5	0 ·028	—	—	—

[Table 4 contd]

		Probability			
	a	0·05	0·025	0·01	0·005
$A=8$ $B=5$	8	2 ·035^{-}	1 ·007	1 ·007	0 ·001
	7	1 ·032	0 ·005^{-}	0 ·005^{-}	0 ·005^{-}
	6	0 ·016	0 ·016	—	—
	5	0 ·044	—	—	—
4	8	1 ·018	1 ·018	0 ·002	0 ·002
	7	0 ·010^{+}	0 ·010^{+}	—	—
	6	0 ·030	—	—	—
3	8	0 ·006	0 ·006	0 ·006	—
	7	0 ·024	0 ·024	—	—
2	8	0 ·022	0 ·022	—	—
$A=9$ $B=9$	9	5 ·041	4 ·015^{-}	3 ·005^{-}	3 ·005^{-}
	8	3 ·025^{-}	3 ·025^{-}	2 ·008	1 ·002
	7	2 ·028	1 ·008	1 ·008	0 ·001
	6	1 ·025^{-}	1 ·025^{-}	0 ·005^{-}	0 ·005^{-}
	5	0 ·015^{-}	0 ·015^{-}	—	—
	4	0 ·041	—	—	—
8	9	4 ·029	3 ·009	3 ·009	2 ·002
	8	3 ·043	2 ·013	1 ·003	1 ·003
	7	2 ·044	1 ·012	0 ·002	0 ·002
	6	1 ·036	0 ·007	0 ·007	—
	5	0 ·020	0 ·020	—	—
7	9	3 ·019	3 ·019	2 ·005^{-}	2 ·005^{-}
	8	2 ·024	2 ·024	1 ·006	0 ·001
	7	1 ·020	1 ·020	0 ·003	0 ·003
	6	0 ·010^{+}	0 ·010^{+}	—	—
	5	0 ·029	—	—	—

		Probability			
	a	0·05	0·025	0·01	0·005
$A=9$ $B=6$	9	3 ·044	2 ·011	1 ·002	1 ·002
	8	2 ·047	1 ·011	0 ·001	0 ·001
	7	1 ·035^{-}	0 ·006	0 ·006	—
	6	0 ·017	0 ·017	—	—
	5	0 ·042	—	—	—
5	9	2 ·027	1 ·005^{-}	1 ·005^{-}	1 ·005^{-}
	8	1 ·023	1 ·023	0 ·003	0 ·003
	7	0 ·010^{+}	0 ·010^{+}	—	—
	6	0 ·028	—	—	—
4	9	1 ·014	1 ·014	0 ·001	0 ·001
	8	0 ·007	0 ·007	0 ·007	—
	7	0 ·021	0 ·021	—	—
	6	0 ·049	—	—	—
3	9	1 ·045^{+}	0 ·005^{-}	0 ·005^{-}	0 ·005^{-}
	8	0 ·018	0 ·018	—	—
	7	0 ·045^{+}	—	—	—
2	9	0 ·018	0 ·018	—	—
$A=10$ $B=10$	10	6 ·043	5 ·016	4 ·005^{+}	3 ·002
	9	4 ·029	3 ·010^{-}	3 ·010^{-}	2 ·003
	8	3 ·035^{-}	2 ·012	1 ·003	1 ·003
	7	2 ·035^{-}	1 ·010^{-}	1 ·010^{-}	0 ·002
	6	1 ·029	0 ·005^{+}	0 ·005^{+}	—
	5	0 ·016	0 ·016	—	—
	4	0 ·043	—	—	—
9	10	5 ·033	4 ·011	3 ·003	3 ·003
	9	4 ·050^{-}	3 ·017	2 ·005^{-}	2 ·005^{-}

A=10 B=9	8	2 ·019	2 ·019	1 ·004	1 ·004
	7	1 ·015^{-}	1 ·015^{-}	0 ·002	0 ·002
	6	1 ·040	0 ·008	0 ·008	—
	5	0 ·022	0 ·022	—	—
8	10	4 ·023	4 ·023	3 ·007	2 ·002
	9	3 ·032	2 ·009	2 ·009	1 ·002
	8	2 ·031	1 ·008	1 ·008	0 ·001
	7	1 ·023	1 ·023	0 ·004	0 ·004
	6	0 ·011	0 ·011	—	—
	5	0 ·029	—	—	—
7	10	3 ·015^{-}	3 ·015^{-}	2 ·003	2 ·003
	9	2 ·018	2 ·018	1 ·004	1 ·004
	8	1 ·013	1 ·013	0 ·002	0 ·002
	7	1 ·036	0 ·006	0 ·006	—
	6	0 ·017	0 ·017	—	—
	5	0 ·041	—	—	—
6	10	3 ·036	2 ·008	2 ·008	1 ·001
	9	2 ·036	1 ·008	1 ·008	0 ·001
	8	1 ·024	1 ·024	0 ·003	0 ·003
	7	0 ·010^{+}	0 ·010^{+}	—	—
	6	0 ·026	—	—	—
5	10	2 ·022	2 ·022	1 ·004	1 ·004
	9	1 ·017	1 ·017	0 ·002	0 ·002
	8	1 ·047	0 ·007	0 ·007	—
	7	0 ·019	0 ·019	—	—
	6	0 ·042	—	—	—
4	10	1 ·011	1 ·011	0 ·001	0 ·001
	9	1 ·041	0 ·005^{-}	0 ·005^{-}	0 ·005^{-}
	8	0 ·015^{-}	0 ·015^{-}	—	—
	7	0 ·035^{-}	—	—	—
3	10	1 ·038	0 ·003	0 ·003	0 ·003
	9	0 ·014	0 ·014	—	—

A=10 B=3	8	0 ·035^{-}	—	—	—
2	10	0 ·015^{+}	0 ·015^{+}	—	—
	9	0 ·045^{+}	—	—	—
A=11 B=11	11	7 ·045^{+}	6 ·018	5 ·006	4 ·002
	10	5 ·032	4 ·012	3 ·004	3 ·004
	9	4 ·040	3 ·015^{-}	2 ·004	2 ·004
	8	3 ·043	2 ·015^{-}	1 ·004	1 ·004
	7	2 ·040	1 ·012	0 ·002	0 ·002
	6	1 ·032	0 ·006	0 ·006	—
	5	0 ·018	0 ·018	—	—
	4	0 ·045^{+}	—	—	—
10	11	6 ·035^{+}	5 ·012	4 ·004	4 ·004
	10	4 ·021	4 ·021	3 ·007	2 ·002
	9	3 ·024	3 ·024	2 ·007	1 ·002
	8	2 ·023	2 ·023	1 ·006	0 ·001
	7	1 ·017	1 ·017	0 ·003	0 ·003
	6	1 ·043	0 ·009	0 ·009	—
	5	0 ·023	0 ·023	—	—
9	11	5 ·026	4 ·008	4 ·008	3 ·002
	10	4 ·038	3 ·012	2 ·003	2 ·003
	9	3 ·040	2 ·012	1 ·003	1 ·003
	8	2 ·035^{-}	1 ·009	1 ·009	0 ·001
	7	1 ·025^{-}	1 ·025^{-}	0 ·004	0 ·004
	6	0 ·012	0 ·012	—	—
	5	0 ·030	—	—	—
8	11	4 ·018	4 ·018	3 ·005^{-}	3 ·005^{-}
	10	3 ·024	3 ·024	2 ·006	1 ·001
	9	2 ·022	2 ·022	1 ·005^{-}	1 ·005^{-}
	8	1 ·015^{-}	1 ·015^{-}	0 ·002	0 ·002
	7	1 ·037	0 ·007	0 ·007	—
	6	0 ·017	0 ·017	—	—
	5	0 ·040	—	—	—

[Table 4 contd]

	a	Probability 0·05	0·025	0·01	0·005
A=11 *B*=7	11	4 ·043	3 ·011	2 ·002	2 ·002
	10	3 ·047	2 ·013	1 ·002	1 ·002
	9	2 ·039	1 ·009	1 ·009	0 ·001
	8	1 ·025⁻	1 ·025⁻	0 ·004	0 ·004
	7	0 ·010⁺	0 ·010⁺	—	—
	6	0 ·025⁻	0 ·025⁻	—	—
6	11	3 ·029	2 ·006	2 ·006	1 ·001
	10	2 ·028	1 ·005⁺	1 ·005⁺	0 ·001
	9	1 ·018	1 ·018	0 ·002	0 ·002
	8	1 ·043	0 ·007	0 ·007	—
	7	0 ·017	0 ·017	—	—
	6	0 ·037	—	—	—
5	11	2 ·018	2 ·018	1 ·003	1 ·003
	10	1 ·013	1 ·013	0 ·001	0 ·001
	9	1 ·036	0 ·005⁻	0 ·005⁻	0 ·005⁻
	8	0 ·013	0 ·013	—	—
	7	0 ·029	—	—	—
4	11	1 ·009	1 ·009	1 ·009	0 ·001
	10	1 ·033	0 ·004	0 ·004	0 ·004
	9	0 ·011	0 ·011	—	—
	8	0 ·026	—	—	—
3	11	1 ·033	0 ·003	0 ·003	0 ·003
	10	0 ·011	0 ·011	—	—
	9	0 ·027	—	—	—
2	11	0 ·013	0 ·013	—	—
	10	0 ·038	—	—	—

	a	Probability 0·05	0·025	0·01	0·005
A=12 *B*=12	12	8 ·047	7 ·019	6 ·007	5 ·002
	11	6 ·034	5 ·014	4 ·005⁻	4 ·005⁻
	10	5 ·045⁻	4 ·018	3 ·006	2 ·002
	9	4 ·050⁻	3 ·020	2 ·006	1 ·001
	8	3 ·050⁻	2 ·018	1 ·005⁻	1 ·005⁻
	7	2 ·045⁻	1 ·014	0 ·002	0 ·002
	6	1 ·034	0 ·007	0 ·007	—
	5	0 ·019	0 ·019	—	—
	4	0 ·047	—	—	—
11	12	7 ·037	6 ·014	5 ·005⁻	5 ·005⁻
	11	5 ·024	5 ·024	4 ·008	3 ·002
	10	4 ·029	3 ·010⁺	2 ·003	2 ·003
	9	3 ·030	2 ·009	2 ·009	1 ·002
	8	2 ·026	1 ·007	1 ·007	0 ·001
	7	1 ·019	1 ·019	0 ·003	0 ·003
	6	1 ·045⁻	0 ·009	0 ·009	—
	5	0 ·024	0 ·024	—	—
10	12	6 ·029	5 ·010⁻	5 ·010⁻	4 ·003
	11	5 ·043	4 ·015⁺	3 ·005⁻	3 ·005⁻
	10	4 ·048	3 ·017	2 ·005⁻	2 ·005⁻
	9	3 ·046	2 ·015⁻	1 ·004	1 ·004
	8	2 ·038	1 ·010⁺	0 ·002	0 ·002
	7	1 ·026	0 ·005⁻	0 ·005⁻	0 ·005⁻
	6	0 ·012	0 ·012	—	—
	5	0 ·030	—	—	—
9	12	5 ·021	5 ·021	4 ·006	3 ·002
	11	4 ·029	3 ·009	3 ·009	2 ·002

A=12 B=9	10	3 ·029	2 ·008	2 ·008	1 ·002
	9	2 ·024	2 ·024	1 ·006	0 ·001
	8	1 ·016	1 ·016	0 ·002	0 ·002
	7	1 ·037	0 ·007	0 ·007	—
	6	0 ·017	0 ·017	—	—
	5	0 ·039	—	—	—
8	12	5 ·049	4 ·014	3 ·004	3 ·004
	11	3 ·018	3 ·018	2 ·004	2 ·004
	10	2 ·015$^+$	2 ·015$^+$	1 ·003	1 ·003
	9	2 ·040	1 ·010$^-$	1 ·010$^-$	0 ·001
	8	1 ·025$^-$	1 ·025$^-$	0 ·004	0 ·004
	7	0 ·010$^+$	0 ·010$^+$	—	—
	6	0 ·024	0 ·024	—	—
7	12	4 ·036	3 ·009	3 ·009	2 ·002
	11	3 ·038	2 ·010$^-$	2 ·010$^-$	1 ·002
	10	2 ·029	1 ·006	1 ·006	0 ·001
	9	1 ·017	1 ·017	0 ·002	0 ·002
	8	1 ·040	0 ·007	0 ·007	—
	7	0 ·016	0 ·016	—	—
	6	0 ·034	—	—	—
6	12	3 ·025$^-$	3 ·025$^-$	2 ·005$^-$	2 ·005$^-$
	11	2 ·022	2 ·022	1 ·004	1 ·004
	10	1 ·013	1 ·013	0 ·002	0 ·002
	9	1 ·032	0 ·005$^-$	0 ·005$^-$	0 ·005$^-$
	8	0 ·011	0 ·011	—	—
	7	0 ·025$^-$	0 ·025$^-$	—	—
	6	0 ·050$^-$	—	—	—
5	12	2 ·015$^-$	2 ·015$^-$	1 ·002	1 ·002
	11	1 ·010$^-$	1 ·010$^-$	1 ·010$^-$	0 ·001
	10	1 ·028	0 ·003	0 ·003	0 ·003
	9	0 ·009	0 ·009	0 ·009	—
	8	0 ·020	0 ·020	—	—

A=12 B=5	7	0 ·041	—	—	—
4	12	2 ·050	1 ·007	1 ·007	0 ·001
	11	1 ·027	0 ·003	0 ·003	0 ·003
	10	0 ·008	0 ·008	0 ·008	—
	9	0 ·019	0 ·019	—	—
	8	0 ·038	—	—	—
3	12	1 ·029	0 ·002	0 ·002	0 ·002
	11	0 ·009	0 ·009	0 ·009	—
	10	0 ·022	0 ·022	—	—
	9	0 ·044	—	—	—
2	12	0 ·011	0 ·011	—	—
	11	0 ·033	—	—	—
A=13 B=13	13	9 ·048	8 ·020	7 ·007	6 ·003
	12	7 ·037	6 ·015$^+$	5 ·006	4 ·002
	11	6 ·048	5 ·021	4 ·008	3 ·002
	10	4 ·024	4 ·024	3 ·008	2 ·002
	9	3 ·024	3 ·024	2 ·008	1 ·002
	8	2 ·021	2 ·021	1 ·006	0 ·001
	7	2 ·048	1 ·015$^+$	0 ·003	0 ·003
	6	1 ·037	0 ·007	0 ·007	—
	5	0 ·020	0 ·020	—	—
	4	0 ·048	—	—	—
12	13	8 ·039	7 ·015$^-$	6 ·005$^+$	5 ·002
	12	6 ·027	5 ·010$^-$	5 ·010$^-$	4 ·003
	11	5 ·033	4 ·013	3 ·004	3 ·004
	10	4 ·036	3 ·013	2 ·004	2 ·004
	9	3 ·034	2 ·011	1 ·003	1 ·003
	8	2 ·029	1 ·008	1 ·008	0 ·001
	7	1 ·020	1 ·020	0 ·004	0 ·004
	6	1 ·046	0 ·010$^-$	0 ·010$^-$	—
	5	0 ·024	0 ·024	—	—

[Table 4 contd]

	a	0·05	0·025	0·01	0·005
		Probability			
A=13 B=11	13	7 ·031	6 ·011	5 ·003	5 ·003
	12	6 ·048	5 ·018	4 ·006	3 ·002
	11	4 ·021	4 ·021	3 ·007	2 ·002
	10	3 ·021	3 ·021	2 ·006	1 ·001
	9	3 ·050$^-$	2 ·017	1 ·004	1 ·004
	8	2 ·040	1 ·011	0 ·002	0 ·002
	7	1 ·027	0 ·005$^-$	0 ·005$^-$	0 ·005$^-$
	6	0 ·013	0 ·013	—	—
	5	0 ·030	—	—	—
10	13	6 ·024	6 ·024	5 ·007	4 ·002
	12	5 ·035$^-$	4 ·012	3 ·003	3 ·003
	11	4 ·037	3 ·012	2 ·003	2 ·003
	10	3 ·033	2 ·010$^+$	1 ·002	1 ·002
	9	2 ·026	1 ·006	1 ·006	0 ·001
	8	1 ·017	1 ·017	0 ·003	0 ·003
	7	1 ·038	0 ·007	0 ·007	—
	6	0 ·017	0 ·017	—	—
	5	0 ·038	—	—	—
9	13	5 ·017	5 ·017	4 ·005$^-$	4 ·005$^-$
	12	4 ·023	4 ·023	3 ·007	2 ·001
	11	3 ·022	3 ·022	2 ·006	1 ·001
	10	2 ·017	2 ·017	1 ·004	1 ·004
	9	2 ·040	1 ·010$^+$	0 ·001	0 ·001
	8	1 ·025$^-$	1 ·025$^-$	0 ·004	0 ·004
	7	0 ·010$^+$	0 ·010$^+$	—	—
	6	0 ·023	0 ·023	—	—
	5	0 ·049	—	—	—

	a	0·05	0·025	0·01	0·005
		Probability			
A=13 B=8	13	5 ·042	4 ·012	3 ·003	3 ·003
	12	4 ·047	3 ·014	2 ·003	2 ·003
	11	3 ·041	2 ·011	1 ·002	1 ·002
	10	2 ·029	1 ·007	1 ·007	0 ·001
	9	1 ·017	1 ·017	0 ·002	0 ·002
	8	1 ·037	0 ·006	0 ·006	—
	7	0 ·015$^-$	0 ·015$^-$	—	—
	6	0 ·032	—	—	—
7	13	4 ·031	3 ·007	3 ·007	2 ·001
	12	3 ·031	2 ·007	2 ·007	1 ·001
	11	2 ·022	2 ·022	1 ·004	1 ·004
	10	1 ·012	1 ·012	0 ·002	0 ·002
	9	1 ·029	0 ·004	0 ·004	0 ·004
	8	0 ·010$^+$	0 ·010$^+$	—	—
	7	0 ·022	0 ·022	—	—
	6	0 ·044	—	—	—
6	13	3 ·021	3 ·021	2 ·004	2 ·004
	12	2 ·017	2 ·017	1 ·003	1 ·003
	11	2 ·046	1 ·010$^-$	1 ·010$^-$	0 ·001
	10	1 ·024	1 ·024	0 ·003	0 ·003
	9	1 ·050$^-$	0 ·008	0 ·008	—
	8	0 ·017	0 ·017	—	—
	7	0 ·034	—	—	—
5	13	2 ·012	2 ·012	1 ·002	1 ·002
	12	2 ·044	1 ·008	1 ·008	0 ·001
	11	1 ·022	1 ·022	0 ·002	0 ·002
	10	1 ·047	0 ·007	0 ·007	—

$A=13$ $B=5$	9	0 ·015⁻	0 ·015⁻	—	—
	8	0 ·029	—	—	—
4	13	2 ·044	1 ·006	1 ·006	0 ·000
	12	1 ·022	1 ·022	0 ·002	0 ·002
	11	0 ·006	0 ·006	0 ·006	—
	10	0 ·015⁻	0 ·015⁻	—	—
	9	0 ·029	—	—	—
3	13	1 ·025	1 ·025	0 ·002	0 ·002
	12	0 ·007	0 ·007	0 ·007	—
	11	0 ·018	0 ·018	—	—
	10	0 ·036	—	—	—
2	13	0 ·010⁻	0 ·010⁻	0 ·010⁻	—
	12	0 ·029	—	—	—
$A=14$ $B=14$	14	10 ·049	9 ·020	8 ·008	7 ·003
	13	8 ·038	7 ·016	6 ·006	5 ·002
	12	6 ·023	6 ·023	5 ·009	4 ·003
	11	5 ·027	4 ·011	3 ·004	3 ·004
	10	4 ·028	3 ·011	2 ·003	2 ·003
	9	3 ·027	2 ·009	2 ·009	1 ·002
	8	2 ·023	2 ·023	1 ·006	0 ·001
	7	1 ·016	1 ·016	0 ·003	0 ·003
	6	1 ·038	0 ·008	0 ·008	—
	5	0 ·020	0 ·020	—	—
	4	0 ·049	—	—	—
13	14	9 ·041	8 ·016	7 ·006	6 ·002
	13	7 ·029	6 ·011	5 ·004	5 ·004
	12	6 ·037	5 ·015⁺	4 ·005⁺	3 ·002
	11	5 ·041	4 ·017	3 ·006	2 ·001
	10	4 ·041	3 ·016	2 ·005⁻	2 ·005⁻
	9	3 ·038	2 ·013	1 ·003	1 ·003
	8	2 ·031	1 ·009	1 ·009	0 ·001

$A=14$ $B=13$	7	1 ·021	1 ·021	0 ·004	0 ·004
	6	1 ·048	0 ·010⁺	—	—
	5	0 ·025⁻	0 ·025⁻	—	—
12	14	8 ·033	7 ·012	6 ·004	6 ·004
	13	6 ·021	6 ·021	5 ·007	4 ·002
	12	5 ·025⁺	4 ·009	4 ·009	3 ·003
	11	4 ·026	3 ·009	3 ·009	2 ·002
	10	3 ·024	3 ·024	2 ·007	1 ·002
	9	2 ·019	2 ·019	1 ·005⁻	1 ·005⁻
	8	2 ·042	1 ·012	0 ·002	0 ·002
	7	1 ·028	0 ·005⁺	0 ·005⁺	—
	6	0 ·013	0 ·013	—	—
	5	0 ·030	—	—	—
11	14	7 ·026	6 ·009	6 ·009	5 ·003
	13	6 ·039	5 ·014	4 ·004	4 ·004
	12	5 ·043	4 ·016	3 ·005⁻	3 ·005⁻
	11	4 ·042	3 ·015⁻	2 ·004	2 ·004
	10	3 ·036	2 ·011	1 ·003	1 ·003
	9	2 ·027	1 ·007	1 ·007	0 ·001
	8	1 ·017	1 ·017	0 ·003	0 ·003
	7	1 ·038	0 ·007	0 ·007	—
	6	0 ·017	0 ·017	—	—
	5	0 ·038	—	—	—
10	14	6 ·020	6 ·020	5 ·006	4 ·002
	13	5 ·028	4 ·009	4 ·009	3 ·002
	12	4 ·028	3 ·009	3 ·009	2 ·002
	11	3 ·024	3 ·024	2 ·007	1 ·001
	10	2 ·018	2 ·018	1 ·004	1 ·004
	9	2 ·040	1 ·011	0 ·002	0 ·002
	8	1 ·024	1 ·024	0 ·004	0 ·004
	7	0 ·010⁻	0 ·010⁻	0 ·010⁻	—
	6	0 ·022	0 ·022	—	—

[Table 4 contd]

	a	Probability 0·05	0·025	0·01	0·005
A=14 *B*=10	5	0 ·047	—	—	—
9	14	6 ·047	5 ·014	4 ·004	4 ·004
	13	4 ·018	4 ·018	3 ·005^{-}	3 ·005^{-}
	12	3 ·017	3 ·017	2 ·004	2 ·004
	11	3 ·042	2 ·012	1 ·002	1 ·002
	10	2 ·029	1 ·007	1 ·007	0 ·001
	9	1 ·017	1 ·017	0 ·002	0 ·002
	8	1 ·036	0 ·006	0 ·006	—
	7	0 ·014	0 ·014	—	—
	6	0 ·030	—	—	—
8	14	5 ·036	4 ·010^{-}	4 ·010^{-}	3 ·002
	13	4 ·039	3 ·011	2 ·002	2 ·002
	12	3 ·032	2 ·008	2 ·008	1 ·001
	11	2 ·022	2 ·022	1 ·005^{-}	1 ·005^{-}
	10	2 ·048	1 ·012	0 ·002	0 ·002
	9	1 ·026	0 ·004	0 ·004	0 ·004
	8	0 ·009	0 ·009	0 ·009	—
	7	0 ·020	0 ·020	—	—
	6	0 ·040	—	—	—
7	14	4 ·026	3 ·006	3 ·006	2 ·001
	13	3 ·025	2 ·006	2 ·006	1 ·001
	12	2 ·017	2 ·017	1 ·003	1 ·003
	11	2 ·041	1 ·009	1 ·009	0 ·001
	10	1 ·021	1 ·021	0 ·003	0 ·003
	9	1 ·043	0 ·007	0 ·007	—
	8	0 ·015^{-}	0 ·015^{-}	—	—
	7	0 ·030	—	—	—

	a	Probability 0·05	0·025	0·01	0·005
A=14 *B*=6	14	3 ·018	3 ·018	2 ·003	2 ·003
	13	2 ·014	2 ·014	1 ·002	1 ·002
	12	2 ·037	1 ·007	1 ·007	0 ·001
	11	1 ·018	1 ·018	0 ·002	0 ·002
	10	1 ·038	0 ·005^{+}	0 ·005^{+}	—
	9	0 ·012	0 ·012	—	—
	8	0 ·024	0 ·024	—	—
	7	0 ·044	—	—	—
5	14	2 ·010^{+}	2 ·010^{+}	1 ·001	1 ·001
	13	2 ·037	1 ·006	1 ·006	0 ·001
	12	1 ·017	1 ·017	0 ·002	0 ·002
	11	1 ·038	0 ·005^{-}	0 ·005^{-}	0 ·005^{-}
	10	0 ·011	0 ·011	—	—
	9	0 ·022	0 ·022	—	—
	8	0 ·040	—	—	—
4	14	2 ·039	1 ·005^{-}	1 ·005^{-}	1 ·005^{-}
	13	1 ·019	1 ·019	0 ·002	0 ·002
	12	1 ·044	0 ·005^{-}	0 ·005^{-}	0 ·005^{-}
	11	0 ·011	0 ·011	—	—
	10	0 ·023	0 ·023	—	—
	9	0 ·041	—	—	—
3	14	1 ·022	1 ·022	0 ·001	0 ·001
	13	0 ·006	0 ·006	0 ·006	—
	12	0 ·015^{-}	0 ·015^{-}	—	—
	11	0 ·029	—	—	—
2	14	0 ·008	0 ·008	0 ·008	—

A=14 B=2	13	0 ·025	0 ·025	—	—
	12	0 ·050	—	—	—
A=15 B=15	15	11 ·050^{-}	10 ·021	9 ·008	8 ·003
	14	9 ·040	8 ·018	7 ·007	6 ·003
	13	7 ·025^{+}	6 ·010^{+}	5 ·004	5 ·004
	12	6 ·030	5 ·013	4 ·005^{-}	4 ·005^{-}
	11	5 ·033	4 ·013	3 ·005^{-}	3 ·005^{-}
	10	4 ·033	3 ·013	2 ·004	2 ·004
	9	3 ·030	2 ·010^{+}	1 ·003	1 ·003
	8	2 ·025^{+}	1 ·007	1 ·007	0 ·001
	7	1 ·018	1 ·018	0 ·003	0 ·003
	6	1 ·040	0 ·008	0 ·008	—
	5	0 ·021	0 ·021	—	—
	4	0 ·050^{-}	—	—	—
14	15	10 ·042	9 ·017	8 ·006	7 ·002
	14	8 ·031	7 ·013	6 ·005^{-}	6 ·005^{-}
	13	7 ·041	6 ·017	5 ·007	4 ·002
	12	6 ·046	5 ·020	4 ·007	3 ·002
	11	5 ·048	4 ·020	3 ·007	2 ·002
	10	4 ·046	3 ·018	2 ·006	1 ·001
	9	3 ·041	2 ·014	1 ·004	1 ·004
	8	2 ·033	1 ·009	1 ·009	0 ·001
	7	1 ·022	1 ·022	0 ·004	0 ·004
	6	1 ·049	0 ·011	—	—
	5	0 ·025^{+}	—	—	—
13	15	9 ·035^{-}	8 ·013	7 ·005^{-}	7 ·005^{-}
	14	7 ·023	7 ·023	6 ·009	5 ·003
	13	6 ·029	5 ·011	4 ·004	4 ·004
	12	5 ·031	4 ·012	3 ·004	3 ·004
	11	4 ·030	3 ·011	2 ·003	2 ·003
	10	3 ·026	2 ·008	2 ·008	1 ·002
A=15 B=13	9	2 ·020	2 ·020	1 ·005^{+}	0 ·001
	8	2 ·043	1 ·013	0 ·002	0 ·002
	7	1 ·029	0 ·005^{+}	0 ·005^{+}	—
	6	0 ·013	0 ·013	—	—
	5	0 ·031	—	—	—
12	15	8 ·028	7 ·010^{-}	7 ·010^{+}	6 ·003
	14	7 ·043	6 ·016	5 ·006	4 ·002
	13	6 ·049	5 ·019	4 ·007	3 ·002
	12	5 ·049	4 ·019	3 ·006	2 ·002
	11	4 ·045^{+}	3 ·017	2 ·005^{-}	2 ·005^{-}
	10	3 ·038	2 ·012	1 ·003	1 ·003
	9	2 ·028	1 ·007	1 ·007	0 ·001
	8	1 ·018	1 ·018	0 ·003	0 ·003
	7	1 ·038	0 ·007	0 ·007	—
	6	0 ·017	0 ·017	—	—
	5	0 ·037	—	—	—
11	15	7 ·022	7 ·022	6 ·007	5 ·002
	14	6 ·032	5 ·011	4 ·003	4 ·003
	13	5 ·034	4 ·012	3 ·003	3 ·003
	12	4 ·032	3 ·010^{+}	2 ·003	2 ·003
	11	3 ·026	2 ·008	2 ·008	1 ·002
	10	2 ·019	2 ·019	1 ·004	1 ·004
	9	2 ·040	1 ·011	0 ·002	0 ·002
	8	1 ·024	1 ·024	0 ·004	0 ·004
	7	1 ·049	0 ·010^{-}	0 ·010^{-}	—
	6	0 ·022	0 ·022	—	—
	5	0 ·046	—	—	—
10	15	6 ·017	6 ·017	5 ·005^{-}	5 ·005^{-}
	14	5 ·023	5 ·023	4 ·007	3 ·002
	13	4 ·022	4 ·022	3 ·007	2 ·001
	12	3 ·018	3 ·018	2 ·005^{-}	2 ·005^{-}

[Table 4 contd]

	a	Probability 0·05	0·025	0·01	0·005
A=15 B=10	11	3 ·042	2 ·013	1 ·003	1 ·003
	10	2 ·029	1 ·007	1 ·007	0 ·001
	9	1 ·016	1 ·016	0 ·002	0 ·002
	8	1 ·034	0 ·006	0 ·006	—
	7	0 ·013	0 ·013	—	—
	6	0 ·028	—	—	—
9	15	6 ·042	5 ·012	4 ·003	4 ·003
	14	5 ·047	4 ·015⁻	3 ·004	3 ·004
	13	4 ·042	3 ·013	2 ·003	2 ·003
	12	3 ·032	2 ·009	2 ·009	1 ·002
	11	2 ·021	2 ·021	1 ·005⁻	1 ·005⁻
	10	2 ·045⁻	1 ·011	0 ·002	0 ·002
	9	1 ·024	1 ·024	0 ·004	0 ·004
	8	1 ·048	0 ·009	0 ·009	—
	7	0 ·019	0 ·019	—	—
	6	0 ·037	—	—	—
8	15	5 ·032	4 ·008	4 ·008	3 ·002
	14	4 ·033	3 ·009	3 ·009	2 ·002
	13	3 ·026	2 ·006	2 ·006	1 ·001
	12	2 ·017	2 ·017	1 ·003	1 ·003
	11	2 ·037	1 ·008	1 ·008	0 ·001
	10	1 ·019	1 ·019	0 ·003	0 ·003
	9	1 ·038	0 ·006	0 ·006	—
	8	0 ·013	0 ·013	—	—
	7	0 ·026	—	—	—
	6	0 ·050⁻	—	—	—
7	15	4 ·023	4 ·023	3 ·005⁻	3 ·005⁻
	14	3 ·021	3 ·021	2 ·004	2 ·004

	a	Probability 0·05	0·025	0·01	0·005
A=15 B=7	13	2 ·014	2 ·014	1 ·002	1 ·002
	12	2 ·032	1 ·007	1 ·007	0 ·001
	11	1 ·015⁺	1 ·015⁺	0 ·002	0 ·002
	10	1 ·032	0 ·005⁻	0 ·005⁻	0 ·005⁻
	9	0 ·010⁺	0 ·010⁺	—	—
	8	0 ·020	0 ·020	—	—
	7	0 ·038	—	—	—
6	15	3 ·015⁺	3 ·015⁺	2 ·003	2 ·003
	14	2 ·011	2 ·011	1 ·002	1 ·002
	13	2 ·031	1 ·006	1 ·006	0 ·001
	12	1 ·014	1 ·014	0 ·002	0 ·002
	11	1 ·029	0 ·004	0 ·004	0 ·004
	10	0 ·009	0 ·009	0 ·009	—
	9	0 ·017	0 ·017	—	—
	8	0 ·032	—	—	—
5	15	2 ·009	2 ·009	2 ·009	1 ·001
	14	2 ·032	1 ·005⁻	1 ·005⁻	1 ·005⁻
	13	1 ·014	1 ·014	0 ·001	0 ·001
	12	1 ·031	0 ·004	0 ·004	0 ·004
	11	0 ·008	0 ·008	0 ·008	—
	10	0 ·016	0 ·016	—	—
	9	0 ·030	—	—	—
4	15	2 ·035⁺	1 ·004	1 ·004	1 ·004
	14	1 ·016	1 ·016	0 ·001	0 ·001
	13	1 ·037	0 ·004	0 ·004	0 ·004
	12	0 ·009	0 ·009	0 ·009	—
	11	0 ·018	0 ·018	—	—
	10	0 ·033	—	—	—

$A=15$ $B=3$	15	1 ·020	1 ·020	0 ·001	0 ·001	$A=15$ $B=2$	15	0 ·007	0 ·007	0 ·007	—
	14	0 ·005⁻	0 ·005⁻	0 ·005⁻	0 ·005⁻		14	0 ·022	0 ·022	—	—
	13	0 ·012	0 ·012	—	—		13	0 ·044	—	—	—
	12	0 ·025⁻	0 ·025⁻	—	—						
	11	0 ·043	—	—	—						

Table 5 Cumulative Poisson probabilities.
Source: J. White, A. Yeats and G. Skipworth, *Tables for statisticians* (London 1974), 12–13.

λ	0·1	0·2	0·3	0·4	0·5	0·6	0·7	0·8	0·9	1·0
$r = 0$	1·00000	1·00000	1·00000	1·00000	1·00000	1·00000	1·00000	1·00000	1·00000	1·00000
1	0·09516	0·18127	0·25918	0·32968	0·39347	0·45119	0·50341	0·55067	0·59343	0·63212
2	0·00468	0·01752	0·03694	0·06155	0·09020	0·12190	0·15580	0·19121	0·22752	0·26424
3	0·00015	0·00115	0·00360	0·00793	0·01439	0·02312	0·03414	0·04742	0·06286	0·08030
4		0·00006	0·00027	0·00078	0·00175	0·00336	0·00575	0·00908	0·01346	0·01899
5			0·00002	0·00006	0·00017	0·00039	0·00079	0·00141	0·00234	0·00366
6					0·00001	0·00004	0·00009	0·00018	0·00034	0·00059
7							0·00001	0·00002	0·00004	0·00008
8										0·00001

λ	1·1	1·2	1·3	1·4	1·5	1·6	1·7	1·8	1·9	2·0
$r = 0$	1·00000	1·00000	1·00000	1·00000	1·00000	1·00000	1·00000	1·00000	1·00000	1·00000
1	0·66713	0·69881	0·72747	0·75340	0·77687	0·79810	0·81732	0·83470	0·85043	0·86466
2	0·30097	0·33737	0·37318	0·40817	0·44217	0·47507	0·50675	0·53716	0·56625	0·59399
3	0·09958	0·12051	0·14289	0·16650	0·19115	0·21664	0·24278	0·26938	0·29628	0·32332
4	0·02574	0·03377	0·04310	0·05373	0·06564	0·07881	0·09319	0·10871	0·12530	0·14288
5	0·00544	0·00775	0·01066	0·01425	0·01858	0·02368	0·02961	0·03641	0·04408	0·05265
6	0·00097	0·00150	0·00223	0·00320	0·00446	0·00604	0·00800	0·01038	0·01322	0·01656
7	0·00015	0·00025	0·00040	0·00062	0·00093	0·00134	0·00188	0·00257	0·00345	0·00453
8	0·00002	0·00004	0·00006	0·00011	0·00017	0·00026	0·00039	0·00056	0·00079	0·00110
9			0·00001	0·00002	0·00003	0·00005	0·00007	0·00011	0·00016	0·00024
10						0·00001	0·00001	0·00002	0·00003	0·00005
11									0·00001	0·00001

The probability that *r or more* random events occur in an interval when the average number of such events per interval is λ is tabulated below.

λ	2·1	2·2	2·3	2·4	2·5	2·6	2·7	2·8	2·9	3·0
$r = 0$	1·00000	1·00000	1·00000	1·00000	1·00000	1·00000	1·00000	1·00000	1·00000	1·00000
1	0·87754	0·88920	0·89974	0·90928	0·91792	0·92573	0·93279	0·93919	0·94498	0·95021
2	0·62039	0·64543	0·66915	0·69156	0·71270	0·73262	0·75134	0·76892	0·78541	0·80085
3	0·35037	0·37729	0·40396	0·43029	0·45619	0·48157	0·50638	0·53055	0·55404	0·57681
4	0·16136	0·18065	0·20065	0·22128	0·24242	0·26400	0·28591	0·30806	0·33038	0·35277
5	0·06213	0·07250	0·08375	0·09587	0·10882	0·12258	0·13709	0·15232	0·16822	0·18474
6	0·02045	0·02491	0·02998	0·03567	0·04202	0·04904	0·05673	0·06511	0·07417	0·08392
7	0·00586	0·00746	0·00936	0·01159	0·01419	0·01717	0·02057	0·02441	0·02872	0·03351
8	0·00149	0·00198	0·00259	0·00334	0·00425	0·00533	0·00662	0·00813	0·00988	0·01190
9	0·00034	0·00047	0·00064	0·00086	0·00114	0·00149	0·00191	0·00243	0·00306	0·00380
10	0·00007	0·00010	0·00014	0·00020	0·00028	0·00038	0·00050	0·00066	0·00086	0·00110
11	0·00001	0·00002	0·00003	0·00004	0·00006	0·00009	0·00012	0·00016	0·00022	0·00029
12			0·00001	0·00001	0·00001	0·00002	0·00003	0·00004	0·00005	0·00007
13							0·00001	0·00001	0·00001	0·00002

λ	3·1	3·2	3·3	3·4	3·5	3·6	3·7	3·8	3·9	4·0
$r = 0$	1·00000	1·00000	1·00000	1·00000	1·00000	1·00000	1·00000	1·00000	1·00000	1·00000
1	0·95495	0·95924	0·96312	0·96663	0·96980	0·97268	0·97528	0·97763	0·97976	0·98168
2	0·81530	0·82880	0·84140	0·85316	0·86411	0·87431	0·88380	0·89262	0·90081	0·90842
3	0·59884	0·62010	0·64057	0·66026	0·67915	0·69725	0·71457	0·73110	0·76487	0·76190
4	0·37516	0·39748	0·41966	0·44164	0·46337	0·48478	0·50585	0·52652	0·54675	0·56653
5	0·20181	0·21939	0·23741	0·25582	0·27456	0·29356	0·31278	0·33216	0·35163	0·37116
6	0·09433	0·10541	0·11712	0·12946	0·14239	0·15588	0·16991	0·18444	0·19944	0·21487
7	0·03880	0·04462	0·05097	0·05785	0·06529	0·07327	0·08181	0·09089	0·10052	0·11067

[Table 5 contd]

λ	3·1	3·2	3·3	3·4	3·5	3·6	3·7	3·8	3·9	4·0
$r = 8$	0·01421	0·01683	0·01978	0·02307	0·02674	0·03079	0·03524	0·04011	0·04540	0·05113
9	0·00468	0·00571	0·00691	0·00829	0·00987	0·01167	0·01370	0·01598	0·01853	0·02136
10	0·00140	0·00176	0·00219	0·00271	0·00331	0·00402	0·00485	0·00580	0·00689	0·00813
11	0·00038	0·00050	0·00064	0·00081	0·00102	0·00127	0·00157	0·00193	0·00235	0·00284
12	0·00010	0·00013	0·00017	0·00022	0·00029	0·00037	0·00047	0·00059	0·00074	0·00092
13	0·00002	0·00003	0·00004	0·00006	0·00008	0·00010	0·00013	0·00017	0·00022	0·00027
14		0·00001	0·00001	0·00001	0·00002	0·00003	0·00003	0·00004	0·00006	0·00008
15						0·00001	0·00001	0·00001	0·00001	0·00002

λ	4·1	4·2	4·3	4·4	4·5	4·6	4·7	4·8	4·9	5·0
$r = 0$	1·00000	1·00000	1·00000	1·00000	1·00000	1·00000	1·00000	1·00000	1·00000	1·00000
1	0·98343	0·98500	0·98643	0·98772	0·98889	0·98995	0·99090	0·99177	0·99255	0·99326
2	0·91548	0·92202	0·92809	0·93370	0·93890	0·94371	0·94816	0·95227	0·95607	0·95957
3	0·77619	0·78976	0·80265	0·81486	0·82642	0·83736	0·84770	0·85746	0·86667	0·87535
4	0·58582	0·60460	0·62285	0·64055	0·65770	0·67429	0·69032	0·70577	0·72066	0·73497
5	0·39069	0·41017	0·42956	0·44882	0·46790	0·48677	0·50539	0·52374	0·54179	0·55951
6	0·23069	0·24686	0·26334	0·28009	0·29707	0·31424	0·33156	0·34899	0·36650	0·38404
7	0·12135	0·13254	0·14421	0·15635	0·16895	0·18197	0·19539	0·20920	0·22335	0·23782
8	0·05731	0·06394	0·07103	0·07858	0·08659	0·09505	0·10397	0·11333	0·12314	0·13337
9	0·02449	0·02793	0·03170	0·03580	0·04026	0·04507	0·05026	0·05582	0·06176	0·06809
10	0·00954	0·01113	0·01291	0·01489	0·01709	0·01953	0·02221	0·02514	0·02834	0·03183
11	0·00341	0·00407	0·00482	0·00569	0·00667	0·00778	0·00902	0·01042	0·01197	0·01370
12	0·00113	0·00137	0·00167	0·00201	0·00240	0·00286	0·00339	0·00399	0·00468	0·00545
13	0·00034	0·00043	0·00053	0·00066	0·00081	0·00098	0·00118	0·00142	0·00170	0·00202
14	0·00010	0·00013	0·00016	0·00020	0·00025	0·00031	0·00039	0·00047	0·00058	0·00070
15	0·00003	0·00003	0·00004	0·00006	0·00007	0·00009	0·00012	0·00015	0·00018	0·00023
16	0·00001	0·00001	0·00001	0·00002	0·00002	0·00003	0·00003	0·00004	0·00005	0·00007
17					0·00001	0·00001	0·00001	0·00001	0·00002	0·00002

Table 6 Critical lower-tail values of W_n for Wilcoxon's rank-sum test.
Source: E. S. Pearson and H. O. Hartley (eds.), *Biometrika tables for statisticians: Vol. 2* (Cambridge 1972), 227–30.

Largest values of W_n which are just significant at the probability level given for each column. $2\overline{W} = 2\,E(W)$ and the upper-tail critical value, $W_n' = 2\overline{W} - W_n$.

	$n = 1$							$n = 2$							
m	0·001	0·005	0·010	0·025	0·05	0·10	$2\overline{W}$	0·001	0·005	0·010	0·025	0·05	0·10	$2\overline{W}$	m
2							4						—	10	2
3							5						3	12	3
4							6					—	3	14	4
5							7					3	4	16	5
6							8					3	4	18	6
7							9				—	3	4	20	7
8						—	10				3	4	5	22	8
9						1	11				3	4	5	24	9
10						1	12				3	4	6	26	10
11						1	13				3	4	6	28	11
12						1	14			—	4	5	7	30	12
13						1	15			3	4	5	7	32	13
14						1	16			3	4	6	8	34	14
15						1	17			3	4	6	8	36	15
16						1	18			3	4	6	8	38	16
17						1	19			3	5	6	9	40	17
18					—	1	20		—	3	5	7	9	42	18
19					1	2	21		3	4	5	7	10	44	19
20					1	2	22		3	4	5	7	10	46	20
21					1	2	23		3	4	6	8	11	48	21
22					1	2	24		3	4	6	8	11	50	22

[Table 6 contd]

	$n = 1$							$n = 2$							
m	0·001	0·005	0·010	0·025	0·05	0·10	$2\overline{W}$	0·001	0·005	0·010	0·025	0·05	0·10	$2\overline{W}$	m
23					1	2	25		3	4	6	8	12	52	23
24					1	2	26		3	4	6	9	12	54	24
25	—	—	—	—	1	2	27	—	3	4	6	9	12	56	25

	$n = 3$							$n = 4$							
m	0·001	0·005	0·010	0·025	0·05	0·10	$2\overline{W}$	0·001	0·005	0·010	0·025	0·05	0·10	$2\overline{W}$	m
3					6	7	21								
4				—	6	7	24			—	10	11	13	36	4
5				6	7	8	27		—	10	11	12	14	40	5
6			—	7	8	9	30		10	11	12	13	15	44	6
7			6	7	8	10	33		10	11	13	14	16	48	7
8		—	6	8	9	11	36		11	12	14	15	17	52	8
9		6	7	8	10	11	39	—	11	13	14	16	19	56	9
10		6	7	9	10	12	42	10	12	13	15	17	20	60	10
11		6	7	9	11	13	45	10	12	14	16	18	21	64	11
12		7	8	10	11	14	48	10	13	15	17	19	22	68	12
13		7	8	10	12	15	51	11	13	15	18	20	23	72	13
14		7	8	11	13	16	54	11	14	16	19	21	25	76	14
15		8	9	11	13	16	57	11	15	17	20	22	26	80	15
16	—	8	9	12	14	17	60	12	15	17	21	24	27	84	16
17	6	8	10	12	15	18	63	12	16	18	21	25	28	88	17
18	6	8	10	13	15	19	66	13	16	19	22	26	30	92	18
19	6	9	10	13	16	20	69	13	17	19	23	27	31	96	19
20	6	9	11	14	17	21	72	13	18	20	24	28	32	100	20
21	7	9	11	14	17	21	75	14	18	21	25	29	33	104	21
22	7	10	12	15	18	22	78	14	19	21	26	30	35	108	22
23	7	10	12	15	19	23	81	14	19	22	27	31	36	112	23
24	7	10	[illegible]	16	19	24	84	15	20	[illegible]	[illegible]	[illegible]	[illegible]	116	24

m	n = 5 0·001	0·005	0·010	0·025	0·05	0·10	$2\overline{W}$	n = 6 0·001	0·005	0·010	0·025	0·05	0·10	$2\overline{W}$	m
5		15	16	17	19	20	55								
6		16	17	18	20	22	60	—	23	24	26	28	30	78	6
7	—	16	18	20	21	23	65	21	24	25	27	29	32	84	7
8	15	17	19	21	23	25	70	22	25	27	29	31	34	90	8
9	16	18	20	22	24	27	75	23	26	28	31	33	36	96	9
10	16	19	21	23	26	28	80	24	27	29	32	35	38	102	10
11	17	20	22	24	27	30	85	25	28	30	34	37	40	108	11
12	17	21	23	26	28	32	90	25	30	32	35	38	42	114	12
13	18	22	24	27	30	33	95	26	31	33	37	40	44	120	13
14	18	22	25	28	31	35	100	27	32	34	38	42	46	126	14
15	19	23	26	29	33	37	105	28	33	36	40	44	48	132	15
16	20	24	27	30	34	38	110	29	34	37	42	46	50	138	16
17	20	25	28	32	35	40	115	30	36	39	43	47	52	144	17
18	21	26	29	33	37	42	120	31	37	40	45	49	55	150	18
19	22	27	30	34	38	43	125	32	38	41	46	51	57	156	19
20	22	28	31	35	40	45	130	33	39	43	48	53	59	162	20
21	23	29	32	37	41	47	135	33	40	44	50	55	61	168	21
22	23	29	33	38	43	48	140	34	42	45	51	57	63	174	22
23	24	30	34	39	44	50	145	35	43	47	53	58	65	180	23
24	25	31	35	40	45	51	150	36	44	48	54	60	67	186	24
25	25	32	36	42	47	53	155	37	45	50	56	62	69	192	25

m	n = 7 0·001	0·005	0·010	0·025	0·05	0·10	$2\overline{W}$	n = 8 0·001	0·005	0·010	0·025	0·05	0·10	$2\overline{W}$	m
7	29	32	34	36	39	41	105								
8	30	34	35	38	41	44	112	40	43	45	49	51	55	136	8
9	31	35	37	40	43	46	119	41	45	47	51	54	58	144	9
10	33	37	39	42	45	49	126	42	47	49	53	56	60	152	10

[Table 6 contd]

	$n = 7$							$n = 8$							
m	0·001	0·005	0·010	0·025	0·05	0·10	$2\overline{W}$	0·001	0·005	0·010	0·025	0·05	0·10	$2\overline{W}$	m
11	34	38	40	44	47	51	133	44	49	51	55	59	63	160	11
12	35	40	42	46	49	54	140	45	51	53	58	62	66	168	12
13	36	41	44	48	52	56	147	47	53	56	60	64	69	176	13
14	37	43	45	50	54	59	154	48	54	58	62	67	72	184	14
15	38	44	47	52	56	61	161	50	56	60	65	69	75	192	15
16	39	46	49	54	58	64	168	51	58	62	67	72	78	200	16
17	41	47	51	56	61	66	175	53	60	64	70	75	81	208	17
18	42	49	52	58	63	69	182	54	62	66	72	77	84	216	18
19	43	50	54	60	65	71	189	56	64	68	74	80	87	224	19
20	44	52	56	62	67	74	196	57	66	70	77	83	90	232	20
21	46	53	58	64	69	76	203	59	68	72	79	85	92	240	21
22	47	55	59	66	72	79	210	60	70	74	81	88	95	248	22
23	48	57	61	68	74	81	217	62	71	76	84	90	98	256	23
24	49	58	63	70	76	84	224	64	73	78	86	93	101	264	24
25	50	60	64	72	78	86	231	65	75	81	89	96	104	272	25

	$n = 9$							$n = 10$							
m	0·001	0·005	0·010	0·025	0·05	0·10	$2\overline{W}$	0·001	0·005	0·010	0·025	0·05	0·10	$2\overline{W}$	m
9	52	56	59	62	66	70	171								
10	53	58	61	65	69	73	180	65	71	74	78	82	87	210	10
11	55	61	63	68	72	76	189	67	73	77	81	86	91	220	11
12	57	63	66	71	75	80	198	69	76	79	84	89	94	230	12
13	59	65	68	73	78	83	207	72	79	82	88	92	98	240	13
14	60	67	71	76	81	86	216	74	81	85	91	96	102	250	14
15	62	69	73	79	84	90	225	76	84	88	94	99	106	260	15

m															m
16	64	72	76	82	87	93	234	78	86	91	97	103	109	270	16
17	66	74	78	84	90	97	243	80	89	93	100	106	113	280	17
18	68	76	81	87	93	100	252	82	92	96	103	110	117	290	18
19	70	78	83	90	96	103	261	84	94	99	107	113	121	300	19
20	71	81	85	93	99	107	270	87	97	102	110	117	125	310	20
21	73	83	88	95	102	110	279	89	99	105	113	120	128	320	21
22	75	85	90	98	105	113	288	91	102	108	116	123	132	330	22
23	77	88	93	101	108	117	297	93	105	110	119	127	136	340	23
24	79	90	95	104	111	120	306	95	107	113	122	130	140	350	24
25	81	92	98	107	114	123	315	98	110	116	126	134	144	360	25

			$n = 11$								$n = 12$				
m	0·001	0·005	0·010	0·025	0·05	0·10	$2\overline{W}$	0·001	0·005	0·010	0·025	0·05	0·10	$2\overline{W}$	m
11	81	87	91	96	100	106	253								
12	83	90	94	99	104	110	264	98	105	109	115	120	127	300	12
13	86	93	97	103	108	114	275	101	109	113	119	125	131	312	13
14	88	96	100	106	112	118	286	103	112	116	123	129	136	324	14
15	90	99	103	110	116	123	297	106	115	120	127	133	141	336	15
16	93	102	107	113	120	127	308	109	119	124	131	138	145	348	16
17	95	105	110	117	123	131	319	112	122	127	135	142	150	360	17
18	98	108	113	121	127	135	330	115	125	131	139	146	155	372	18
19	100	111	116	124	131	139	341	118	129	134	143	150	159	384	19
20	103	114	119	128	135	144	352	120	132	138	147	155	164	396	20
21	106	117	123	131	139	148	363	123	136	142	151	159	169	408	21
22	108	120	126	135	143	152	374	126	139	145	155	163	173	420	22
23	111	123	129	139	147	156	385	129	142	149	159	168	178	434	23
24	113	126	132	142	151	161	396	132	146	153	163	172	183	444	24
25	116	129	136	146	155	165	407	135	149	156	167	176	187	456	25

[Table 6 contd]

m	$n = 13$ 0·001	0·005	0·010	0·025	0·05	0·10	$2\overline{W}$	$n = 14$ 0·001	0·005	0·010	0·025	0·05	0·10	$2\overline{W}$	m
13	117	125	130	136	142	149	351								
14	120	129	134	141	147	154	364	137	147	152	160	166	174	406	14
15	123	133	138	145	152	159	377	141	151	156	164	171	179	420	15
16	126	136	142	150	156	165	390	144	155	161	169	176	185	434	16
17	129	140	146	154	161	170	403	148	159	165	174	182	190	448	17
18	133	144	150	158	166	175	416	151	163	170	179	187	196	462	18
19	136	148	154	163	171	180	429	155	168	174	183	192	202	476	19
20	139	151	158	167	175	185	442	159	172	178	188	197	207	490	20
21	142	155	162	171	180	190	455	162	176	183	193	202	213	504	21
22	145	159	166	176	185	195	468	166	180	187	198	207	218	518	22
23	149	163	170	180	189	200	481	169	184	192	203	212	224	532	23
24	152	166	174	185	194	205	494	173	188	196	207	218	229	546	24
25	155	170	179	189	199	211	507	177	192	200	212	223	235	560	25

m	$n = 15$ 0·001	0·005	0·010	0·025	0·05	0·10	$2\overline{W}$	$n = 16$ 0·001	0·005	0·010	0·025	0·05	0·10	$2\overline{W}$	m
15	160	171	176	184	192	200	465								
16	163	175	181	190	197	206	480	184	196	202	211	219	229	528	16
17	167	180	186	195	203	212	495	188	201	207	217	225	235	544	17
18	171	184	190	200	208	218	510	192	206	212	222	231	242	560	18
19	175	189	195	205	214	224	525	196	210	218	228	237	248	576	19
20	179	193	200	210	220	230	540	201	215	223	234	243	255	592	20
21	183	198	205	216	225	236	555	205	220	228	239	249	261	608	21
22	187	202	210	221	231	242	570	209	225	233	245	255	267	624	22
23	191	207	214	226	236	248	585	214	230	238	251	261	274	640	23
24	195	211	219	231	242	254	600	218	235	244	256	267	280	656	24
25	199	216	224	237	248	260	615	222	240	249	262	273	287	672	25

	$n = 17$							$n = 18$							
m	0·001	0·005	0·010	0·025	0·05	0·10	$2\overline{W}$	0·001	0·005	0·010	0·025	0·05	0·10	$2\overline{W}$	m
17	210	223	230	240	249	259	595								
18	214	228	235	246	255	266	612	237	252	259	270	280	291	666	18
19	219	234	241	252	262	273	629	242	258	265	277	287	299	684	19
20	223	239	246	258	268	280	646	247	263	271	283	294	306	702	20
21	228	244	252	264	274	287	663	252	269	277	290	301	313	720	21
22	233	249	258	270	281	294	680	257	275	283	296	307	321	738	22
23	238	255	263	276	287	300	697	262	280	289	303	314	328	756	23
24	242	260	269	282	294	307	714	267	286	295	309	321	335	774	24
25	247	265	275	288	300	314	731	273	292	301	316	328	343	792	25

	$n = 19$							$n = 20$							
m	0·001	0·005	0·010	0·025	0·05	0·10	$2\overline{W}$	0·001	0·005	0·010	0·025	0·05	0·10	$2\overline{W}$	m
19	267	283	291	303	313	325	741								
20	272	289	297	309	320	333	760	298	315	324	337	348	361	820	20
21	277	295	303	316	328	341	779	304	322	331	344	356	370	840	21
22	283	301	310	323	335	349	798	309	328	337	351	364	378	860	22
23	288	307	316	330	342	357	817	315	335	344	359	371	386	880	23
24	294	313	323	337	350	364	836	321	341	351	366	379	394	900	24
25	299	319	329	344	357	372	855	327	348	358	373	387	403	920	25

	$n = 21$							$n = 22$							
m	0·001	0·005	0·010	0·025	0·05	0·10	$2\overline{W}$	0·001	0·005	0·010	0·025	0·05	0·10	$2\overline{W}$	m
21	331	349	359	373	385	399	903								
22	337	356	366	381	393	408	924	365	386	396	411	424	439	990	22
23	343	363	373	388	401	417	945	372	393	403	419	432	448	1012	23
24	349	370	381	396	410	425	966	379	400	411	427	441	457	1034	24
25	356	377	388	404	418	434	987	385	408	419	435	450	467	1056	25

[Table 6 contd]

	n = 23							n = 24							
m	0·001	0·005	0·010	0·025	0·05	0·10	$2\overline{W}$	0·001	0·005	0·010	0·025	0·05	0·10	$2\overline{W}$	m
23	402	424	434	451	465	481	1081								
24	409	431	443	459	474	491	1104	440	464	475	492	507	525	1176	24
25	416	439	451	468	483	500	1127	448	472	484	501	517	535	1200	25

	n = 25						
m	0·001	0·005	0·010	0·025	0·05	0·10	$2\overline{W}$
25	480	505	517	536	552	570	1275

Table 7 Mann-Whitney's *U* test.
Source: Adapted from L. R. Verdooren, 'Extended tables of critical values for Wilcoxon's test statistic', *Biometrika* 50 (1963), 177–86, table 1.

The entries in this table are the quantiles w_p of the Mann-Whitney test statistic T, given for selected values of p. Note that $P(T < w_p) \leq p$. Upper quantiles may be found from the equation

$$w_{1-p} = nm - w_p$$

Critical regions correspond to values less than (or greater than) but not including the appropriate quantile.

For n or m greater than 20, the pth quantile w_p of the Mann-Whitney test statistic may be approximated by

$$w_p = \frac{nm}{2} + x_p \frac{\sqrt{nm(n+m+1)}}{12}$$

where x_p is the pth quantile of a standard normal random variable, obtained from table 1.

n	*p*	*m* = 2	3	4	5	6	7	8	9	10	11	12	13	14	15	16	17	18	19	20
2	·01	0	0	0	0	0	0	0	0	0	0	0	1	1	1	1	1	1	2	2
	·025	0	0	0	0	0	0	1	1	1	1	2	2	2	2	2	3	3	3	3
	·05	0	0	0	1	1	1	2	2	2	2	3	3	4	4	4	4	5	5	5
	·10	0	1	1	2	2	2	3	3	4	4	5	5	5	6	6	7	7	8	8
3	·01	0	0	0	0	0	1	1	2	2	2	3	3	3	4	4	5	5	5	6
	·025	0	0	0	1	2	2	3	3	4	4	5	5	6	6	7	7	8	8	9
	·05	0	1	1	2	3	3	4	5	5	6	6	7	8	8	9	10	10	11	12
	·10	1	2	2	3	4	5	6	6	7	8	9	10	11	11	12	13	14	15	16
4	·01	0	0	0	1	2	2	3	4	4	5	6	6	7	9	8	9	10	10	11
	·025	0	0	1	2	3	4	5	5	6	7	8	9	10	11	12	12	13	14	15
	·05	0	1	2	3	4	5	6	7	8	9	10	11	12	13	15	16	17	18	19
	·10	1	2	4	5	6	7	8	10	11	12	13	14	16	17	18	19	21	22	23

[Table 7 contd]

n	p	$m = 2$	3	4	5	6	7	8	9	10	11	12	13	14	15	16	17	18	19	20
5	·01	0	0	1	2	3	4	5	6	7	8	9	10	11	12	13	14	15	16	17
	·025	0	1	2	3	4	6	7	8	9	10	12	13	14	15	16	18	19	20	21
	·05	1	2	3	5	6	7	9	10	12	13	14	16	17	19	20	21	23	24	26
	·10	2	3	5	6	8	9	11	13	14	16	18	19	21	23	24	26	28	29	31
6	·01	0	0	2	3	4	5	7	8	9	10	12	13	14	16	17	19	20	21	23
	·025	0	2	3	4	6	7	9	11	12	14	15	17	18	20	22	23	25	26	28
	·05	1	3	4	6	8	9	11	13	15	17	18	20	22	24	26	27	29	31	33
	·10	2	4	6	8	10	12	14	16	18	20	22	24	26	28	30	32	35	37	39
7	·01	0	1	2	4	5	7	8	10	12	13	15	17	18	20	22	24	25	27	29
	·025	0	2	4	6	7	9	11	13	15	17	19	21	23	25	27	29	31	33	35
	·05	1	3	5	7	9	12	14	16	18	20	22	25	27	29	31	34	36	38	40
	·10	2	5	7	9	12	14	17	19	22	24	27	29	32	34	37	39	42	44	47
8	·01	0	1	3	5	7	8	10	12	14	16	18	21	23	25	27	29	31	33	35
	·025	1	3	5	7	9	11	14	16	18	20	23	25	27	30	32	35	37	39	42
	·05	2	4	6	9	11	14	16	19	21	24	27	29	32	34	37	40	42	45	48
	·10	3	6	8	11	14	17	20	23	25	28	31	34	37	40	43	46	49	52	55
9	·01	0	2	4	6	8	10	12	15	17	19	22	24	27	29	32	34	37	39	41
	·025	1	3	5	8	11	13	16	18	21	24	27	29	32	35	38	40	43	46	49
	·05	2	5	7	10	13	16	19	22	25	28	31	34	37	40	43	46	49	52	55
	·10	3	6	10	13	16	19	23	26	29	32	36	39	42	46	49	53	56	59	63
10	·01	0	2	4	7	9	12	14	17	20	23	25	28	31	34	37	39	42	45	48
	·025	1	4	6	9	12	15	18	21	24	27	30	34	37	40	43	46	49	53	56
	·05	2	5	8	12	15	18	21	25	28	32	35	38	42	45	49	52	56	59	63
	·10	4	7	11	14	18	22	25	29	33	37	40	44	48	52	55	59	63	67	71
11	·01	0	2	5	8	10	13	16	19	23	26	29	32	35	38	42	45	48	51	54
	·025	1	4	7	10	14	17	20	24	27	31	34	38	41	45	48	52	56	59	63
	·05	2	6	9	13	17	20	24	28	32	35	39	43	47	51	55	58	62	66	70
	·10	4	8	12	16	20	24	28	32	37	41	45	49	53	58	62	66	70	74	79

n	p	$m = 2$	3	4	5	6	7	8	9	10	11	12	13	14	15	16	17	18	19	20
12	·01	0	3	6	9	12	15	18	22	25	29	32	36	39	43	47	50	54	57	61
	·025	2	5	8	12	15	19	23	27	30	34	38	42	46	50	54	58	62	66	70
	·05	3	6	10	14	18	22	27	31	35	39	43	48	52	56	61	65	69	73	78
	·10	5	9	13	18	22	27	31	36	40	45	50	54	59	64	68	73	78	82	87
13	·01	1	3	6	10	13	17	21	24	28	32	36	40	44	48	52	56	60	64	68
	·025	2	5	9	13	17	21	25	29	34	38	42	46	51	55	60	64	68	73	77
	·05	3	7	11	16	20	25	29	34	38	43	48	52	57	62	66	71	76	81	85
	·10	5	10	14	19	24	29	34	39	44	49	54	59	64	69	75	80	85	90	95
14	·01	1	3	7	11	14	18	23	27	31	35	39	44	48	52	57	61	66	70	74
	·025	2	6	10	14	18	23	27	32	37	41	46	51	56	60	65	70	75	79	84
	·05	4	8	12	17	22	27	32	37	42	47	52	57	62	67	72	78	83	88	93
	·10	5	11	16	21	26	32	37	42	48	53	59	64	70	75	81	86	92	98	103
15	·01	1	4	8	12	16	20	25	29	34	38	43	48	52	57	62	67	71	76	81
	·025	2	6	11	15	20	25	30	35	40	45	50	55	60	65	71	76	81	86	91
	·05	4	8	13	19	24	29	34	40	45	51	56	62	67	73	78	84	89	95	101
	·10	6	11	17	23	28	34	40	46	52	58	64	69	75	81	87	93	99	105	111
16	·01	1	4	8	13	17	22	27	32	37	42	47	52	57	62	67	72	77	83	88
	·025	2	7	12	16	22	27	32	38	43	48	54	60	65	71	76	82	87	93	99
	·05	4	9	15	20	26	31	37	43	49	55	61	66	72	78	84	90	96	102	108
	·10	6	12	18	24	30	37	43	49	55	62	68	75	81	87	94	100	107	113	120
17	·01	1	5	9	14	19	24	29	34	39	45	50	56	61	67	72	78	83	89	94
	·025	3	7	12	18	23	29	35	40	46	52	58	64	70	76	82	88	94	100	106
	·05	4	10	16	21	27	34	40	46	52	58	65	71	78	84	90	97	103	110	116
	·10	7	13	19	26	32	39	46	53	59	66	73	80	86	93	100	107	114	121	128
18	·01	1	5	10	15	20	25	31	37	42	48	54	60	66	71	77	83	89	95	101
	·025	3	8	13	19	25	31	37	43	49	56	62	68	75	81	87	94	100	107	113
	·05	5	10	17	23	29	36	42	49	56	62	69	76	83	89	96	103	110	117	124
	·10	7	14	21	28	35	42	49	56	63	70	78	85	92	99	107	114	121	129	136

[Table 7 contd]

n	p	$m = 2$	3	4	5	6	7	8	9	10	11	12	13	14	15	16	17	18	19	20
19	·01	2	5	10	16	21	27	33	39	45	51	57	64	70	76	83	89	95	102	108
	·025	3	8	14	20	26	33	39	46	53	59	66	73	79	86	93	100	107	114	120
	·05	5	11	18	24	31	38	45	52	59	66	73	81	88	95	102	110	117	124	131
	·10	8	15	22	29	37	44	52	59	67	74	82	90	98	105	113	121	129	136	144
20	·01	2	6	11	17	23	29	35	41	48	54	61	68	74	81	88	94	101	108	115
	·025	3	9	15	21	28	35	42	49	56	63	70	77	84	91	99	106	113	120	128
	·05	5	12	19	26	33	40	48	55	63	70	78	85	93	101	108	116	124	131	139
	·10	8	16	23	31	39	47	55	63	71	79	87	95	103	111	120	128	136	144	152

Table 8 Critical values for the Kruskal-Wallis test statistic for three samples and small sample sizes.
Source: W. H. Kruskal and W. A. Wallis, 'Use of ranks in one-criterion variance analysis', *J. Am. Stat. Assoc.* 47 (1952), 614–17, and 48 (1953), 907–11 (corrections).

The null hypothesis may be rejected at the probability shown in the column α if the Kruskal-Wallis test statistic is greater than or equal to the critical value shown.

Sample sizes			Critical value for KW	α
n_1	n_2	n_3		
2	1	1	2·7000	0·500
2	2	1	3·6000	0·200
2	2	2	4·5714	0·067
			3·7143	0·200
3	1	1	3·2000	0·300
3	2	1	4·2857	0·100
			3·8571	0·133
3	2	2	5·3572	0·029
			4·7143	0·048
			4·5000	0·067
			4·4643	0·105
3	3	1	5·1429	0·043
			4·5714	0·100
			4·0000	0·129
3	3	2	6·2500	0·011
			5·3611	0·032
			5·1389	0·061
			4·5556	0·100
			4·2500	0·121
3	3	3	7·2000	0·004
			6·4889	0·011
			5·6889	0·029
			5·6000	0·050
			5·0667	0·086
			4·6222	0·100
4	1	1	3·5714	0·200
4	2	1	4·8214	0·057
			4·5000	0·076
			4·0179	0·114
4	2	2	6·0000	0·014
			5·3333	0·033
			5·1250	0·052
			4·4583	0·100
			4·1667	0·105
4	3	1	5·8333	0·021
			5·2083	0·050
			5·0000	0·057
			4·0556	0·093
			3·8889	0·129
4	3	2	6·4444	0·008
			6·3000	0·011
			5·4444	0·046
			5·4000	0·051
			4·5111	0·098
			4·4444	0·102
4	3	3	6·7455	0·010
			6·7091	0·013
			5·7909	0·046
			5·7273	0·050
			4·7091	0·092
			4·7000	0·101
4	4	1	6·6667	0·010
			6·1667	0·022
			4·9667	0·048
			4·8667	0·054
			4·1667	0·082
			4·0667	0·102
4	4	2	7·0364	0·006
			6·8727	0·011
			5·4545	0·046

[Table 8 contd]

Sample sizes n_1	n_2	n_3	Critical value for KW	α
4	4	2	5·2364	0·052
			4·5545	0·098
			4·4455	0·103
4	4	3	7·1439	0·010
			7·1364	0·011
			5·5985	0·049
			5·5758	0·051
			4·5455	0·099
			4·4773	0·102
4	4	4	7·6538	0·008
			7·5385	0·011
			5·6923	0·049
			5·6538	0·054
			4·6539	0·097
			4·5001	0·104
5	1	1	3·8571	0·143
5	2	1	5·2500	0·036
			5·0000	0·048
			4·4500	0·071
			4·2000	0·095
			4·0500	0·119
5	2	2	6·5333	0·008
			6·1333	0·013
			5·1600	0·034
			5·0400	0·056
			4·3733	0·090
			4·2933	0·122
5	3	1	6·4000	0·012
			4·9600	0·048
			4·8711	0·052
			4·0178	0·095
			3·8400	0·123
5	3	2	6·9091	0·009
			6·8218	0·010
			5·2509	0·049
			5·1055	0·052
			4·6509	0·091
			4·4945	0·101
5	3	3	7·0788	0·009
			6·9818	0·011
			5·6485	0·049
			5·5152	0·051
			4·5333	0·097
			4·4121	0·109
5	4	1	6·9545	0·008
			6·8400	0·011
			4·9855	0·044
			4·8600	0·056
			3·9873	0·098
			3·9600	0·102
5	4	2	7·2045	0·009
			7·1182	0·010
			5·2727	0·049
			5·2682	0·050
			4·5409	0·098
			4·5182	0·101
5	4	3	7·4449	0·010
			7·3949	0·011
			5·6564	0·049
			5·6308	0·050
			4·5487	0·099
			4·5231	0·103
5	4	4	7·7604	0·009
			7·7440	0·011
			5·6571	0·049
			5·6176	0·050
			4·6187	0·100
			4·5527	0·102
5	5	1	7·3091	0·009
			6·8364	0·011
			5·1273	0·046
			4·9091	0·053
			4·1091	0·086
			4·0364	0·105
5	5	2	7·3385	0·010
			7·2692	0·010

Sample sizes n_1	n_2	n_3	Critical value for KW	α
5	5	2	5·3385	0·047
			5·2462	0·051
			4·6231	0·097
			4·5077	0·100
5	5	3	7·5780	0·010
			7·5429	0·010
			5·7055	0·046
			5·6264	0·051
			4·5451	0·100
			4·5363	0·102
5	5	4	7·8229	0·010
			7·7914	0·010
			5·6657	0·049
			5·6429	0·050
			4·5229	0·099
			4·5200	0·101
5	5	5	8·0000	0·009
			7·9800	0·010
			5·7800	0·049
			5·6600	0·051
			4·5600	0·100
			4·5000	0·102

Table 9 Critical values for Friedman's related samples test statistic.
Source: Adapted from D. B. Owen, *Handbook of statistical tables* (Reading, Mass. 1962), table 14.1.

The null hypothesis may be rejected at the probability shown in the column α if the Friedman test statistic is greater than or equal to the critical value shown.

	Significance level, α			
R	0·10	0·05	0·01	0·001
		$C = 3$		
2	—	—	—	—
3	18 0·028	18 0·028	—	—
4	24 0·069	26 0·042	32 0·0046	—
5	26 0·093	32 0·039	42 0·0085	50 0·00077
6	32 0·072	42 0·029	54 0·0081	72 0·00013
7	38 0·085	50 0·027	62 0·0084	86 0·00032
8	42 0·079	50 0·047	72 0·0099	98 0·00086

[Table 9 contd]

R	0·10	0·05	0·01	0·001
		Significance level, α		
9	50 0·069	56 0·048	78 0·010	114 0·00072
10	50 0·092	62 0·046	96 0·0073	122 0·0010
11	54 0·100	72 0·043	104 0·0066	146 0·00067
12	62 0·080	74 0·050	114 0·0080	150 0·00087
13	62 0·098	78 0·050	122 0·0076	168 0·00081
14	72 0·089	86 0·049	126 0·010	186 0·00058
15	74 0·096	96 0·047	134 0·010	194 0·00095
		$C = 4$		
2	20 0·042	20 0·042	–	–
3	33 0·075	37 0·026	45 0·0017	–
4	42 0·093	52 0·036	64 0·0056	74 0·00094
5	53 0·089	65 0·049	83 0·0092	105 0·00040
6	64 0·088	76 0·043	102 0·0097	128 0·0010
7	75 0·093	91 0·040	121 0·0091	161 0·00084
8	84 0·098	102 0·049	138 0·010	184 0·00100

Table 10 Critical values of R for total number of runs test.
Source: Adapted from F. S. Swed and C. Eisenhart, 'Tables for testing randomness of grouping in a sequence of alternatives', *Ann. Math. Stat.* 14 (1943), 66–87, table II.

The null hypothesis may be rejected at the α level shown if the value of R is smaller than or equal to the critical value shown. The label n_1 is given to the larger sample, n_2 to the smaller sample.

	n_2																		
	2	3	4	5	6	7	8	9	10	11	12	13	14	15	16	17	18	19	20
n_1	$\alpha = 0{\cdot}05$																		
4			2																
5		2	2	3															
6		2	3	3	3														
7		2	3	3	4	4													
8	2	2	3	3	4	4	5												
9	2	2	3	4	4	5	5	6											
10	2	3	3	4	5	5	6	6	6										
11	2	3	3	4	5	5	6	6	7	7									
12	2	3	4	4	5	6	6	7	7	8	8								
13	2	3	4	4	5	6	6	7	8	8	9	9							
14	2	3	4	5	5	6	7	7	8	8	9	9	10						
15	2	3	4	5	6	6	7	8	8	9	9	10	10	11					
16	2	3	4	5	6	6	7	8	8	9	10	10	11	11	11				
17	2	3	4	5	6	7	7	8	9	9	10	10	11	11	12	12			
18	2	3	4	5	6	7	8	8	9	10	10	11	11	12	12	13	13		
19	2	3	4	5	6	7	8	8	9	10	10	11	12	12	13	13	14	14	
20	2	3	4	5	6	7	8	9	9	10	11	11	12	12	13	13	14	14	15
n_1	$\alpha = 0{\cdot}01$																		
5				2															
6			2	2	2														
7			2	2	3	3													
8			2	2	3	3	4												
9		2	2	3	3	4	4	4											
10		2	2	3	3	4	4	5	5										
11		2	2	3	4	4	5	5	5	6									
12		2	3	3	4	4	5	5	6	6	7								
13		2	3	3	4	5	5	6	6	6	7	7							
14		2	3	3	4	5	5	6	6	7	7	8	8						
15		2	3	4	4	5	5	6	7	7	8	8	8	9					
16		2	3	4	4	5	6	6	7	7	8	8	9	9	10				
17		2	3	4	5	5	6	7	7	8	8	9	9	10	10	10			
18		2	3	4	5	5	6	7	7	8	8	9	9	10	10	11	11		
19	2	2	3	4	5	6	6	7	8	8	9	9	10	10	11	11	12	12	
20	2	2	3	4	5	6	6	7	8	8	9	10	10	11	11	11	12	12	13

Table 11 Cumulative probabilities for Edgington's total number of runs up and down test.
Source: E. S. Edgington, 'Probability table for number of runs of signs of first differences in ordered series', *J. Am. Stat. Assoc.* 56 (1961), 156–9, table 1.

The probabilities of r or fewer runs of signs of first differences for n observations.

Number of runs	Number of observations n												
r	1	2	3	4	5	6	7	8	9	10	11	12	13
1		1·0000	0·3333	0·0833	0·0167	0·0028	0·0004	0·0000	0·0000	0·0000	0·0000	0·0000	0·0000
2			1·0000	0·5833	0·2500	0·0861	0·0250	0·0063	0·0014	0·0003	0·0001	0·0000	0·0000
3				1·0000	0·7333	0·4139	0·1909	0·0749	0·0257	0·0079	0·0022	0·0005	0·0001
4					1·0000	0·8306	0·5583	0·3124	0·1500	0·0633	0·0239	0·0082	0·0026
5						1·0000	0·8921	0·6750	0·4347	0·2427	0·1196	0·0529	0·0213
6							1·0000	0·9313	0·7653	0·5476	0·3438	0·1918	0·0964
7								1·0000	0·9563	0·8329	0·6460	0·4453	0·2749
8									1·0000	0·9722	0·8823	0·7280	0·5413
9										1·0000	0·9823	0·9179	0·7942
10											1·0000	0·9887	0·9432
11												1·0000	0·9928
12													1·0000
13													
14													
15													
16													
17													
18													
19													
20													
21													
22													
23													
24													

Number of runs	Number of observations n											
r	14	15	16	17	18	19	20	21	22	23	24	25
1	0·0000	0·0000	0·0000	0·0000	0·0000	0·0000	0·0000	0·0000	0·0000	0·0000	0·0000	0·0000
2	0·0000	0·0000	0·0000	0·0000	0·0000	0·0000	0·0000	0·0000	0·0000	0·0000	0·0000	0·0000
3	0·0000	0·0000	0·0000	0·0000	0·0000	0·0000	0·0000	0·0000	0·0000	0·0000	0·0000	0·0000
4	0·0007	0·0002	0·0001	0·0000	0·0000	0·0000	0·0000	0·0000	0·0000	0·0000	0·0000	0·0000
5	0·0079	0·0027	0·0009	0·0003	0·0001	0·0000	0·0000	0·0000	0·0000	0·0000	0·0000	0·0000
6	0·0441	0·0186	0·0072	0·0026	0·0009	0·0003	0·0001	0·0000	0·0000	0·0000	0·0000	0·0000
7	0·1534	0·0782	0·0367	0·0160	0·0065	0·0025	0·0009	0·0003	0·0001	0·0000	0·0000	0·0000
8	0·3633	0·2216	0·1238	0·0638	0·0306	0·0137	0·0058	0·0023	0·0009	0·0003	0·0001	0·0000
9	0·6278	0·4520	0·2975	0·1799	0·1006	0·0523	0·0255	0·0117	0·0050	0·0021	0·0008	0·0003
10	0·8464	0·7030	0·5369	0·3770	0·2443	0·1467	0·0821	0·0431	0·0213	0·0099	0·0044	0·0018
11	0·9609	0·8866	0·7665	0·6150	0·4568	0·3144	0·2012	0·1202	0·0674	0·0356	0·0177	0·0084
12	0·9954	0·9733	0·9172	0·8188	0·6848	0·5337	0·3873	0·2622	0·1661	0·0988	0·0554	0·0294
13	1·0000	0·9971	0·9818	0·9400	0·8611	0·7454	0·6055	0·4603	0·3276	0·2188	0·1374	0·0815
14		1·0000	0·9981	0·9877	0·9569	0·8945	0·7969	0·6707	0·5312	0·3953	0·2768	0·1827
15			1·0000	0·9988	0·9917	0·9692	0·9207	0·8398	0·7286	0·5980	0·4631	0·3384
16				1·0000	0·9992	0·9944	0·9782	0·9409	0·8749	0·7789	0·6595	0·5292
17					1·0000	0·9995	0·9962	0·9846	0·9563	0·9032	0·8217	0·7148
18						1·0000	0·9997	0·9975	0·9892	0·9679	0·9258	0·8577
19							1·0000	0·9998	0·9983	0·9924	0·9765	0·9436
20								1·0000	0·9999	0·9989	0·9947	0·9830
21									1·0000	0·9999	0·9993	0·9963
22										1·0000	1·0000	0·9995
23											1·0000	1·0000
24												1·0000

Table 12 Critical upper-tail values of Spearman's rank-order correlation test statistic.
Source: Adapted from G. J. Glasser and R. F. Winter, 'Critical values of the coefficient of rank correlation for testing the hypothesis of independence', *Biometrika* 48 (1961), 444–8.

The null hypothesis may be rejected at the selected α in favour of the one-sided alternative for calculated values of the test statistic which are as large or larger than the tabulated values.

n	$p = 0{\cdot}900$	0·950	0·975	0·990	0·995	0·999
4	0·8000	0·8000				
5	0·7000	0·8000	0·9000	0·9000		
6	0·6000	0·7714	0·8286	0·8857	0·9429	
7	0·5357	0·6786	0·7450	0·8571	0·8929	0·9643
8	0·5000	0·6190	0·7143	0·8095	0·8571	0·9286
9	0·4667	0·5833	0·6833	0·7667	0·8167	0·9000
10	0·4424	0·5515	0·6364	0·7333	0·7818	0·8667
11	0·4182	0·5273	0·6091	0·7000	0·7455	0·8364
12	0·3986	0·4965	0·5804	0·6713	0·7273	0·8182
13	0·3791	0·4780	0·5549	0·6429	0·6978	0·7912
14	0·3626	0·4593	0·5341	0·6220	0·6747	0·7670
15	0·3500	0·4429	0·5179	0·6000	0·6536	0·7464
16	0·3382	0·4265	0·5000	0·5824	0·6324	0·7265
17	0·3260	0·4118	0·4853	0·5637	0·6152	0·7083
18	0·3148	0·3994	0·4716	0·5480	0·5975	0·6904
19	0·3070	0·3895	0·4579	0·5333	0·5825	0·6737
20	0·2977	0·3789	0·4451	0·5203	0·5684	0·6586
21	0·2909	0·3688	0·4351	0·5078	0·5545	0·6455
22	0·2829	0·3597	0·4241	0·4963	0·5426	0·6318
23	0·2767	0·3518	0·4150	0·4852	0·5306	0·6186
24	0·2704	0·3435	0·4061	0·4748	0·5200	0·6070
25	0·2646	0·3362	0·3977	0·4654	0·5100	0·5962
26	0·2588	0·3299	0·3894	0·4564	0·5002	0·5856
27	0·2540	0·3236	0·3822	0·4481	0·4915	0·5757
28	0·2490	0·3175	0·3749	0·4401	0·4828	0·5660
29	0·2443	0·3113	0·3685	0·4320	0·4744	0·5567
30	0·2400	0·3059	0·3620	0·4251	0·4665	0·5479

For n greater than 30 the approximate quantiles of ρ may be obtained from

$$w_p \cong \frac{x_p}{\sqrt{n-1}}$$

where x_p is the p quantile of a standard normal random variable obtained from table 1.

Table 13 Critical upper-tail values of Kendall's rank-order correlation test statistic. Source: L. Kaarsemaker and A. van Wijngaarden, 'Tables for use in rank correlation', *Statistica Neerlandica* 7 (1953), 41–54, table III.

The null hypothesis may be rejected at the selected α in favour of the one-sided alternative for calculated values of the test statistic which are as large or larger than the tabulated values.

n	$\alpha = 0{\cdot}005$	$\alpha = 0{\cdot}010$	$\alpha = 0{\cdot}025$	$\alpha = 0{\cdot}050$	$\alpha = 0{\cdot}100$
4	8	8	8	6	6
5	12	10	10	8	8
6	15	13	13	11	9
7	19	17	15	13	11
8	22	20	18	16	12
9	26	24	20	18	14
10	29	27	23	21	17
11	33	31	27	23	19
12	38	36	30	26	20
13	44	40	34	28	24
14	47	43	37	33	25
15	53	49	41	35	29
16	58	52	46	38	30
17	64	58	50	42	34
18	69	63	53	45	37
19	75	67	57	49	39
20	80	72	62	52	42
21	86	78	66	56	44
22	91	83	71	61	47
23	99	89	75	65	51
24	104	94	80	68	54
25	110	100	86	72	58
26	117	107	91	77	61
27	125	113	95	81	63
28	130	118	100	86	68
29	138	126	106	90	70
30	145	131	111	95	75
31	151	137	117	99	77
32	160	144	122	104	82
33	166	152	128	108	86
34	175	157	133	113	89
35	181	165	139	117	93
36	190	172	146	122	96
37	198	178	152	128	100
38	205	185	157	133	105
39	213	193	163	139	109
40	222	200	170	144	112

Table 14 Critical lower-tail values of R^{+} for Wilcoxon's signed-rank test. Source: F. Wilcoxon, S. K. Katti and R. A. Wilcox, *Critical values and probability levels for the Wilcoxon rank-sum test and the Wilcoxon signed-rank test* (American Cyanamid Company, New York/Florida State Univ., Tallahassee, 1963), table II.

The null hypothesis may be rejected at the probability shown in the column α if the obtained value of R^{+} is less than or equal to the tabulated value. The exact probability is also given for each value.

	$\alpha = 0{\cdot}05$		$\alpha = 0{\cdot}025$		$\alpha = 0{\cdot}01$		$\alpha = 0{\cdot}005$	
5	0	0·0313						
	1	0·0625						
6	2	0·0469	0	0·0156				
	3	0·0781	1	0·0313				
7	3	0·0391	2	0·0234	0	0·0078		
	4	0·0547	3	0·0391	1	0·0156		
8	5	0·0391	3	0·0195	1	0·0078	0	0·0039
	6	0·0547	4	0·0273	2	0·0117	1	0·0078
9	8	0·0488	5	0·0195	3	0·0098	1	0·0039
	9	0·0645	6	0·0273	4	0·0137	2	0·0059
10	10	0·0420	8	0·0244	5	0·0098	3	0·0049
	11	0·0527	9	0·0322	6	0·0137	4	0·0068
11	13	0·0415	10	0·0210	7	0·0093	5	0·0049
	14	0·0508	11	0·0269	8	0·0122	6	0·0068
12	17	0·0461	13	0·0212	9	0·0081	7	0·0046
	18	0·0549	14	0·0261	10	0·0105	8	0·0061
13	21	0·0471	17	0·0239	12	0·0085	9	0·0040
	22	0·0549	18	0·0287	13	0·0107	10	0·0052
14	25	0·0453	21	0·0247	15	0·0083	12	0·0043
	26	0·0520	22	0·0290	16	0·0101	13	0·0054
15	30	0·0473	25	0·0240	19	0·0090	15	0·0042
	31	0·0535	26	0·0277	20	0·0108	16	0·0051
16	35	0·0467	29	0·0222	23	0·0091	19	0·0046
	36	0·0523	30	0·0253	24	0·0107	20	0·0055
17	41	0·0492	34	0·0224	27	0·0087	23	0·0047
	42	0·0544	35	0·0253	28	0·0101	24	0·0055
18	47	0·0494	40	0·0241	32	0·0091	27	0·0045
	48	0·0542	41	0·0269	33	0·0104	28	0·0052
19	53	0·0478	46	0·0247	37	0·0090	32	0·0047
	54	0·0521	47	0·0273	38	0·0102	33	0·0054
20	60	0·0487	52	0·0242	43	0·0096	37	0·0047
	61	0·0527	53	0·0266	44	0·0107	38	0·0053

Table 15 Critical values of the Smirnov test statistic for two samples of different size n and m.
Source: F. J. Massey, 'Distribution table for the deviation between two sample cumulatives', *Ann. Math. Stat.* 23 (1952), 435–41.

With $n < m$, the null hypothesis may be rejected at α if the observed value exceeds the value given in the table. For n, m greater than covered by the table use the large sample approximation.

One-sided test:		$p = 0{\cdot}90$	0·95	0·975	0·99	0·995
Two-sided test:		$p = 0{\cdot}80$	0·90	0·95	0·98	0·99
$N_1 = 1$	$N_2 =$ 9	17/18				
	10	9/10				
$N_1 = 2$	$N_2 =$ 3	5/6				
	4	3/4				
	5	4/5	4/5			
	6	5/6	5/6			
	7	5/7	6/7			
	8	3/4	7/8	7/8		
	9	7/9	8/9	8/9		
	10	7/10	4/5	9/10		
$N_1 = 3$	$N_2 =$ 4	3/4	3/4			
	5	2/3	4/5	4/5		
	6	2/3	2/3	5/6		
	7	2/3	5/7	6/7	6/7	
	8	5/8	3/4	3/4	7/8	
	9	2/3	2/3	7/9	8/9	8/9
	10	3/5	7/10	4/5	9/10	9/10
	12	7/12	2/3	3/4	5/6	11/12
$N_1 = 4$	$N_2 =$ 5	3/5	3/4	4/5	4/5	
	6	7/12	2/3	3/4	5/6	5/6
	7	17/28	5/7	3/4	6/7	6/7
	8	5/8	5/8	3/4	7/8	7/8
	9	5/9	2/3	3/4	7/9	8/9
	10	11/20	13/20	7/10	4/5	4/5
	12	7/12	2/3	2/3	3/4	5/6
	16	9/16	5/8	11/16	3/4	13/16
$N_1 = 5$	$N_2 =$ 6	3/5	2/3	2/3	5/6	5/6
	7	4/7	23/35	5/7	29/35	6/7
	8	11/20	5/8	27/40	4/5	4/5
	9	5/9	3/5	31/45	7/9	4/5
	10	1/2	3/5	7/10	7/10	4/5
	15	8/15	3/5	2/3	11/15	11/15
	20	1/2	11/20	3/5	7/10	3/4
$N_1 = 6$	$N_2 =$ 7	23/42	4/7	29/42	5/7	5/6
	8	1/2	7/12	2/3	3/4	3/4
	9	1/2	5/9	2/3	13/18	7/9
	10	1/2	17/30	19/30	7/10	11/15
	12	1/2	7/12	7/12	2/3	3/4
	18	4/9	5/9	11/18	2/3	13/18
	24	11/24	1/2	7/12	5/8	2/3

[Table 15 contd]

One-sided test:		$p = 0{\cdot}90$	0·95	0·975	0·99	0·995
Two-sided test:		$p = 0{\cdot}80$	0·90	0·95	0·98	0·99
$N_1 = 7$	$N_2 = 8$	27/56	33/56	5/8	41/56	3/4
	9	31/63	5/9	40/63	5/7	47/63
	10	33/70	39/70	43/70	7/10	5/7
	14	3/7	1/2	4/7	9/14	5/7
	28	3/7	13/28	15/28	17/28	9/14
$N_1 = 8$	$N_2 = 9$	4/9	13/24	5/8	2/3	3/4
	10	19/40	21/40	23/40	27/40	7/10
	12	11/24	1/2	7/12	5/8	2/3
	16	7/16	1/2	9/16	5/8	5/8
	32	13/32	7/16	1/2	9/16	19/32
$N_1 = 9$	$N_2 = 10$	7/15	1/2	26/45	2/3	31/45
	12	4/9	1/2	5/9	11/18	2/3
	15	19/45	22/45	8/15	3/5	29/45
	18	7/18	4/9	1/2	5/9	11/18
	36	13/36	5/12	17/36	19/36	5/9
$N_1 = 10$	$N_2 = 15$	2/5	7/15	1/2	17/30	19/30
	20	2/5	9/20	1/2	11/20	3/5
	40	7/20	2/5	9/20	1/2	
$N_1 = 12$	$N_2 = 15$	23/60	9/20	1/2	11/20	7/12
	16	3/8	7/16	23/48	13/24	7/12
	18	13/36	5/12	17/36	19/36	5/9
	20	11/30	5/12	7/15	31/60	17/30
$N_1 = 15$	$N_2 = 20$	7/20	2/5	13/30	29/60	31/60
$N_1 = 16$	$N_2 = 20$	27/80	31/80	17/40	19/40	41/80
Large sample approximation		$1{\cdot}07\sqrt{\frac{m+n}{mn}}$	$1{\cdot}22\sqrt{\frac{m+n}{mn}}$	$1{\cdot}36\sqrt{\frac{m+n}{mn}}$	$1{\cdot}52\sqrt{\frac{m+n}{mn}}$	$1{\cdot}63\sqrt{\frac{m+n}{mn}}$

Table 16 Kolmogorov's goodness-of-fit test.
Source: L. H. Miller, 'Table of percentage points of Kolmogorov statistics', *J. Am. Stat. Assoc.* 51 (1956), 111–21.

The null hypothesis may be rejected at the selected probability if the observed value exceeds the critical value given in the table. For $n > 40$ use the approximation.

One-sided test											
	p = 0·90	0·95	0·975	0·99	0·995		*p* = 0·90	0·95	0·975	0·99	0·995
Two-sided test											
	p = 0·80	0·90	0·95	0·98	0·99		*p* = 0·80	0·90	0·95	0·98	0·99
$n = 1$	0·900	0·950	0·975	0·990	0·995	$n = 21$	0·226	0·259	0·287	0·321	0·344
2	0·684	0·776	0·842	0·900	0·929	22	0·221	0·253	0·281	0·314	0·337
3	0·565	0·636	0·708	0·785	0·829	23	0·216	0·247	0·275	0·307	0·330
4	0·493	0·565	0·624	0·689	0·734	24	0·212	0·242	0·269	0·301	0·323
5	0·447	0·509	0·563	0·627	0·669	25	0·208	0·238	0·264	0·295	0·317
6	0·410	0·468	0·519	0·577	0·617	26	0·204	0·233	0·259	0·290	0·311
7	0·381	0·436	0·483	0·538	0·576	27	0·200	0·229	0·254	0·284	0·305
8	0·358	0·410	0·454	0·507	0·542	28	0·197	0·225	0·250	0·279	0·300
9	0·339	0·387	0·430	0·480	0·513	29	0·193	0·221	0·246	0·275	0·295
10	0·323	0·369	0·409	0·457	0·489	30	0·190	0·218	0·242	0·270	0·290
11	0·308	0·352	0·391	0·437	0·468	31	0·187	0·214	0·238	0·266	0·285
12	0·296	0·338	0·375	0·419	0·449	32	0·184	0·211	0·234	0·262	0·281
13	0·285	0·325	0·361	0·404	0·432	33	0·182	0·208	0·231	0·258	0·277
14	0·275	0·314	0·349	0·390	0·418	34	0·179	0·205	0·227	0·254	0·273
15	0·266	0·304	0·338	0·377	0·404	35	0·177	0·202	0·224	0·251	0·269
16	0·258	0·295	0·327	0·366	0·392	36	0·174	0·199	0·221	0·247	0·265
17	0·250	0·286	0·318	0·355	0·381	37	0·172	0·196	0·218	0·244	0·262
18	0·244	0·279	0·309	0·346	0·371	38	0·170	0·194	0·215	0·241	0·258
19	0·237	0·271	0·301	0·337	0·361	39	0·168	0·191	0·213	0·238	0·255
20	0·232	0·265	0·294	0·329	0·352	40	0·165	0·189	0·210	0·235	0·252
				Approximation for $n > 40$			$\frac{1{\cdot}07}{\sqrt{n}}$	$\frac{1{\cdot}22}{\sqrt{n}}$	$\frac{1{\cdot}36}{\sqrt{n}}$	$\frac{1{\cdot}52}{\sqrt{n}}$	$\frac{1{\cdot}63}{\sqrt{n}}$

Table 17 Critical values for Kuiper's test statistic.
Source: K. V. Mardia, *Statistics of directional data* (London 1972), 308.

The null hypothesis may be rejected at a selected α for observed values of the test statistic as large or larger than the tabulated values.

n	$\alpha \rightarrow$ 0·10	0·05	0·01	0·005
5	1·46	1·57	1·76	1·84
6	1·47	1·58	1·79	1·87
7	1·48	1·60	1·81	1·89
8	1·49	1·61	1·83	1·91
9	1·50	1·62	1·84	1·93
10	1·51	1·63	1·85	1·94
11	1·51	1·63	1·86	1·95
12	1·52	1·64	1·87	1·96
13	1·52	1·64	1·88	1·96
14	1·53	1·65	1·88	1·97
15	1·53	1·65	1·89	1·98
16	1·53	1·66	1·89	1·98
17	1·54	1·66	1·90	1·99
18	1·54	1·66	1·90	1·99
19	1·54	1·66	1·90	2·00
20	1·55	1·67	1·91	2·00
25	1·55	1·67	1·92	2·01
30	1·56	1·68	1·93	2·02
40	1·57	1·69	1·94	2·03
50	1·58	1·70	1·95	2·04
100	1·59	1·72	1·97	2·06
∞	1·62	1·75	2·00	2·10

Table 18 Critical values for the two-sample uniform scores test, with n, m in the samples.
Source: K. V. Mardia, *Statistics of directional data* (London 1972), 312–13.

The null hypothesis may be rejected at the selected α if the observed value of the test statistic is as large or larger than the tabulated critical values.

n	n_1	$\alpha \rightarrow$ 0·001	0·01	0·05	0·10
8	4				6·83
9	3			8·29	6·41
	4				4·88
10	3				6·85
	4			9·47	6·24
	5			10·47	6·85

n	n_1	$\alpha \rightarrow$ 0·001	0·01	0·05	0·10
11	3			7·20	5·23
	4			10·42	7·43
	5		12·34	8·74	6·60
12	3			7·46	5·73
	4		11·20	8·46	7·46
	5		13·93	10·46	7·46
	6		14·93	11·20	7·46
13	3			7·68	6·15
	4		11·83	9·35	7·03
	5		15·26	10·15	7·39
	6		17·31	10·42	8·04
14	3			7·85	6·49
	4		12·34	9·30	7·60
	5		16·39	10·30	7·85
	6	19·20	15·59	12·21	7·94
	7	20·20	16·39	11·65	8·85
15	3			7·99	6·78
	4		12·78	8·74	7·91
	5	17·35	14·52	10·36	7·91
	6	20·92	17·48	11·61	9·12
	7	22·88	16·14	11·57	9·06
16	3			8·11	5·83
	4		13·14	9·44	7·38
	5	18·16	15·55	10·44	9·03
	6	22·43	16·98	11·54	9·11
	7	25·27	18·16	12·66	9·78
17	3		8·21	7·23	6·14
	4	13·44	11·76	9·74	7·64
	5	18·86	16·44	11·03	8·76
	6	23·73	17·76	12·21	9·41
	7	27·40	17·98	12·63	10·11
	8	29·37	19·11	13·36	10·15
18	2				3·88
	3		8·29	7·41	6·41
	4	13·70	12·17	9·94	8·06
	5	19·46	16·05	11·45	8·76
	6	24·87	17·40	12·25	9·94
	7	29·28	19·46	13·41	10·29
	8	28·40	20·11	13·82	10·60
	9	29·28	20·23	13·99	11·04
19	2				3·89
	3		8·36	7·56	6·48
	4	13·93	12·52	9·69	7·54
	5	19·98	15·88	11·29	8·96
	6	25·87	18·19	12·57	9·87
	7	27·71	19·34	13·54	10·55

[Table 18 contd]

n	n_1	$\alpha \rightarrow$ 0·001	0·01	0·05	0·10
	8	31·04	21·12	14·29	11·12
	9	29·46	21·07	14·58	11·37
20	2				3·90
	3		8·42	7·70	6·70
	4	14·12	12·83	9·87	7·80
	5	20·43	16·29	11·49	9·08
	6	26·75	18·64	12·93	9·98
	7	29·36	20·43	14·05	11·03
	8	30·08	21·77	14·77	11·47
	9	32·44	22·99	15·45	11·97
	10	33·26	22·67	15·39	12·19
	$R = \chi_2^2$	13·816	9·210	5·991	4·605

Answers to exercises

1.1 6, 24, 120, 40 320.

1.2 336, 6720.

1.3 21, 56, 1, 1, 55, 55.

1.4 n, $n(n-1)$.

1.5 Expand and cancel.

1.6 Expand both sides.

1.7 Expand both sides.

1.8 $35 = 35$, expand left-hand side (lhs) to give

$k!\,(k-r+1) + k!\,(r)/r!\,(k-r+1)! \; = \; k!\,(k+1)/r!\,(k-r+1)!$

then expand right-hand side and equate.

1.10 35.

1.11 11 760.

1.12 7!

1.13 25.

1.14 26.

1.15 50/4.

1.16 0·6310 = 460/729.

1.17c) 6/4, 18/16; 18/4, 18/16.

1.18 5/2 . $E(X)$ does not have to be a realizable value.

1.19 24 arrangements, $X_0 = 9$, $X_1 = 8$, $X_2 = 6$, $X_4 = 1$.

1.20 7.

1.21 Expand and cancel to give p.

1.22 36, 64.

1.23 H_0 : matching is random. H_1 : matching is better than under random expectation. Probability of 4 matches is $\frac{1}{24}$, and this fixes α and X_i. Probability of no matches is $\frac{3}{8}$. For n maps and n assertions the probability of no matches tends to 0·368 as n increases.

1.24 1. Replace 'does not' by 'does'.
2. Replace 'does not' by 'does'.
3. Replace 'are' by 'are not'.
To direct the hypotheses to imply a one-tailed test for the alternative put in the H_1, the direction preferred.
1. Increases/decreases.
2. Increases/decreases.
3. Clustered.

1.25 1.(a) Is not changing or (b) decreasing.
2.(a) Are not or (b) are raw-material oriented.
3.(a) Are not different from the expected frequency of the random variable implied in the H_0. This will be dealt with in section 2.
In all cases the measurement and the property need careful definitions.

1.26 Nominal. Non-parametric.

1.27 Discrete-quantity.

1.28 It cannot.

1.29 (a) Not identically. (b) This is a question of what is an acceptable level of similarity (see article 3.3.2). c) Yes, but no more of the order information is usually used. d) Join and count the number of inversions; in article 3.2.2 a measure of relationship is introduced using concordances and discordances.

1.30 a) 6, 17, 0·5 dinches.

1.31 $\{d, f\}$ $\{a, b, c, e, g, h, i\}$ at nominal level.
A discrete-quantity at least is then possible providing an ordering relationship for the objects.

1.32 a) H_1.
b) H_0 : students with experience in statistical inference get marks in geography examinations which are lower or no different from the marks obtained by students without such experience.
c) Let the r.v. X be students with experience in statistical inference. Let Y be students without such experience. Let $(X > Y) = +$, let $(X < Y) = -$, let $(X = Y) = 0$. Then

$$H_0 : P(X > Y) \leqslant P(X < Y)$$
$$H_1 : P(X > Y) > P(X < Y)$$

d) One-tailed.

2.1 a) None, b) bijective, c) surjective, d) none, e) injective.

2.2 Surjective.

2.3 Bijective. No.

2.4 Because there is an infinite number of random scatters of points for classes of random variable. Yes.

2.5 Point probability is $P(X = 3) = \binom{10}{3}\left(\frac{6}{20}\right)^3\left(\frac{14}{20}\right)^7 = 0{\cdot}266814$

2.6 a) $P(1,8,1) = \frac{10!}{1!8!1!} \cdot (0{\cdot}25)^1 . (0{\cdot}20)^8 (0{\cdot}55)^1 = 0{\cdot}00003168.$

b) The probability of the following more extreme results is 0·061154. This is derived from $P(8,1,1)$, $P(1,1,8)$, $P(9,1,0)$, $P(1,9,0)$, $P(9,0,1)$, $P(0,9,1)$, $P(0,1,9)$, $P(1,0,9)$, $P(10,0,0)$, $P(0,10,0)$, $P(0,0,10)$. You may consider the six arrangements of (8,2,0) should be included too.

c) Yes.

2.7 a) H_0 : There is no selective occupation with respect to aquifer.
H_1 : Occupation is selective against aquifers yielding water deleterious to high quality paper.

b) Hypergeometric with the following outcomes

		Aquifer		
		Favourable	Deleterious	Total
Occupation of site	Continues	7	1	8
	Rejected	5	6	11
	Total	12	7	19

c) The probability distribution is asymmetric and the point probability of the given table is 0·023577 and there is evidence of selective occupation.

2.8 $P(0;\lambda{=}2) = 0{\cdot}1353$, $P(1;\lambda{=}2) = 0{\cdot}2707$, $P(2;\lambda{=}2) = 0{\cdot}2707$, $P(3;\lambda{=}2) = 0{\cdot}1804$, $P(4;\lambda{=}2) = 0{\cdot}0902$, $P(5;\lambda{=}2) = 0{\cdot}0361$, $P(6;\lambda{=}2) = 0{\cdot}0120$. The cumulative probability is then 0·9955.

2.9 $P(0;\lambda{=}1{\cdot}68) = 0{\cdot}1864$, $P(1;\lambda{=}1{\cdot}68) = 0{\cdot}3131$, $P(2;\lambda{=}1{\cdot}68) = 0{\cdot}2630$, $P(3;\lambda{=}3) = 0{\cdot}1473$, $P(4;\lambda{=}1{\cdot}68) = 0{\cdot}0619$, $P(5;\lambda{=}1{\cdot}68) = 0{\cdot}0208$. The cumulative probability is then 0·9924.

2.10 $\lambda = 2{\cdot}30136$. $P(0;\lambda) = 0{\cdot}1001$, $P(1;\lambda) = 0{\cdot}2304$, $P(2;\lambda) = 0{\cdot}2651$, $P(3;\lambda) = 0{\cdot}2034$, $P(4;\lambda) = 0{\cdot}1170$, $P(\geqslant 5;\lambda) = 0{\cdot}0839$.

2.11 $X = 32$, $P(X = 32) = 0{\cdot}9^{31} . 0{\cdot}1 = 0{\cdot}004$.

2.12 $$P(X = 3) = \binom{15-1}{3-1}(0{\cdot}8)^{12}(0{\cdot}2)^3$$
$$= 0{\cdot}05003.$$

Better here to discuss the probability than to accept H_0 on the basis of $\alpha : 0{\cdot}05$.

2.14 a) H_1 : Parks in areas of high residential density show a greater intensity of usage.
H_0 : Usage is independent of residential density.

2.14 b) One-tailed test at $\alpha : 0{\cdot}05$, $P(X \geqslant 11) = 0{\cdot}0673$. $P(X = 11) = 0{\cdot}05846$.
c) The use of χ^2 as an approximating distribution would be rejected because $N = 24$ is smaller than preferred, and the cell-expected values are small.

2.15 $2R = N; x = n - r \Rightarrow n - r = \dfrac{nN - Nr}{N} = \dfrac{N(n-r)}{N}$

$2n = N; x = R - r \Rightarrow R - r = \dfrac{NR - Nr}{N} = \dfrac{N(R-r)}{N}$

2.16 If H_1 had not been directed in exercise 2.14 we should have had to allow for the table that was an extreme as the given table but in the opposite direction. The most extreme table in the opposite direction is

5	10
9	0

. The appropriate table is

7	8
7	2

.

2.17 Yes. Ian Fleming contains good social geography! $X^2 = 6{\cdot}394$ with 1 *df*. X_1^2 at $\alpha : 0{\cdot}05 = 3{\cdot}841$.

2.18 H_0 : Nationality and immigrant preference are independent.
H_1 is justified at any reasonable α.
$X^2 = 17{\cdot}524$. Use values coded 10^{-3}.

2.19 $X^2 = 2{\cdot}3$. Accept H_0. Again use values coded 10^{-3}. As coding affects the decision it is justified? Without coding $X^2 = 2327{\cdot}7$. $df = 1$.

2.21 Reject hypothesis of independence. χ^2 with 4 $df = 9{\cdot}488$ at $\alpha : 0{\cdot}05$. $X^2 > \chi^2$.

2.22 a) Single-sample model.
b) H_0 : probability that a visitor will come from a particular distance class is independent of the sort of user he is.
c) Summer bank-holiday: $X^2 = 11{\cdot}991$, $df = 3$.
Non-bank holiday: $X^2 = 18{\cdot}705$, $df = 3$.
Combined: $X^2 = 17{\cdot}618$, $df = 3$.
χ^2 $df = 3, \alpha : 0{\cdot}05 = 7{\cdot}815$.

2.23 a) $X^2 = 15{\cdot}736$, $df = 4$, reject H_0 at $\alpha : 0{\cdot}05$, infer that size and distance class to market are related.
b) $X^2 = 28{\cdot}215$, $df = 4$, reject H_0 at $\alpha : 0{\cdot}05$, infer that size and distance class to raw material are related.

2.24 Four subtables are

25	8		33	32		39	19		58	40
14	11		25	8		24	16		40	26

$X^2 = 2{\cdot}521$ $\quad X^2 = 5{\cdot}658$ $\quad X^2 = 0{\cdot}541$ $\quad X^2 = 0{\cdot}33$

χ^2 $df = 1$, $\alpha : 0{\cdot}05 = 3{\cdot}841$.

2.25 Market subtables. 1 *df* in each case.

25	18	43	40	42	27	69	62
17	9	26	22	50	21	71	21

$X^2 = 0{\cdot}3164$ $X^2 = 0{\cdot}0723$ $X^2 = 1{\cdot}4616$ $X^2 = 13{\cdot}8857$

Raw material subtables. 1 *df* in each case.

19	19	38	45	36	32	68	18
17	13	30	18	55	16	71	21

$X^2 = 0{\cdot}377$ $X^2 = 3{\cdot}619$ $X^2 = 9{\cdot}526$ $X^2 = 14{\cdot}693$

2.26 a) Yes. $X^2 = 39{\cdot}34$, reject H_0.
b) Yes. $X^2 = \hat{L}^2/S^2 = 0{\cdot}08187$, accept H_0.

$$\hat{L} = 0{\cdot}28768;\ S^2 = 1{\cdot}0109.$$

2.27 a) $X^2 = 8{\cdot}777$, $df = 4$. $< \chi^2_4\ \alpha : 0{\cdot}05$.
b) Subsets $[-2,-1]\ [0,1,2]$ $X^2 = 2{\cdot}252$, $df = 1$.
$[-2]\ [-1]$ $X^2 = 0{\cdot}153$, $df = 1$.
$[0]\ [1]\ [2]$ $X^2 = 6{\cdot}372$, $df = 2$.

Subsets $[+2,+1]\ [0,-1,-2]$ $X^2 = 0{\cdot}526$, $df = 1$.
$[+2]\ [+1]$ $X^2 = 6{\cdot}247$, $df = 1$.
$[0]\ [-1]\ [-2]$ $X^2 = 2{\cdot}004$, $df = 2$.

c) X^2 component due to linear regression $= 3{\cdot}204$, 1 *df*.
X^2 component due to non-linear regression $= 5{\cdot}573$, 3 *df*.
e) For $Y =$ survival, $X =$ distance class we have

$$Y = 0{\cdot}63813 - 0{\cdot}06794\,X$$

2.28 Response is independent of size given the level of location. Try a joint variable.

2.29 a) $B \otimes C \cap A = \Phi | BC$.
b) If we know the joint variable, category/location, then category is independent of location/size. The H_0 implies $f_{ijk} = \mu + \lambda^B + \lambda^C$ and no interaction terms are required.
c) Simply those for category and location.

C_1 134 L_1 174
C_2 111 L_2 272

2.30 a) $X^2 = 0{\cdot}041$, $df = 1$.
b) $X^2 = 181{\cdot}069$, $df = 4$.
c) $X^2 = 153{\cdot}311$, $df = 2$.

2.31 a) If we set $\alpha : 0{\cdot}05$ such that we accept H_0 that there is no difference in income, if probability H_0 is $> 0{\cdot}05$ and reject if $H_0 \leqslant 0{\cdot}05$, then we reject H_0, because the Wilcoxon rank-sum is $W_n = 24$ and for $n = 5$, $m = 9$, $\alpha : 0{\cdot}05 = W_n = 24$.
b) Initial size, capital, return to scale of freezing-plant.

2.31 c) 2002.

d) As $W_n = 25$ is exactly at $\alpha : 0{\cdot}05$ then the number of such sums, x, is

$$\frac{x}{2002} = \frac{5}{100}, \quad x = \frac{5\,.\,2002}{100} \simeq 100.$$

e) Consult the diagram. Yes I believe so as diagram (ii) shows the information actually used in the test.

2.32 For H_0 given, propose H_1 that scheduled buildings lead to lower values. If H_1 is true then we expect lower sum of values. Give rank 1 to lowest value. Then we expect H_1 to lead to lower sum of ranks for scheduled buildings. Obtained $W_n = 66$. With $n = 8$, $m = 14$, $W_n = 67$ at $\alpha : 0{\cdot}05$. Accept H_1.

2.33 Ex. 2.31: $U = 9$.
Ex. 2.32: $U = 30$.

2.34 If seeding increases rainfall then we expect higher values. If we give low rank to high values we expect W_n to be unusually low.
Calculated $W_n = 55$. Probability distribution for $n = 6$, $m = 9$ is $W_n = 33$ at $\alpha : 0{\cdot}05$. We accept H_0 of no difference in yield of precipitation.

2.35 a) Accept H_0.

b) Accept H_0.

c) Reject H_0. Again we expect more litter and higher organic content under oak than under pine. If we give low rank to high value of organic content, and if H_1 is true, we expect a smaller sum of ranks than expected under H_0. The obtained $W_n = 48$, the expected W_n at $\alpha : 0{\cdot}025$ is 49, and at $\alpha : 0{\cdot}05 = 51$. We reject H_0.

2.36 Because tied values in this case have so little effect on the sum of ranks.

2.37 a) Accept H_0 at $\alpha : 0{\cdot}05$. $F = 0{\cdot}5$.

b) Reject H_0 at $\alpha : 0{\cdot}05$. $F = 91{\cdot}5 \simeq$ prob $0{\cdot}0222$ and for the preferred order we accept H_1 at any reasonable α as $(2!\alpha)/3! = \frac{1}{3}\, 0{\cdot}0222 = 0{\cdot}0074$.

3.1 a) Accept.

b) Accept. The appropriate table is

2	15
4	0

for $P(r = 4)$.

$P(r = 4) = 0{\cdot}000104$.

3.2 Strong evidence. For 17 runs, $n_1 =$ dry days $= 32$, $n_2 =$ wet days $= 18$, we find mean $= 24{\cdot}04$, variance $= 10{\cdot}3633$ and $Z = 2{\cdot}186$. Using the normal approximation equation we reject H_0.

3.3 For Noether's test we find

a) BOD 1965/67 2 monotone runs in 5 runs of triples giving

$$\binom{5}{2}\left(\frac{1}{3}\right)^2\left(\frac{2}{3}\right)^3 = 80/243 \text{ as the probability;}$$

b) BOD 1959/61 1 monotone run in 5 runs of triples giving

$$\binom{5}{1}\left(\frac{1}{3}\right)^1\left(\frac{2}{3}\right)^4 = 80/243 = 0{\cdot}329.$$

Each is likely to occur under randomization about $\frac{1}{3}$ of the time and we accept H_0 in each case.

3.4 All give evidence of association. In a sense the figure is a graph of concordances and discordances.

3.5 Yes. rho $= -0{\cdot}373$, tau $= -0{\cdot}255$.

3.6 In exercise 2.31 we have $n_1 = 5$, $n_2 = 9$, $r = 8$. Wald-Wolfowitz at $\alpha : 0{\cdot}05$ gives 4 runs. So accept H_0.
In exercise 2.32 we have $n_1 = 8$, $n_2 = 14$, $r = 9$. Wald-Wolfowitz at $\alpha : 0{\cdot}05$ gives 7 runs. So accept H_0.

3.7 Change in land use is used in this test. Of 16 changes, 13 have the same sign, i.e. have changed in the same way. We take $n = 16, p = (1-p) = 0{\cdot}5$ as the likelihood of a change.
For a two-tailed test of H_0 we look at tables of the cumulative binomial distribution with $\alpha/2 = 0{\cdot}025$. The value for rejection is $x \leqslant 3$ or $x \geqslant 12$ when x is either + or −. We reject H_0 in this case.
With $T = (13 - 3)^2/16 = 100/16 = 6{\cdot}25$ again reject with χ^2.

3.8 Let 1965/7 < 1959/61 be (−). Then $n = 18$, $(+) = 1$, $p = 0{\cdot}0001$.

3.9 a) Accept H_0. There are 1482 paths less extreme, with $p = 0{\cdot}8636$, thus $1 - 0{\cdot}8636 = 0{\cdot}1364$ is the probability of a result as extreme or more extreme. Using the method of fig. 3.31 we find $22/42 < 24/42$ which is the value of the fraction at $\alpha : 0{\cdot}05$. Again accept H_0.
c) 1716.
d) Wilcoxon at $\alpha : 0{\cdot}05$, $W_n = 29$; actual $W_n = 33$, $n = 6$, $m = 7$.

3.10 Exercise 2.31: Smirnov value $25/45 < 27/45$ which is critical value. Accept H_0.
Exercise 2.32: Smirnov value $29/56 > 1/2$ which is critical value. Reject H_0.
Exercise 2.34: Smirnov value $7/18 < 5/9$ which is critical value. Accept H_0.
Exercise 2.35: Smirnov value $5/8 > 4/8$ which is critical value. Reject H_0.

3.11 a) Accept H_0 as $D_{max} = 0{\cdot}2428 <$ critical $0{\cdot}338$ for $\alpha : 0{\cdot}05$.
b) Accept H_0 as $D_{max} = 0{\cdot}1329$.

3.12 The values expected under Poisson probability are $n_0 = 7{\cdot}309$, $n_1 = 16{\cdot}821$, $n_2 = 19{\cdot}355$, $n_3 = 14{\cdot}848$, $n_4 = 8{\cdot}542$, $n_5 = 6{\cdot}125$.
$X^2 = 0{\cdot}352$. χ^2 with $(6-2) = 4$ *df*, $\alpha : 0{\cdot}05 = 9{\cdot}488$.

3.13 $R^2 = 2{\cdot}308 <$ the critical value for $n = 14$, $n_1 = 6$ at $\alpha : 0{\cdot}05$ which is $12{\cdot}21$ and so accept H_0 of no difference.

3.14 $r_1 = 8 \Rightarrow r_2 = 7 \Rightarrow r = r_1 + r_2 = 15$.

Bibliography and further reading

AITCHISON, J. (1968) *Statistics*, Vols. I, II. Edinburgh.

ALONSO, W. (1964) *Location and land use: toward a general theory of land rent.* Cambridge, Mass.

ARMITAGE, P. (1955) Tests for linear trends in proportions and frequencies. *Biom.* 11, 375–86.

ARTHURS, A. M. (1965) *Probability theory.* London.

BACHI, R. (1968) *Graphical rational patterns. A new approach to graphical presentation of statistics.* Jerusalem.

BAILEY, N. T. J. (1964) *The elements of stochastic processes with applications to the natural sciences.* New York.

BARTHOLOMEW, D. J. (1973) *Stochastic models for social processes.* New York.

BARTLETT, M. S. (1935) Contingency table interactions. *J. Roy. Stat. Soc.*, Ser. B, 2, 248–52.

BERKSON, J. (1938) Some difficulties of interpretation. *J. Am. Stat. Assoc.* 33, 526–36.

BERKSON, J. (1939) A note on the chi-square test, the Poisson and the binomial. *J. Am. Stat. Assoc.* 34, 362–7.

BIRCH, M. M. (1963) Maximum likelihood in three-way contingency tables. *J. Roy. Stat. Soc.*, Ser. B, 25, 220–33.

BISHOP, Y. M. M. (1969) Full contingency tables, logits and split contingency tables. *Biom.* 25, 383–400.

BISHOP, Y. and FIENBERG, S. E. (1969) Incomplete two-dimensional contingency tables. *Biom.* 25, 119–28.

BRADLEY, J. V. (1968) *Distribution-free statistical tests.* Englewood Cliffs, N.J.

CAMP, B. H. (1938) Further interpretations of the chi-square test. *J. Am. Stat. Assoc.* 33, 537–42.

CAMP, B. H. (1939) Further comments on Berkson's problem. *J. Am. Stat. Assoc.* 34, 368–76. (See BERKSON 1938, 1939.)

CHRISTALLER, W. (1966) *Central places in southern Germany.* Translated by C. BASKIN. Englewood Cliffs, N.J.

COCHRAN, W. G. (1954) Some methods for strengthening the common χ^2 tests. *Biom.* 10, 417–51.

COCHRAN, W. G. (1955) A test of linear function of the deviations between observed and expected numbers. *J. Am. Stat. Assoc.* 50, 377–97.
CONOVER, W. J. (1971) *Practical non-parametric statistics.* New York.
COX, D. R. and LAUH, E. (1967) A note on the graphical analysis of multidimensional contingency tables. *Technometrics* 9, 481–8.
DALEY, D. J. and VERE-JONES, D. (1972) A summary of the theory of point processes, in P. A. W. LEWIS (ed.) *Stochastic point processes: statistical analysis, theory and applications*, 299–383. New York.
DE FINETTI, B. (1972) *Probability, induction and statistics.* New York.
DUNN, J. E. (1969) A compounded multiple runs distribution. *J. Am. Stat. Assoc.* 64, 1415–23.
DUNN, O. J. (1964) Multiple comparisons using rank sums. *Technometrics* 6, 241–52.
DYKE, G. V. and PATTERSON, H. D. (1952) Analysis of factorial arrangements when data are proportions. *Biom.* 8, 1–12.
EDGINGTON, E. S. (1961) Probability table for number of runs of signs of first differences in ordered series. *J. Am. Stat. Assoc.* 56, 156–9.
ELLIS, B. (1966) *Basic concepts of measurement.* Cambridge.
FIENBERG, S. E. and GILBERT, J. P. (1969) The geometry of a two by two contingency table. *J. Am. Stat. Assoc.* 64, 694–701.
FISHER, L. (1972) A survey of the mathematical theory of multi-dimensional point processes, in P. A. W. LEWIS (ed.) *Stochastic point processes: statistical analysis, theory and applications*, 468–513. New York.
FRENCH, H. M. (1972) Quantitative methods and non-parametric statistics, in H. M. FRENCH and J. B. RACINE (eds.) *Quantitative and qualitative geography*, 119–28. Univ. of Ottawa, Dept. of Geogr., Occas. Pap. 1.
GANI, J. (1972) Point processes in epidemiology, in P. A. W. LEWIS (ed.) *Stochastic point processes: statistical analysis, theory and applications*, 756–73. New York.
GASKING, D. A. T. (1960) Clusters. *Aust. J. Phil.* 38 (1), 1–36.
GERIG, T. M. (1969) A multivariate extension of Friedman's test. *J. Am. Stat. Assoc.* 64, 1595–608.
GIBBONS, J. D. (1971) *Non-parametric statistical inference.* New York.
GLASSER, G. J. (1962) A distribution-free test of independence with a sample of paired observations. *J. Am. Stat. Assoc.*, 57, 116–33.
GOOD, I. J., GOVER, T. N. and MITCHELL, G. J. (1970) Exact distribution for χ^2 and for the likelihood-ratio statistic for the equiprobable multinomial distribution. *J. Am. Stat. Assoc.* 65, 267–83.
GOODMAN, L. A. (1958) Simplified runs tests and likelihood ratio tests for Markoff chains. *Biometrika* 45, 181–97.
GOODMAN, L. A. (1963) On Plackett's test for contingency table interactions. *J. Roy. Stat. Soc.*, Ser. B, 25, 179–88.
GOODMAN, L. A. (1964) Simple methods for analysing three-factor interaction in contingency tables. *J. Am. Stat. Assoc.* 59, 319–52.
GOODMAN, L. A. (1968) The analysis of cross-classified data: independence, quasi-independence, and interactions in contingency tables with or without missing entries. *J. Am. Stat. Assoc.* 63, 1091–131.
GOODMAN, L. A. (1969) How to ransack social mobility tables and other kinds of cross-classification tables. *Am. J. Sociol.* 75, 1–40.
GOODMAN, L. A. (1970) The multivariate analysis of qualitative data: interactions among multiple classifications. *J. Am. Stat. Assoc.* 65, 226–56.

GOODMAN, L. A. (1971) The analysis of multidimensional contingency tables: stepwise procedures and direct estimation methods for building models for multiple classifications. *Technometrics* 13, 33–61.

GOODMAN, L. A. and KRUSKAL, W. H. (1954) Measures of association for cross classifications. *J. Am. Stat. Assoc.* 49, 732–64.

GOODMAN, L. A. and KRUSKAL, W. H. (1959) Measures of association for cross classifications. II: Further discussions and references. *J. Am. Stat. Assoc.* 54, 124–63.

GOODMAN, L. A. and KRUSKAL, W. H. (1963) Measures of association for cross classification. III: Approximate sampling theory. *J. Am. Stat. Assoc.* 58, 310–64.

GREIG-SMITH, P. (1964) *Quantitative plant ecology.* London.

HABERMAN, S. J. (1972) Log-linear fit for contingency tables algorithm AS 51. *Appl. Stat.* 21, 218–25.

HAGERSTRAND, T. (1957) Migration and area, in *Migration in Sweden.* Lund Stud. in *Geogr.*, Ser. B, 13.

HEAP, B. R. (1972) *Algorithms for the production of contour maps over an irregular triangular mesh.* Nation. Phys. Labor. Pap. NAC 10.

HOLGATE, P. (1964) Estimation for the bivariate Poisson distribution. *Biometrika* 51 (1), 241–5.

HOLGATE, P. (1972) The use of distance methods for the analysis of spatial distribution of points, in P. A. W. LEWIS (ed.) *Stochastic point processes: statistical analysis, theory and applications*, 122–35. New York.

HOLLANDER, M. (1967) Asymptotic efficiency of two non-parametric competitors of Wilcoxon's two sample test. *J. Am. Stat. Assoc.* 62, 939–49.

IRWIN, J. O. (1949) A note on the subdivision of χ^2 into components. *Biometrika* 36, 130–4.

ISHII, G. (1960) Intra-class contingency tables. *Ann. Inst. of Stat. Maths* 12, 161–207.

JACOBSON, J. E. (1963) The Wilcoxon two-sample statistic: tables and bibliography. *J. Am. Stat. Assoc.* 58, 1086–103.

JOHNSON, E. M. (1972) The Fisher-Yates exact test and unequal sample sizes. *Psychometrika* 37, 103–6.

JONES, P. N. (1969) Some aspects of immigration into the Glamorgan coalfield between 1881 and 1911. *Trans. Hon. Soc. Cymmrodorion*, Part I, 82–98.

KENDALL, M. G. (1949) *Rank correlation methods.* London.

KIMBALL, A. W. (1954) Short-cut formulas for the exact partition of χ^2 in contingency tables. *Biom.* 10, 452–8.

KRUSKAL, W. H. (1957) Historical notes on the Wilcoxon unpaired two-sample test. *J. Am. Stat. Assoc.* 52, 356–60.

KRUSKAL, W. H. and WALLIS, W. A. (1952) Use of ranks in one-criterion variance analysis. *J. Am. Stat. Assoc.* 47, 583–621.

KU, H. H. and KULLBACK, S. (1968) Interaction in multidimensional contingency tables: an information theoretic approach. *J. Res. Nat. Bur. Stud.* 72 B, 159–99.

KULLBACK, S., KUPPERMAN, M. and KU, H. H. (1962) Tests for contingency tables and Markov chains. *Technometrics* 4, 573–608.

LANCASTER, H. O. (1969) *The chi-squared distribution.* New York.

LEWIS, B. (1962) On the analysis of interaction in multidimensional contingency tables. *J. Roy. Stat. Soc.*, Ser. A, 125, 88–117.

LEWIS, P. A. W. ed. (1972) *Stochastic point processes: statistical analysis, theory and applications.* New York.

LEWIS, P. W. (1969) *A numerical approach to the location of industry.* Hull.
LEWIS, P. W. (1970) Measuring spatial interaction. *Geografiska Ann.* 52, Ser. B, 22–39.
LEWIS, P. W. (1974) *Papermaking in the expanded EEC.* Paper.
LILLIEFORS, H. W. (1967) On the Kolmogorov-Smirnov test for normality with mean and variance unknown. *J. Am. Stat. Assoc.* 62, 399–402.
McCORNACK, R. L. (1965) Extended tables of the Wilcoxon matched pair signed rank statistic. *J. Am. Stat. Assoc.* 60, 864–71.
MARCUS, A. (1972) Some point process models of lunar and planetary surfaces, in P. A. W. LEWIS (ed.) *Stochastic point processes: statistical analysis, theory and applications*, 682–99. New York.
MARDIA, K. V. (1972) *Statistics of directional data.* London and New York.
MASSEY, F. J. (1951) The Kolmogorov-Smirnov test for goodness of fit. *J. Am. Stat. Assoc.* 46, 68–78.
MAXWELL, A. E. (1961) *Analysing qualitative data.* London.
MEDHURST, F. and PARRY-LEWIS, J. (1969) *Urban decay: an analysis and policy.* London.
MEYER, P. L. (1970) *Introductory probability and statistical applications.* Reading, Mass.
MILTON, R. C. (1964) An extended table of critical values for the Mann Whitney (Wilcoxon) two-sample statistic. *J. Am. Stat. Assoc.* 59, 925–34.
MONKHOUSE, F. J. and WILKINSON, H. R. (1971) *Maps and diagrams.* 3rd edn. London.
MOSTELLER, F. (1968) Association and estimation in contingency tables. *J. Am. Stat. Assoc.* 63, 1–27.
NAUS, J. I. (1965) The distribution of the size of the maximum cluster of points on a line. *J. Am. Stat. Assoc.* 60, 532–8.
NOETHER, G. E. (1956) Two sequential tests against trend. *J. Am. Stat. Assoc.* 51, 440–50.
NOETHER, G. E. (1967) Wilcoxon confidence intervals for location parameters in the discrete case. *J. Am. Stat. Assoc.* 62, 184–8.
OKAMOTO, M. and ISHII, G. (1961) Test of independence in interclass 2 × 2 tables. *Biometrika* 48, 181–90.
OWEN, D. B. (1962) *Handbook of statistical tables.* London.
PEARSON, K. (1930) On the theory of contingency. *J. Am. Stat. Assoc.* 25, 320–3, 327.
PIELOU, E. C. (1969) *Introduction to mathematical ecology.* New York.
PLACKETT, R. L. (1962) A note on interactions in contingency tables. *J. Roy. Stat. Soc.*, Ser. B, 24, 162–6.
PLACKETT, R. L. (1969) Multidimensional contingency tables: a survey of models and methods. Paper presented at Roy. Stat. Soc. Conf., Newcastle.
PLACKETT, R. L. (1971) Multivariate categorical data. Paper presented at Roy. Stat. Soc. Conf., Newcastle.
PLACKETT, R. L. (1972) *Introduction to the theory of statistics.* London.
PRATT, J. W. (1959) Remarks on zeros and ties in the Wilcoxon signed rank procedures. *J. Am. Stat. Assoc.* 54, 655–67.
QUADE, D. (1967) Rank analysis of covariance. *J. Am. Stat. Assoc.* 62, 1187–200.
RIEDWYL, H. (1967) Goodness of fit (class of distribution-free statistics with tables). *J. Am. Stat. Assoc.* 62, 390–8.
ROBINSON, A. H. (1953) *Elements of cartography.* New York.

ROY, S. N. and KASTENBAUM, M. A. (1956) On the hypothesis of no interaction in a multiway contingency table. *Ann. Math. Stat.* 27, 749–57.

SAVAGE, I. R. (1953) Bibliography of nonparametric statistics and related topics. *J. Am. Stat. Assoc.* 48, 844–912.

SAVAGE, L. J. *et al.* (1962) *The foundations of statistical inference.* London.

SIMPSON, E. H. (1951) The interpretation of interaction in contingency tables. *J. Roy. Stat. Soc.*, Ser. B, 13, 238–41.

SLAKTER, M. J. (1965) A comparison of the Pearson chi-square and Kolmogorov goodness-of-fit tests with respect to validity. *J. Am. Stat. Assoc.* 60, 854–8.

STEWART, J. Q. (1950) The development of social physics. *Am. J. Phys.* 18 (5), 239–43.

STOUFFER, S. A. (1940) Intervening opportunities: a theory relating mobility and distance. *Am. Sociol. Rev.* 5, 845–67.

TOBLER, W. (1961) *Map transformations of geographic space.* University microfilms. Ann Arbor, Mich.

TRENT RIVER AUTHORITY (1969) Fourth annual report.

VON THUNEN, J. H. (1966) *The isolated state.* Translated by P. HALL. Oxford.

WARNTZ, W. (1959) *Toward a geography of price.* Philadelphia, Penn.

WEBER, A. (1929) *Theory of the location of industries.* Translated by C. J. FRIEDRICH. Chicago.

WHITE, M. R. (1973) An investigation of the nature of the 'zone in transition' of E. W. Burgess for a sample area in northeast London. Univ. of London dissertation.

YATES, F. (1934) Contingency tables involving small numbers and the χ^2 test. *J. Roy. Stat. Soc. Suppl.* 1, 217–35.

Index

Note: Page numbers printed in bold type refer to exercises.